Multifield Problems

Springer-Verlag Berlin Heidelberg GmbH

A.-M. Sändig W. Schiehlen
W.L. Wendland (Eds.)

Multifield Problems

State of the Art

With 93 Figures

Springer

Professor Dr.
Anna-Margarete Sändig
Mathematisches Institut A
Universität Stuttgart
Pfaffenwaldring 57
70569 Stuttgart
Germany
E-mail:
saendig@mathematik.uni-stuttgart.de

Professor Dr. Dr. h.c.
Wolfgang L. Wendland
Mathematisches Institut A
Universität Stuttgart
Pfaffenwaldring 57
70569 Stuttgart
Germany
E-mail:
wendland@mathematik.uni-stuttgart.de

Professor Dr. Dr. h.c.
Werner Schiehlen
Institut B für Mathematik
Universität Stuttgart
Pfaffenwaldring 9
70569 Stuttgart
Germany
E-mail:
wos@mechb.uni-stuttgart.de

Library of Congress Cataloging-in-Publication Data applied for.

Die Deutsche Bibliothek - CIP-Einheitsaufnahme

Multifield problems : state of the art / S.-M. Sändig ... (ed.).
- Berlin ; Heidelberg ; New York ; Barcelona ; Hong
Kong ; London ; Milan ; Paris ; Singapore ; Tokyo : Springer, 2000

ISBN 978-3-642-08693-9 ISBN 978-3-662-04015-7 (eBook)
DOI 10.1007/978-3-662-04015-7

Originally published by Springer-Verlag Berlin Heidelberg New York in 2000.

Softcover reprint of the hardcover 1st edition 2000

Typesetting: Camera-ready copy by the authors/editors
Cover design: de'blik, Berlin

Printed on acid-free paper SPIN 10768587 62/3144/tr 5 4 3 2 1 0

Preface

This book on the state of the art in "Multifield Problems" consists of selected articles based on a conference on this topic at the University of Stuttgart in 1999. The first two articles are contributions to the general modelling of multifield problems. S.S. Antman presents the important rôle of viscoelastic dissipation in the mathematical modelling of bifurcation analysis of nonlinear elasticity for large deformations. G.A. Maugin presents the basic theoretical foundations for the combination of three scales – the microscopic lattice of crystals, the mesoscopic thermomechanical model and the macroscopic continuum mechanics model for describing the propagation of phase transition fronts in terms of the Landau-Ginzburg theory and the modelling of nonlinear waves.

The other contributions are associated with five main areas of multifield modelling such as two and multiphase flows, the mechanics of materials in terms of multiscaling, the interaction of solids and fluids, efficient solution methods of the discrete equations including adaptivity, and the modelling of contact and fracture.

Within Chapter 1 on two- and multiphase flows we present three articles on the modelling of the sedimentation of particles: First, a measure-valued stochastic model which is able to justify macroscopic constitutive hindered settling models, and second, computer simulations of the sedimentation of particles in comparison with experiments, which also provide constitutive laws for the phenomenological theory of sedimentation-consolidation.This phenomenological theory, which assumes that the material behaviour of the mixture is modelled by constitutive laws for the hindered settling function and the effective solid stress, is topic of the third contribution on sedimentatin theory. The model arising from this theory is an initial-boundary value problem of parabolic-hyperbolic type, where a free surface in the solution separates the parabolic and hyperbolic regions. This free surface separates the hindered settling zone, where the equation is hyperbolic, from the compression zone, where it is parabolic. Another field of applications consists of multiphase flows through porous media. In this volume, we present a model which is based on Lattice-Boltzmann simulations and includes hysteresis effects. Here gas particle and gas-water flows are analyzed and simulated. The fifth contribution is on front tracking for the accurate simulation of interfaces between different fluids; a very interesting example shows the simulation of rising bubbles in fluids.

Chapter 2 is devoted to the mechanics of materials where homogenization and multiscales play a dominant rôle in the corresponding modelling, discretization and algorithmic implementation of complex inelastic response of solid materials. The mathematical foundation of these models is currently an intensive research field. Microstructures form if oscillating sequences infimize a given non-convex energy functional. Different forms of convexification allow macroscopic approximations and the microscopic quantities can be removed from relaxed models. Corresponding finite element approximations including a priori and a posteriori error bounds allow numerical computations which simulate, e.g., phase transitions in elastic solids. A very important rôle in the transition from micro to macroscales play homogenization techniques. Here the mathematical analysis as well as numerical homogenization strategies allow to model and to simulate numerically the behaviour of strongly heterogeneous elastic materials. Multigrid discretizations with local finite elements and patchwise homogenization provide efficient numerical simulation techniques for these rather complex materials. For the transition from the micro to macroscales one needs a reliable analysis of the mixture of the behaviour in multiphase models for alloys in order to get reliable models on the macroscopic scale. The two-scale micro-macro approach allows also to model the behaviour of materials with finite crystal plasticity of the grains for large strain problems.

Solid-fluid interaction is considered in Chapter 3. Here models for the interaction of fluid flow and mechanical deformation mechanics are of central interest, and we present two fundamentally different approaches: The classical coupling of fluid flow with moving, perhaps elastic and vibrating bodies; and on the other hand, modelling of interface conditions for the complex material behaviour of two-phase continua with fluid-solid interaction. For the first type of problems we present the modelling and numerical computation of the vibrational response of ships to the fluid forces generated by the ship's movement in the fluid. Here coupling of finite and boundary elements and the application of model adaption, homogenization and mesh refinement allows the numerical simulation of complete ship vibrations. Also coupling of existing computational fluid dynamics methods based on the Navier-Stokes equations and finite element structural dynamics can be used to compute e.g. the behaviour of a pendulum in a channel flow. Fluid-structure interaction like fluids sloshing in a tank, wind induced vibrations of a bridge and vibrating off-shore structures are rather complex and therefore complexity has to be taken into account when modelling the different fields as well as their interaction. Here a finite element Lagrangian-Eulerian formulation (ALE) leads to a three field model with moving mesh and a robust iterative substructuring scheme. For the second type of problems we present the derivation of interface conditions for the sliding between deformable porous media involving viscosity of the interstitial fluid. In a boundary layer, the viscous drag competes with the resistive drag between solid and fluid; a set of boundary conditions like solid to solid frictional fraction interaction is derived. Local concentration of plastic strains in frictional geomaterials yields the development of sheerbands which is simulated with a time and space adaptive finite element method. The complex material behaviour of viscoelastic two-phase models with fluid-solid interaction and intrinsic skeleton viscoelasticity leads to a macroscopic theory of porous media with an appropriate viscoelastic constitutive law. Here finite elements can be used for the simulation of flow-dependent consolidation processes.

Chapter 4 is devoted to efficient solvers and adaptivity. Due to the complexity of multifield problems, the construction of effective solution algorithms for the discretized equations is of extraordinary importance. Here we present efficient finite and boundary element methods. For multifield problems, domain decomposition methods provide a natural tool for solving complex problems and also for the application of hierarchical finite element analysis. For geometrical domain decomposition involving non-matching grids on the interfaces between the individual subdomains, the mortar methods based on an additional Lagrangian field on the interface boundaries provide an efficient method for relaxed interface linkage of adjacent subdomains. Here such methods are applied to the design of high power electrical devices as, e.g., in high-speed train engines. The corresponding electrothermal mechanical coupling with Joule selfheating is modelled by a combination of finite elements, mortars on the interfaces and curl-conforming edge-elements for the Maxwell equations. The corresponding mixed saddlepoint equations then are solved with an efficient preconditioned iterative solver. A posteriori error estimators drive the adaption of the grids. Geometric domain decomposition also underlies the finite element tearing and interconnecting method for solving elasticity problems for compressible elastic materials; and a very efficient preconditioned conjugate gradient method is obtained by using approximate Neumann subdomain solvers and Dirichlet preconditioners. Parallel finite element multigrid methods can be used to perform efficient elastoplastic simulations where the radial return algorithm is applied to viscoplasticity with nonlinear isotropic and kinematic hardening. Boundary element methods can very efficiently be used for nonlocal equations such as in the modelling of radiation heat transfer applied to the modelling of industrial exhaust pipes and catalytic converters which is needed in the automobile industry. Here a degenerate kernel approximation of the boundary integral operators in combination with adaptive compression reduces the originally fully populated matrices to a sparse representation in terms of low-rank matrices which allow an extremely fast matrix times vector multiplication. In combination with preconditioned conjugate gradient methods, boundary integral equations in this form can be solved with almost optimal efficiency. For the boundary integral equations, efficient and reliable error estimators are still an important task. Based on an appropriate approximation of the classical Neumann series for solving a Fredholm integration of the second kind for the error approximately, one obtains efficient residual-based error estimators that drive adaptive boundary element methods.

In Chapter 5 we consider contact and fracture. A permanent challenge in structural mechanics is the design of engineering multistructures in the presence of stress singularities, and in contact and fracture mechanics. For dynamic problems involving contact and Coulomb friction, sensitivity and optimal control is formulated in terms of the adjoint dynamical system. The dynamic evolution of contact problems for viscoelastic bodies without and with heat production from friction can be formulated in terms of an evolution based on variational inequalities. Here new existence and regularity results are obtained. The longitudinal wave propagation in conical rods can be formulated as a wave equation on a one-dimensional interval. D'Alembert's solution formula yields with the reflection at the boundaries recurrence relations and and explicit solution algorithm which can be used with computer algebra for computing the wave's propagation. The acoustic radiation of noise generated by the tire-road interaction can numerically be simulated by using e sequential finite element-boundary element formulation based on an Eulerian-Lagrangian model. Quasistatic contact problems with elastic plates can be treated by an augmented Lagrangian method. For these non-convex problems existence results and an iterative solution algorithm is presented. For mildy nonlinear elastic deformation of composites, where stress singularities develop at crossing points of interfaces between different materials, a regularity analysis based on the Mellin technique yields explicit asymptotic characterizations of the stress fields which in this case behave similar to corresponding solutions of linearized problems.

As these examples show, multifield problems need careful investigations with respect to the efficiency of the individual models and solution algorithms. The successful performance requires a close multi-disciplinary interaction of engineers from applied mechanics with mathematicians who work on the analysis and numerical solution of partial differential equations; hence, scientists from different disciplines have created this work on multifield problems.

The authors and editors express their sincere gratitude to the German Research Foundation DFG which supported this activity. Special thanks are due to Dr. Werner Kolbe who did all the technical work in preparing this final form. We also thank the Springer-Verlag for its patience and the opportunity to publish this book.

Stuttgart, March 2000 Wolfgang Wendland

Table of Contents

The Many Roles of Viscosity in Solid Mechanics
Stuart S. Antman 1

1. Introduction 1
2. A two-dimensional shearing problem 1
3. Self-sustained oscillations of a viscoelastic layer 4
4. Longitudinal motions 5
5. Longitudinal motions of incompressible rods 6
6. The motion of rods in space 8
7. Shocks and viscosity 9

References 10

Multiscale Approach to a Basic Problem of Materials Mechanics (Propagation of Phase-Transition Fronts)
Gérard A. Maugin 11

1. Introduction 11
2. Microscopic condensed-matter-physics approach: solitonics 12
3. Macroscopic engineering approach: singular surface and thermodynamic criterion 13
4. Mesoscopic applied-mathematics approach: structured front 16
5. Theoretical-physics approach: quasi-particle and transient motion .. 16
6. Numerics: from finite-differences to continuous cellular automata ... 17
   6.1 Finite-Difference Method 17
   6.2 Finite-Element Method 18
   6.3 Finite-Volume Method and cellular automaton 18
7. Conclusion 19

References 21

Section 1: Two and Multiphase Flows
Edited by H. Herrmann and S. Schwarzer 23

Analysing Particle Sedimentation in Fluids by Measure-Valued Stochastic Processes
Christian Hesse and Armin Dunz 25

1. Introduction 25
2. System analysis by measure-valued stochastic processes 27
3. Sedimentation in 1-d 30

References 33

Phenomenological Model of Sedimentation-Consolidation Processes
R. Bürger 34

1. Introduction 34
2. Phenomenological sedimentation-consolidation model 34
3. Analysis of the initial-boundary value problem 37
4. Comparison with experimental data 38
5. Conclusions and directions of future research 40

References 40

The Structure of Bidisperse Suspensions at Low Reynolds Numbers
K. Höfler and S. Schwarzer 42

1. Introduction 42
2. Numerical method 42
3. Bidisperse suspensions 44
4. An advection-diffusion model for polydisperse suspensions 46

References 49

Lattice Boltzmann Simulations of Complex Multiphase Flows
M. Krafczyk, P. Lehmann, O. Filippova, D. Hänel and U. Lantermann . 50

1. Lattice-Boltzmann simulation of gas-particle flow in complex geometry 50
   1.1 Introduction 50
   1.2 Method and results 51
2. Direct calculation of hysteresis by Lattice-Boltzmann simulations .. 54
   2.1 Introduction 54
   2.2 Hysteresis 54
   2.3 Method and results 55
   2.4 Discussion 56
   2.5 Acknowledgements 57

References 57

Interface Tracking in Multiphase Flows
A.-K. Tornberg and B. Engquist ... 58

1. Introduction ... 58
2. The multiphase flow equations ... 59
   2.1 Discretization ... 60
3. The segment projection method ... 62
4. Comparison of three methods for a buoyant bubble ... 63

References ... 65

Section 2: Mechanics of Materials and Multiscaling
Edited by C. Miehe ... 67

Finite Element Computation of Macroscopic Quantities in Nonconvex Minimisation Problems and Applications in Materials Science
Sören Bartels, Carsten Carstensen and Petr Plecháč ... 69

1. Introduction ... 69
2. Scalar double-well problem and its numerical solution ... 71
3. Numerical analysis of linearised phase transitions in elastic solids .. 76

References ... 79

Homogenization of an Initial-Boundary Value Problem Describing Evolving Microstructure
Hans-Dieter Alber ... 80

1. Introduction ... 80
2. A mathematical model with sharp phase interfaces ... 80
   2.1 Equations for the stress, displacement and internal variables ... 80
   2.2 Evolution equation for the phase interface, dissipation inequality 82
3. Homogenization of the equations for materials with evolving microstructure ... 85

References ... 87

Aspects of Homogenization Techniques and Multigrid Solving
C. G. Bayreuther, C. Miehe and J. Schröder ... 88

1. Introduction ... 88
2. The basic multigrid concept ... 88
3. A numerical homogenization procedure ... 90
4. Transfer operators for heterogeneous structures ... 91
   4.1 Smoothing adapted transfer operators ... 91
   4.2 Transfer operators associated with unit displacements ... 92
   4.3 Algebraical motivated transfer operators ... 92
   4.4 Homogenization based transfer operators ... 93

5. Comparative study of model problems 93
6. Conclusion 95
References 95

Estimates on the Mixture Function for Multiphase Problems in Elasticity
Alexander Mielke 96

1. Introduction 96
2. Definition of the mixture function 97
3. The $\mathbb{H}$-measure associated with χ 99
4. A lower estimate 100
5. Upper bounds 101
6. A special case 102
References 103

A Two–Scale Micro–Macro–Approach to Anisotropic Finite Plasticity of Polycrystals
C. Miehe and J. Schotte 104

1. Introduction 104
2. A class of mirco-macro-transitions at large strains 105
3. A computational model of finite crystal plasticity 108
4. Texture evolution in homogeneous compression test 110
References 111

Section 3: Solid–Fluid–Interaction
Edited by W. Ehlers 113

Boundary Conditions at the Viscous Sliding Interface of Incompressible Porous Deformable Media
G.A. Ateshian and X. Wang 115

1. Introduction 115
2. Formulation 116
   2.1 Kinematic conditions 117
   2.2 Interface tractions 118
3. Limiting cases 120
   3.1 Solid-solid interface 121
   3.2 Fluid-fluid interface 121
   3.3 Solid-fluid interface 121
   3.4 Biphasic-fluid interface 121
   3.5 Biphasic-solid interface 121
   3.6 Inviscid fluid phase 122
4. Governing equations 122
5. Discussion 123
References 123

Computational Experience from the Solution of Coupled Problems in Ship Dynamics
Frank Ihlenburg ... 125

1. Introduction ... 125
2. Physical assumptions and governing equations ... 125
3. Numerical method and computational implementation ... 127
4. Computational results ... 127
   4.1 Fluid-structure interaction ... 128
   4.2 Structural model adaptation ... 130
   4.3 Structural mesh refinement ... 132
5. Conclusions ... 132

References ... 134

On the Adaptive Computation of Shear Bands in Frictional Geomaterials
Wolfgang Ehlers and Peter Ellsiepen ... 135

1. Introduction ... 135
2. Governing equations ... 136
3. Spatial discretization and numerical tools ... 138
4. Numerical example ... 141

References ... 142

Intrinsic Viscoelasticity of Porous Materials
Wolfgang Ehlers and Bernd Markert ... 143

1. Introduction ... 143
2. Governing equations ... 144
3. Finite viscoelasticity law ... 145
4. Examples ... 147
5. Conclusions ... 149

References ... 150

Numerical Simulation of Fluids Interacting with Moving Rigid Bodies
S. Meynen and M. Schäfer ... 151

1. Introduction ... 151
2. Governing equations and solution methods ... 152
   2.1 Navier-Stokes equations for moving grids ... 152
   2.2 Linear rigid body dynamics ... 153
3. Coupling algorithm ... 154
4. Numerical example ... 155
5. Conclusions ... 157

References ... 158

Partitioned Analysis of Transient Nonlinear Fluid Structure Interaction Problems Including Free Surface Effects
W.A. Wall, D.P. Mok, J. Schmidt and E. Ramm 159

1. Introduction 159
2. CFD - Computational Fluid Dynamics 159
   2.1 Arbitrary Lagrangean Eulerian (ALE) formulation 160
   2.2 Stabilized finite element method 160
   2.3 Description of the free surface 161
3. CSD - Computational Structural Dynamics 161
4. CMD - Computational Mesh Dynamics 162
5. Partitioned staggered analysis approach 162
   5.1 Algorithmic setup 163
   5.2 A robust iterative substructuring scheme 164
6. Numerical examples 165
   6.1 Floating vertical cylinder in viscous fluid 165
   6.2 Cavity with oscillating top plate and flexible bottom 165

References 166

Section 4: Efficient Solvers and Adaptivity
Edited by O. Steinbach and C. Wieners 167

Domain Decomposition Methods in the Design of High Power Electronic Devices
R. Hoppe, Y. Iliash, and G. Mazurkevitch 169

1. Introduction 169
2. High power electronic devices and systems 170
   2.1 Electrothermomechanical couplings in IHV-Modules 170
   2.2 Minimization of parasitic effects in Converter Modules 172
3. The mortar element methods 173
   3.1 Mortar methods for Lagrangian finite elements 173
   3.2 Mortar methods for curl-conforming edge elements 176
4. Numerical results 178

References 181

Matrix Compression for the Radiation Heat Transfer in Exhaust Pipes
M. Bebendorf and S. Rjasanow 183

1. Introduction 183
2. Mathematical model 184
3. Approximation of the full dense matrices 187
4. Adaptive cross approximation 188
   4.1 Matrix partitioning 188
   4.2 Low-rank approximation 189

5. Numerical examples 190
References 191

An Iterative Substructuring Method with Lagrange Multipliers for Elasticity Problems Using Approximate Neumann Subdomain Solvers
A. Klawonn 193

1. Introduction 193
2. The FETI method 194
3. The new domain decomposition method with approximate subdomain solves 196
4. Numerical results 198
References 200

A New a Posteriori Error Estimator in Adaptive Direct Boundary Element Methods. The Neumann Problem
H. Schulz and O. Steinbach 201

1. Introduction 201
2. An a posteriori error estimator 203
3. Implementation 205
4. Numerical example 207
References 208

Efficient Elasto-Plastic Simulation
C. Wieners 209

1. Introduction 209
2. A general framework for quasi-static plasticity 210
3. Return mapping algorithms for incremental plasticity 211
4. Standard materials 212
5. An example with nonlinear hardening 213
6. The computation of the return parameter 214
7. A numerical experiment 215
8. Conclusion 216
References 216

Section 5: Contact and Fracture
Edited by W. L. Wendland 217

Sensitivity and Optimal Control in Contact Mechanics
G. Szefer 219

1. Introduction 219
2. Setting of the optimal control and sensitivity problem 219
3. Adjoint system 225

4. Conclusions 228
References 228

On the Treatment of Contact Problems in Elasto-Plasticity
M. Bach, C. Eck and M. Schulz 229

1. Introduction 229
2. Problem setting 230
3. Boundary element method 234
4. Numerical example 236
References 236

On Contact Problems for Linear and Nonlinear Elastic Plates: Existence of Solutions and Application of Augmented Lagrangian Method
W. Bielski, A. Gałka, and J.J. Telega 237

1. Geometrically linear plates 237
   1.1 Thin plates 237
   1.2 Reissner's plate model 239
2. Von Kármán plates 240
3. Augmented Lagrangians methods for a class of nonconvex problems 241
4. Example 243
References 245

Longitudinal Wave Propagation in Conical Rods Subject to Impacts
P. Eberhard, B. Hu and W. Schiehlen 246

1. Introduction 246
2. Symbolical computation for a conical rod 247
3. Experimental investigation of longitudinal waves 250
4. Conclusions 252
References 253

A Survey on Dynamic Contact Problems with Coulomb Friction
Ch. Eck and J. Jarušek 254

1. Introduction 254
2. The dynamic contact problem with Coulomb friction 254
3. Frictional contact problems with heat transfer 257
References 261

Numerical Simulation of Noise Radiation from Rolling Tires
L. Gaul, U. Nackenhorst and B. Nolte 262

1. Introduction 262

2. The simulation procedure 263
3. Numerical simulation of stationary rolling wheels 264
4. Numerical analysis of sound radiation 265
5. Example: A simple wheel 267
6. Conclusions 267
References 269

Stress Singularities in Composites
A.-M. Sändig 270

1. Introduction 270
2. Linear examples: Laplacian and Lamé system 271
3. Linear elliptic boundary transmission problems in Sobolev spaces with detached asymptotics 274
4. Semilinear problems 275
5. Continuity and Fréchet–differentiability of the semilinear operator 276
6. Asymptotic behaviour of the solution of the semilinear problem 276
References 277

List of Coeditors

Prof. Dr. W. Ehlers
Universität Stuttgart, Institut für Mechanik (Bauwesen),
Pfaffenwaldring 7, 70569 Stuttgart, Germany.

Prof. Dr. H. Herrmann
Universität Stuttgart, Institut für Computerphysik,
Pfaffenwaldring 27, 70569 Stuttgart, Germany.

Prof. Dr. C. Miehe
Universität Stuttgart, Institut für Mechanik (Bauwesen),
Pfaffenwaldring 7, 70569 Stuttgart, Germany.

Dr. S. Schwarzer
Universität Stuttgart, Institut für Computerphysik,
Pfaffenwaldring 27, 70569 Stuttgart, Germany.

Dr. O. Steinbach
Universität Stuttgart, Mathematisches Institut A,
Pfaffenwaldring 57, 70569 Stuttgart, Germany.

Dr. C. Wieners
Universität Heidelberg,
Interdisziplinäres Zentrum für Wissenschaftliches Rechnen,
Im Neuenheimer Feld 368, 69120 Heidelberg.

List of Contributors

Prof. Dr. H.-D. Alber
Technische Universität Darmstadt, Fachbereich Mathematik,
Schlossgartenstr. 7, 64289 Darmstadt, Germany.

Prof. Dr. S. S. Antman
University of Maryland, Department of Mathematics,
College Park, Maryland 20742-4015, USA.

Prof. Dr. G. A. Ateshian
Columbia University, Department of Mechanical Engineering,
500 West 120th Street, Mail Code 4703, New York, N.Y. 10027-6699, USA.

Dr. M. Bach
Universität Stuttgart, Mathematisches Institut A,
Pfaffenwaldring 57, 70569 Stuttgart, Germany.

M.Sc. S. Bartels
Christian-Albrechts-Universität Kiel, Lehrstuhl für Wissenschaftl. Rechnen,
Ludewig-Meyn-Str. 4, 24098 Kiel, Germany.

Dipl.–Math. M. Bebendorf
Universität des Saarlandes, Fachbereich 9 - Mathematik,
Postfach 151150, 66041 Saarbrücken, Germany.

Dipl.–Ing. K. Bayreuther
Universität Stuttgart, Institut für Mechanik (Bauwesen),
Pfaffenwaldring 7, 70569 Stuttgart, Germany.

Dr. W. R. Bielski
Polish Aademy of Sciences, Institute of Geophysics,
ul. Ksiecies Janusza 64,PL–01– 452 Warsaw, Poland.

Dr. R. Bürger
Universität Stuttgart, Mathematisches Institut A
Pfaffenwaldring 57, 70569 Stuttgart, Germany.

Prof. Dr. C. Carstensen
Christian-Albrechts-Universität Kiel, Lehrstuhl für Wissenschaftl. Rechnen,
Ludewig-Meyn-Str. 4, 24098 Kiel, Germany.

Dr. S. Diebels
Universität Stuttgart, Institut für Mechanik (Bauwesen),
Pfaffenwaldring 7, 70569 Stuttgart, Germany.

Dipl.–Math. A. Dunz
Universität Stuttgart, Mathematisches Institut A
Pfaffenwaldring 57, 70569 Stuttgart, Germany.

Dr. P. Eberhard
Universität Stuttgart, Institut B für Mechanik,
Pfaffenwaldring 9, 70569 Stuttgart, Germany.

Dr. C. Eck
Universität Erlangen - Nürnberg, Institut für Angewandte Mathematik,
Martensstr. 3, 91058 Erlangen, Germany.

Prof. Dr. W. Ehlers
Universität Stuttgart, Institut für Mechanik (Bauwesen),
Pfaffenwaldring 7, 70569 Stuttgart, Germany.

Dr. P. Ellsiepen
Universität Stuttgart, Institut für Mechanik (Bauwesen),
Pfaffenwaldring 9, 70569 Stuttgart, Germany.

Prof. Dr. B. Engquist
Royal Institute of Technology,
Department of Numerical Analysis and Computing Science,
S–10044 Stockholm, Sweden.

Dr. O. Filippova
Gerhard Mercator Universität Duisburg, Inst. f. Verbrennung & Gasdynamik,
47048 Duisburg, Germany.

Dr. A. Gałka
Polish Academy of Sciences, Institut of Fundamental Technological Research,
ul. Swietokrzyska 21, PL–00 – 049 Warszawa, Poland.

Prof. Dr. L. Gaul
Universität Stuttgart, Institut A für Mechanik,
Pfaffenwaldring 9, 70569 Stuttgart, Germany.

Prof. Dr. D. Hänel
Gerhard Mercator Universität Duisburg, Inst. f. Verbrennung & Gasdynamik,
47048 Duisburg, Germany.

Prof. Dr. C. Hesse
Universität Stuttgart, Mathematisches Institut A,
Pfaffenwaldring 57, 70569 Stuttgart, Germany.

Dr K. Höfler
Universität Stuttgart, Institut für Computerphysik,
Pfaffenwaldring 27, 70569 Stuttgart, Germany.

Prof. Dr. R.H.W. Hoppe
Universität Augsburg, Institut für Mathematik,
Universitätsstr. 14, 86159 Augsburg, Germany.

Dr. B. Hu
Universität Stuttgart, Institut B für Mechanik,
Pfaffenwaldring 9, 70569 Stuttgart, Germany.

Dr. F. Ihlenburg
Germanischer Lloyd,
Vorsetzen 32, 20459 Hamburg, Germany.

Dr. Y. Iliash
Universität Augsburg, Institut für Mathematik,
Universitätsstr. 14, 86159 Augsburg, Germany.

Dr. J. Jarušek
Acad. Sci. of Czech. Republic, Mathematical Insitute,
Zitná 25, CZ–11567 Prag, Czech Republic.

Dr. A. Klawonn
Westfälische Wilhelms - Universität,
Inst. f. Numerische und Instrumentelle Mathematik,
Einsteinstr. 82, 48149 Münster, Germany.

Dr. M. Krafczyk
TU München, LS Bauinformatik,
Arcisstr. 21, 80290 München, Germany.

U. Lantermann
Gerhard Mercator Universität Duisburg, Inst. f. Verbrennung & Gasdynamik,
47048 Duisburg, Germany.

P. Lehmann
ETH Zürich, Institute of Terrestrial Ecology,
Grabenstr. 3, CH–8952 Schlieren, Switzerland.

Dipl.–Ing. B. Markert
Universität Stuttgart, Institut für Mechanik (Bauwesen),
Pfaffenwaldring 7, 70569 Stuttgart, Germany.

Prof. Dr. G.Z. Maugin
Université Pierre et Marie Curie, Laboratoire de Modélisation en Mécanique,
4 place Jussieu (Case 162), F–75252 Paris, France.

Dr. G. Mazurkevitch
Universität Augsburg, Institut für Mathematik,
Universitätsstr. 14, 86159 Augsburg, Germany.

Dr. S. Meynen
TU Darmstadt, FB Numerische Berechnungsverfahren im Maschinenbau,
Petersenstr. 30, 64287 Darmstadt, Germany.

Prof. Dr. C. Miehe
Universität Stuttgart, Institut für Mechanik (Bauwesen),
Pfaffenwaldring 7, 70569 Stuttgart, Germany.

Prof. Dr. A. Mielke
Universität Stuttgart, Mathematisches Institut A,
Pfaffenwaldring 57, 70569 Stuttgart, Germany.

Dipl.–Ing. D. P. Mok
Universität Stuttgart, Institut für Baustatik,
Pfaffenwaldring 7, 70569 Stuttgart, Germany.

Dr. U. Nackenhorst
Universität der Bundeswehr, Institut für Mechanik,
Holstenhofweg 85, 22043 Hamburg, Germany.

Dr. B. Nolte
Universität der Bundeswehr, Professur für Mathematik,
Holstenhofweg 85, 22043 Hamburg, Germany.

Prof. Dr. P. Plecháč
Mathematical Sciences Department, University of Delaware,
Newark, DE 19716-2553, USA.

Prof. Dr. E. Ramm
Universität Stuttgart, Institut für Baustatik,
Pfaffenwaldring 7, 70569 Stuttgart, Germany.

Prof. Dr. S. Rjasanow
Universität des Saarlandes, Fachbereich 9 - Mathematik,
Postfach 151150, 66041 Saarbrücken, Germany.

Prof. Dr. A.M. Sändig
Universität Stuttgart, Mathematisches Institut A,
Pfaffenwaldring 57, 70569 Stuttgart, Germany.

Dipl.–Ing. J. Schmidt
Universität Stuttgart, Institut für Baustatik,
Pfaffenwaldring 7, 70569 Stuttgart, Germany.

Prof. Dr. M. Schäfer
TU Darmstadt, FB Numerische Berechnungsverfahren im Maschinenbau,
Petersenstr. 30, 64287 Darmstadt, Germany.

Prof. Dr. W. Schiehlen
Universität Stuttgart, Institut B für Mechanik,
Pfaffenwaldring 9, 70569 Stuttgart, Germany.

Dipl.–Ing. J. Schotte
Universität Stuttgart, Institut für Mechanik (Bauwesen),
Pfaffenwaldring 7, 70569 Stuttgart, Germany.

Dr. J. Schröder
Universität Stuttgart, Institut für Mechanik(Bauwesen),
Pfaffenwaldring 7, 70569 Stuttgart, Germany.

Dr. H. Schulz
Universität Stuttgart, Mathematisches Institut A,
Pfaffenwaldring 57, 70569 Stuttgart, Germany.

Dr. M. Schulz
Universität Stuttgart, Mathematisches Institut A,
Pfaffenwaldring 57, 70569 Stuttgart, Germany.

Dr. S. Schwarzer
Universität Stuttgart, Institut für Computerphysik,
Pfaffenwaldring 27, 70569 Stuttgart, Germany.

Dr. O. Steinbach
Universität Stuttgart, Mathematisches Institut A,
Pfaffenwaldring 57, 70569 Stuttgart, Germany.

Prof. Dr. G. Szefer
Cracow Universtity of Technology, Institute of Structurel Mechanics,
Warszawska 24, PL–31155 Krakow, Poland.

Prof. Dr. J.J. Telega
Polish Academy of Sciences, Institut of Fundamental Technological Research,
ul. Swietokrzyska 21, PL–00 – 049 Warszawa, Poland.

Dipl.–Ing. A.K. Tornberg
Royal Institute of Technology,
Department of Numerical Analysis and Computing Science,
S–10044 Stockholm, Sweden.

Dr. W. A. Wall
Universität Stuttgart, Institut für Baustatik,
Pfaffenwaldring 7, 70569 Stuttgart, Germany.

Dr. X. Wang
Lehman Brothers, Inc.,
New York, NY 10285, USA

Dr. C. Wieners
Universität Heidelberg,
Interdisziplinäres Zentrum für Wissenschaftliches Rechnen,
Im Neuenheimer Feld 368, 69120 Heidelberg.

The Many Roles of Viscosity in Solid Mechanics

Stuart S. Antman

Department of Mathematics, Institute for Physical Science and Technology, and Institute for Systems Research, University of Maryland, College Park, MD 20742-4015, USA

Summary. This paper describes how the strong dissipative mechanism in viscoelastic solids of strain-rate type plays a crucial role in a wide variety of problems.

1. Introduction

Nonlinear elasticity is conceptually, although not mathematically, the simplest theory for the large deformations of solids. Because it lacks any dissipative mechanism whatsoever, many concrete problems for it are not mathematically well-posed, a difficulty that is often remedied by the introduction of the strong dissipative mechanism from viscoelasticity of strain-rate type. (The use of this dissipative mechanism does not preclude the use of other weaker mechanisms accounting for memory effects, although we do not discuss these in any detail.) In this paper we examine the critical role of viscous dissipation in bounding strains and preventing total compressions, in justifying the Hopf Bifurcation Theorem, in influencing the qualitative nature of Hopf bifurcations, in justifying asymptotic expansions, in showing that various initial-boundary-value problems are well-set, and in designing invariant numerical schemes for shocks. We also discuss the limited role of viscous dissipation for problems for incompressible rods, and we discuss its ineffectiveness for determining shock structure in certain problems.

2. A two-dimensional shearing problem

Let $\{\mathbf{i}_1, \mathbf{i}_2, \mathbf{k}\}$ be a fixed right-handed orthonormal basis for Euclidean 3-space. We identify a typical material point in a body by its position $x_1\mathbf{i}_1 + x_2\mathbf{i}_2 + s\mathbf{k}$ in a reference configuration of the body. We suppose that the body is homogeneous and incompressible, and covers the whole (x_1, x_2)-plane, so that $-\infty < x_1 < \infty$, $-\infty < x_2 < \infty$. We study shearing motions in which the position $\mathbf{p}(x_1, x_2, s, t)$ of the material point $x_1\mathbf{i}_1 + x_2\mathbf{i}_2 + s\mathbf{k}$ at time t has the form

$$\mathbf{p}(x_1, x_2, s, t) = [x_1 + u_1(s,t)]\mathbf{i}_1 + [x_2 + u_2(s,t)]\mathbf{i}_2 + s\mathbf{k}. \tag{2.1}$$

If the tractions on any plane $s = \text{const}$ (possibly at infinity) are independent of x_1 and x_2 and if there are no body forces, then the equations of motion for $\mathbf{u} := u_1\mathbf{i}_1 + u_2\mathbf{i}_2$ reduce to

$$\rho\mathbf{u}_{tt} = \mathbf{n}_s, \tag{2.2}$$

where ρ is the density and where $\mathbf{n} = n_1\mathbf{i}_1 + n_2\mathbf{i}_2$ with n_1 and n_2 the shear components of the the first Piola-Kirchhoff stress tensor. For a viscoelastic material of strain-rate type, $\mathbf{n}$ has the form

$$\mathbf{n}(s,t) = \frac{\partial W}{\partial \mathbf{u}_s}(\mathbf{u}_s(s,t)) + \mathbf{n}^{\mathrm{D}}(\mathbf{u}_s(s,t), \mathbf{u}_{st}(s,t)), \tag{2.3}$$

where $\mathbf{u}_s \mapsto W(\mathbf{u}_s)$ is the stored-energy function for the equilibrium response and where the dissipative part $\mathbf{n}^{\mathrm{D}}$ of this stress satisfies $\mathbf{n}^{\mathrm{D}}(\mathbf{u}_s, \mathbf{0}) = \mathbf{0}$. For an isotropic material, (2.3) has the form

$$\mathbf{n} = \mu(\boldsymbol{\eta})\mathbf{u}_s + \nu(\boldsymbol{\eta})\mathbf{u}_{st} \tag{2.4}$$

where $\boldsymbol{\eta} := (\eta_0, \eta_1, \eta_2)$ is the collection of invariants

$$\eta_0 := |\mathbf{u}_s|^2, \quad \eta_1 := 2\mathbf{u}_s \cdot \mathbf{u}_{st}, \quad \eta_2 := |\mathbf{u}_{st}|^2.$$

If s ranges over the unit interval $[0,1]$, then the energy equation for (2.2), (2.3) is

$$\int_0^1 [\tfrac{1}{2}\rho|\mathbf{u}_t|^2 + W(\mathbf{u}_s)]\, ds + \textstyle\int_0^t \int_0^1 \mathbf{n}^{\mathrm{D}} \cdot \mathbf{u}_t \, ds\, dt = \text{const}. \tag{2.5}$$

We assume that there is a positive constant c such that

$$W(\mathbf{u}_s) \geq c\,|\mathbf{u}_s|. \tag{2.6}$$

We require that our material be uniformly dissipative: There is a positive constant c such that

$$\mathbf{a} \cdot \frac{\partial \mathbf{n}^{\mathrm{D}}}{\partial \mathbf{u}_{st}}(\mathbf{u}_s, \mathbf{u}_{st}) \cdot \mathbf{a} \geq c\,|\mathbf{a}|^2 \tag{2.7}$$

for all $\mathbf{u}_s, \mathbf{u}_{st}$ and for all $\mathbf{a} \in \mathbb{E}^2$. Thus the double integral in (2.5) is non-negative. Under the mild assumption that $W(\mathbf{u}_s) \to \infty$ as $|\mathbf{u}_s| \to \infty$, the energy equation merely implies that for any fixed time, $|\mathbf{u}_s|$ is finite for almost all s. It is remarkable that the far stronger and far more useful statement that $|\mathbf{u}_s|$ is bounded on any finite interval of time is an easy consequence of an assumption that the viscosity is mildly strain-dependent (as in the constitutive equation for a Navier-Stokes fluid in the material formulation): There is a continuously differentiable function ψ with $\psi(y) \to \infty$ as $y \to \infty$ and there are non-negative numbers c and C such that

$$\frac{\mathbf{n}^{\mathrm{D}}(\mathbf{u}_s, \mathbf{u}_{st}) \cdot \mathbf{u}_s}{|\mathbf{u}_s|} \geq \psi'(|\mathbf{u}_s|)\frac{\mathbf{u}_s \cdot \mathbf{u}_{st}}{|\mathbf{u}_s|} - C\psi(|\mathbf{u}_s|) + c\,|\mathbf{u}_s||\mathbf{u}_{st}| \tag{2.8}$$

for all $\mathbf{u}_s, \mathbf{u}_{st}$. This inequality is inspired by (2.4). For mathematical simplicity, let us take $C = 0$ in (2.8).

In the ensuing analysis, based on [2], we assume that all the functions that are exhibited are continuous. (This analysis gives us bounds telling us where to look for solutions. One then has to show that solutions can be found there and that these solutions are regular.) We set $\kappa := |\mathbf{u}_s|$ and write

$$\mathbf{u}_s = \kappa\, \mathbf{e}_1, \tag{2.9}$$

$$\mathbf{e}_1 := \cos\theta\, \mathbf{i}_1 + \sin\theta\, \mathbf{i}_2, \quad \mathbf{e}_2 := -\sin\theta\, \mathbf{i}_1 + \cos\theta\, \mathbf{i}_2. \tag{2.10}$$

Thus

$$\partial_t\, \mathbf{e}_1 = \theta_t\, \mathbf{e}_2, \quad \mathbf{u}_{st} = \kappa_t\, \mathbf{e}_1 + \kappa\, \theta_t\, \mathbf{e}_2. \tag{2.11}$$

For $\nu = 1, 2$, we define

$$g_\nu(s,t) = \mathbf{e}_\nu(s,t) \cdot \int_s^1 \rho \mathbf{u}_t(\xi,t)\, d\xi. \tag{2.12}$$

The integration of (2.2) yields

$$\mathbf{e}_1(s,t) \cdot \mathbf{n}(s,t) = \mathbf{e}_1(s,t) \cdot \mathbf{n}(1,t) - \partial_t g_1(s,t) + \theta_t(s,t)\, g_2(s,t). \tag{2.13}$$

Throughout this paper we let Γ, sometimes bearing indices, denote any continuous function that we can compute. Its meaning may change in each appearance. If $\mathbf{n}(1,t)$ is a prescribed continuous function, then (2.8) and (2.13) imply that there is a Γ such that

$$\psi'(\kappa)\, \kappa_t \le \Gamma - \partial_t g_1 + |\theta_t\, g_2| - c\,\kappa\, |\mathbf{u}_{st}|. \tag{2.14}$$

From (2.5) it follows that there are functions Γ_ν such that

$$|g_\nu(s,t)| \le \Gamma_\nu(t), \quad \nu = 1, 2. \tag{2.15}$$

Since $|\mathbf{u}_{st}| = \sqrt{\kappa_t^2 + \kappa^2\theta_t^2}$, we obtain

$$|\theta_t(s,t)\, g_2(s,t)| - c\,\kappa(s,t)\, |\mathbf{u}_{st}(s,t)| \le |\theta_t(s,t)|\, [\Gamma_2(t) - c\,\kappa(s,t)^2]. \tag{2.16}$$

Thus, wherever $\kappa(s,t)^2 \ge \Gamma_2(t)/c$, inequality (2.14) reduces to

$$\psi'(\kappa)\, \kappa_t \le \Gamma - \partial_t g_1. \tag{2.17}$$

Let $\kappa(\cdot, 0)$ be bounded and let Γ_3 be bounded below by $\sqrt{\Gamma_2/c} + \max_s \kappa(s,0)$. We now show that κ is bounded by a continuous function of t. We suppose that there is a ξ in $[0,1]$ and a $\tau_2 > 0$ such that $\kappa(\xi,\tau_2) > \Gamma_3(\tau_2)$. (Were there no such (ξ,τ_2), there would be nothing to prove.) Since κ is continuous and since $\kappa(\xi,0) < \Gamma_3(\tau_2)$, there exists a last time $\tau_1 < \tau_2$ at which $\kappa(\xi,\tau_1) = \Gamma_3(\tau_2)$. We integrate (2.17) for $s = \xi$ from τ_1 to τ_2 and use (2.15) to obtain

$$\psi(\kappa(\xi,\tau_2)) - \psi(\Gamma_3(\tau_1)) \le \Gamma(\tau_2) - g_1(\xi,\tau_2) + g_1(\xi,\tau_1) \le \Gamma(\tau_2). \tag{2.18}$$

Thus, in general,

$$|\mathbf{u}_s| \leq \Gamma. \tag{2.19}$$

A generalization of this argument handles the analogous problem for the large deformation of a visco*plastic* material in which the memory is introduced by means of the evolution of an internal variable [2]. Perhaps the most important application of this method is to longitudinal motions in which the presence of a viscosity that becomes infinite at total compression prevents the total compression everywhere [8] (whereas the unboundedness of energy at a total compression falls to deliver anything so strong). Another application is to the deformation in space of rods that can suffer flexure, torsion, extension, and shear. Here, a strong viscosity prevents cross sections from intersecting within the rod [9]. This problem is discussed in Sect. 6.

One can argue, as is traditional in mechanics, that bodies are not observed to undergo deformations as severe as total compressions, and therefore models need not account for them. But dynamical problems in continuum mechanics can have badly behaved solutions. By accounting for total compressions in the constitutive equations, we can show (for problems with just one spatial variable) that they cannot occur. By ignoring total compressions in the constitutive equations, we could have equations that exhibit not only total compressions, but also changes of orientation, a phenomenon that might be very hard to detect in beautiful computer printouts.

3. Self-sustained oscillations of a viscoelastic layer

We now consider special shearing motions of the form (2.1) in which $u_2 = 0$. Thus the equation for $u := u_1$ has the form

$$\rho u_{tt} = n(u_s, u_{st})_s, \quad 0 < s < 1. \tag{3.1}$$

We assume that the body is fixed to the surface $s = 0$ and that the surface $s = 1$ is subjected to a stick-slip frictional force. Then

$$u(0,t) = 0, \quad n(u_s(1,t), u_{st}(1,t)) = f(c - u_t(1,t)). \tag{3.2}$$

We assume that f is an odd function, continuously differentiable on $(0,\infty)$, with $f(v) > 0$ for $v > 0$, and that there is a $v_0 > 0$ such that $f'(v)(v - v_0) > 0$ for $v > 0$ and $v \neq v_0$.

There is a trivial steady-state solution u° satisfying

$$n(u_s^\circ(s), 0) = f(c). \tag{3.3}$$

The linearization of this nonlinear problem about this solution has the form

$$\rho v_{tt} = A(c)v_{ss} + B(c)v_{sst}, \tag{3.4}$$

$$v(0,t) = 0, \quad A(c)v_s(1,t) + B(c)v_{st}(1,t) = -f'(c)v_t(1,t). \tag{3.5}$$

Proceeding formally, we seek solutions of this linear system in the form

$$v(x,t) = z(x)e^{\lambda t}, \tag{3.6}$$

obtaining the eigenvalue problem

$$\lambda^2 \rho z = [A(c) + \lambda B(c)] z_{ss}, \tag{3.7}$$

$$z(0) = 0, \quad [A(c) + \lambda B(c)] z_s(1) = -\lambda f'(c) z(1). \tag{3.8}$$

The determination of how the eigenvalues depend on the control parameter is tricky [3]. When the viscosity is large, a single complex conjugate pair of eigenvalues, corresponding to a fundamental oscillatory mode, cross the imaginary axis as c is reduced to 0 from a starting value above v_0, and stay thereafter in the right half plane. All other eigenvalues stay in the left half plane. If the viscosity is small, then the eigenvalues for the fundamental mode have the same behavior, but several other pairs of eigenvalues cross the imaginary axis from left to right and then cross back again from right to left, staying thereafter in the left half plane. That each such crossing is accompanied by a Hopf bifurcation leading to a periodic solution of the nonlinear governing equations is ensured by a new Hopf Bifurcation Theorem for fully nonlinear parabolic-hyperbolic partial differential equations subject to nonlinear boundary conditions. The requisite proof [14] relies on the presence of strong viscosity. Thus, when the viscosity is small, solutions of initial-value problems could exhibit very complicated behavior.

4. Longitudinal motions

Purely longitudinal motions of a straight viscoelastic rod of strain-rate type are also governed by (3.1), with $u(s,t)$ now representing the position at time t along an axis of the material point with reference position s. The requirement that the local ratio of deformed to reference length at s never be reduced to zero is that $u_s(s,t) > 0$. This condition restricts the domain of the constitutive function n. That u_s has a positive lower bound on any finite interval of time, proved by the methods of Sect. 2, is crucial in proving that initial-boundary-value problems have regular solutions for all time.

Here we first study the motion of such a rod of scaled length 1, when its density per unit reference length, denoted $\varepsilon\rho A$ is small, when its end $s = 0$ is fixed, and when its other end $s = 1$ carries a particle of mass m. (Thus the small parameter ε can be regarded as the ratio of the mass of the rod to that of the particle.) The governing equations are

$$\varepsilon(\rho A)(s) u_{tt} = n(u_s, u_{st}, s)_s, \tag{4.1}$$

$$u(0,t) = 0, \quad m u_{tt}(1,t) + n(u_s(1,t), u_{st}(1,t), 1) = 0. \tag{4.2}$$

Here we allow the rod to be non-uniform. For simplicity of exposition, we assume that there are no external loads applied to the rod or to the particle. Except for the s derivatives, the second boundary condition looks like the equation of motion for a mass on a massless spring, which is taught in every elementary course in particle mechanics. To make this boundary condition into the standard ordinary differential equation, we would have to strip off the s-derivatives, in the limit as $\varepsilon \to 0$, taking advantage of the reduced problem ($\varepsilon = 0$). But this reduced problem is equivalent to $n(u_s(s,t), u_{st}(s,t), s) = n(u_s(1,t), u_{st}(1,t), 1)$, which is a family of ordinary differential equations for u_s parametrized by s. An analysis of this equation [1] shows that for $\varepsilon = 0$, the motion of the particle is typically not governed by a standard ordinary differential equation for the motion of a particle on a massless spring, but rather by an equation with memory. (Wilber [19] has proved that forced versions of the reduced equations for a *uniform* rod have a global attractor that is also the attractor for an associated ordinary differential equation for a particle.)

To show that this reduced problem with its surprising memory has mathematical significance, we must analyze its role as an approximation to the full initial-boundary-value problem. The full problem is analogous to initial-value problems for singularly perturbed systems of ordinary differential equations, which are treated in detail by [17, 18] inter al. Such work on ordinary differential equations suggests seeking solutions in the form of a regular and an initial-layer expansion in the parameter ε:

$$u(s,t) = \sum_{j=0}^{k} u_j(s,t)\frac{\varepsilon^j}{j!} + \sum_{j=0}^{k} U_j(s,t/\varepsilon)\frac{\varepsilon^j}{j!} + O(\varepsilon^{k+1}). \tag{4.3}$$

A rigorous justification of this expansion consists in showing that the error between the exact solution (the existence of which follows from [8]) and the two sums in (4.3) is indeed of order ε^{k+1}. For this purpose, many of the methods for treating ordinary differential equations are of little avail for our quasilinear partial differential equation. A key step in the justification is a demonstration that there are positive constants C and μ such that many derivatives of the U_j are bounded in absolute value by $Ce^{-\mu t/\varepsilon}$. The uniform viscosity plays a crucial role in this analysis [20], which is quite delicate.

The reduced problem for the motion of rods in space carrying rigid bodies at their ends has solutions with a very rich structure [6]. There is at yet no justification of the asymptotic status of such solutions.

5. Longitudinal motions of incompressible rods

We give u the same interpretation as in the last section. The simplest model for the purely longitudinal motions of an incompressible viscoelastic rod, in which the thickness is proportional to $1/u_s$, is governed by an equation of motion of the form [7]

$$\begin{aligned} -\partial_{ss} n_{(2)}(u_s, u_{ss}, u_{st}, u_{sst}) + \partial_s n_{(1)}(u_s, u_{ss}, u_{st}, u_{sst}) \\ = \rho A u_{tt} + \rho J \partial_s [2 u_s^{-5} {u_{st}}^2 - u_s^{-4} u_{stt}]. \end{aligned} \tag{5.1}$$

where (with a slight change of notation) ρA and ρJ are the mass density and the mass moment of inertia of each cross section per unit reference length, and where the constitutive equations are

$$n_{(1)} = \frac{\partial W}{\partial u_s}(u_s, u_{ss}) + n^{\mathrm{D}}_{(1)}(u_s, u_{ss}, u_{st}, u_{sst}), \tag{5.2}$$

$$n_{(2)} = \frac{\partial W}{\partial u_{ss}}(u_s, u_{ss}) + n^{\mathrm{D}}_{(2)}(u_s, u_{ss}, u_{st}, u_{sst}), \tag{5.3}$$

subject to the very mild dissipativity condition $n^{\mathrm{D}}_{(1)} u_{st} + n^{\mathrm{D}}_{(2)} u_{sst} \geq 0$.

In contrast to compressible bodies, here a sufficiently strong penalization of the energy prevents total compression pointwise: The energy inequality for initial-boundary-value problems with powerless boundary conditions is

$$\int_0^1 \left[W + \frac{\rho A}{2} {u_t}^2 + \frac{\rho J}{2} \frac{{u_{st}}^2}{{u_s}^4} \right] ds < \Gamma(t). \tag{5.4}$$

Suppose that

$$W \geq \text{const} \left[u_s + \frac{1}{u_s} + \frac{|u_{ss}|}{{u_s}^2} \right]. \tag{5.5}$$

Then

$$\frac{1}{u_s(s,t)} = \frac{1}{u_s(\xi,t)} - \int_\xi^s \frac{u_{ss}(\sigma,t)}{u_s(\sigma,t)^2}\, d\sigma \leq \frac{1}{u_s(\xi,t)} + \Gamma(t). \tag{5.6}$$

We integrate this inequality with respect to ξ from 0 to 1 to get the desired bound:

$$\frac{1}{u_s(s,t)} \leq \Gamma(t). \tag{5.7}$$

To get a pointwise upper bound on u_s without using the viscosity, we strengthen (5.5): There are numbers $p, q > 1$ with $1/p + 1/q < 1$ such that

$$W \geq \text{const} \left[{u_s}^q + \frac{1}{u_s} + \left| \frac{u_{ss}}{{u_s}^2} \right|^p \right]. \tag{5.8}$$

We write $q = (1+\alpha)p/(p-1)$ where α is a positive number. Then the Fundamental Theorem of Calculus and the Hölder inequality imply that

$$\begin{aligned} u_s(s,t)^\alpha &= u_s(\xi,t)^\alpha + \alpha \int_\xi^s u_s(\sigma,t)^{\alpha+1} \frac{u_{ss}(\sigma,t)}{u_s(\sigma,t)^2}\, d\sigma \\ &\leq u_s(\xi,t)^\alpha + \alpha \left[\int_\xi^s u_s(\sigma,t)^q d\sigma \right]^{\frac{p-1}{p}} \left[\int_\xi^s \left| \frac{u_{ss}(\sigma,t)}{u_s(\sigma,t)^2} \right|^p d\sigma \right]^{\frac{1}{p}} \\ &\leq u_s(\xi,t)^\alpha + \Gamma(t). \end{aligned} \tag{5.9}$$

We integrate this inequality with respect to ξ from 0 to 1 and again apply the Hölder inequality to get

$$u_s(s,t) \leq \Gamma(t). \tag{5.10}$$

These regularity results, which do not require the presence of viscosity, are no doubt related to that of [13].

When (5.10) holds, (5.4) yields $\int_0^1 {u_{st}}^2\, ds < \Gamma(t)$, which is very useful for further analysis. Note that this inequality relies on the positivity of the small number ρJ (just as the analysis of [10] depends on rotatory inertia). Saxton [16] treated the more difficult and less natural problem in which W does not depend on u_{ss}. The results in this section are new. To my knowledge, there is no analysis of flexural motions for incompressible rods.

6. The motion of rods in space

We define the configuration at time t of a rod that can bend, twist, stretch, and shear in 3-space by a function $s \mapsto \mathbf{r}(s,t)$ and by an orthonormal pair of functions $s \mapsto \mathbf{d}_1(s,t), \mathbf{d}_2(s,t)$. The function $\mathbf{r}(\cdot,t)$ is interpreted as the configuration at time t of some material curve (e.g., the curve of centroids in the reference configuration), and $\mathbf{d}_1(s,t), \mathbf{d}_2(s,t)$ are interpreted as characterizing the configuration at time t of the material cross section at s. We define $\mathbf{d}_3 := \mathbf{d}_1 \times \mathbf{d}_2$. Since the $\mathbf{d}_k$ are orthonormal, there are vectors $\mathbf{u}$ and $\mathbf{w}$ such that

$$\partial_s \mathbf{d}_k = \mathbf{u} \times \mathbf{d}_k, \qquad \partial_t \mathbf{d}_k = \mathbf{w} \times \mathbf{d}_k. \tag{6.1}$$

The functions $u_k := \mathbf{u} \cdot \mathbf{d}_k$ and $v_k := \mathbf{r}_s \cdot \mathbf{d}_k$ are the *strains* for our theory. An inequality relating v_3, u_1, u_2, which depends on the cross-sectional shape, ensures that material cross sections do not intersect. It is handled by the methods of Sect. 2.

Let $\mathbf{n}(s,t)$ and $\mathbf{m}(s,t)$ denote the contact force and contact couple acting across section s at time t. The equations of motion for zero body force and couple have the form

$$\rho A \mathbf{r}_{tt} = \mathbf{n}_s, \tag{6.2}$$

$$\rho J_{pq} \partial_t (w_q \mathbf{d}_p) \equiv \rho J_{pq} [(\partial_t w_q) \mathbf{d}_p + w_q \mathbf{w} \times \mathbf{d}_p] = \mathbf{m}_s + \mathbf{r}_s \times \mathbf{n}, \tag{6.3}$$

where ρJ_{pq} are the components of a positive-definite matrix determined by the mass moments of inertia of the cross sections, $w_q := \mathbf{w} \cdot \mathbf{d}_q$, and we use the summation convention.

The substitution of invariant constitutive equations for viscoelastic rods of strain-rate type into (6.2), (6.3) converts these equations into an eighteenth-order quasilinear system of parabolic-hyperbolic equations. Here we encounter new difficulties not associated with the size of the system: Eq (6.3) has two nonlinearities: $\mathbf{r}_s \times \mathbf{n}$ and the acceleration term. In analyzing these equations we have to take their time derivative to derive a sort of energy equation for higher derivatives. In doing so, we obtain a cubic nonlinearity in $\mathbf{w}$ in the acceleration terms, and similar complications in the derivative of $\mathbf{r}_s \times \mathbf{n}$. It is a pleasant surprise that the uniform dissipativity is able to dominate these cubic nonlinearities and thereby support a proof that initial-boundary-value problems for these equations have regular solutions that exist for all time [9].

7. Shocks and viscosity

If the constitutive equations giving the components of $\mathbf{n}$ and $\mathbf{m}$ with respect to the basis $\{\mathbf{d}_k\}$ depend only on the v_k and u_k and form a monotone mapping of these strains, then (6.2), (6.3) becomes a quasilinear hyperbolic system susceptible to shocks. All the standard numerical methods for capturing shocks employ some numerical dissipation. The most transparent method is that of Lax and Friedrichs [15]. One version of it consists in writing (6.2), (6.3) as a system of first-order equations, replacing each equation in the system of the form $f_t = g_s + h$ with $f_t = g_s + h + \alpha f_{ss}$ where α is a small positive number, and then treating the resulting system by difference methods. The final system, which has a sort of viscous dissipation not only in the momentum equations, but also in the compatibility equations, is not invariant under rigid motions, and appears to give rise to serious numerical errors for rapidly rotating bodies. It is a straightforward exercise to modify the momentum equations by making the dissipation have a properly invariant strain-rate form, but the treatment of the compatibility equations is trickier. The basic idea is to reconstitute the Lax-Friedrichs equations as a system of momentum equations with each equation having two time derivatives. The invariance criterion is that these momentum equations correspond to materials with invariant constitutive equations. This process (which is not unique) leads to compatibility equations of parabolic-hyperbolic type, rather than the purely parabolic type of the standard Lax-Friedrichs scheme, and leads to momentum equations in which there are four s-derivatives, corresponding to materials of strain-gradient type [4].

Shock structure can often be determined by travelling-wave solutions of dissipative versions of the underlying hyperbolic equations. The travelling wave should represent a unique trajectory connecting singular points in its phase space that correspond to states on each side of the shock satisfying the Rankine-Hugoniot relations. It is shown in [5] that travelling waves for (2.2), (2.4) have trajectories joining singular points that typically are not unique. It is therefore tempting to augment the viscous dissipation with dissipation of the strain-gradient type, which arises both in invariant versions of the Lax-Friedrichs scheme and in the models for incompressible rods of Sect. 5. The corresponding travelling waves are far more difficult to analyze, however, and the general theory is beset by subtle difficulties [11, 12].

Acknowledgement. This article was prepared while I was a visitor at the Max-Planck Institute for Mathematics in the Sciences, Leipzig. The research leading to it was supported in part by an NSF Grant and by an ARO-MURI Grant to the Center for Dynamics and Control of Smart Structures.

References

1. S. S. Antman: The paradoxical asymptotic status of massless springs, *SIAM J. Appl. Math.* **48** (1988), 1319–1334.
2. S. S. Antman and F. Klaus: The shearing of nonlinearly viscoplastic slabs, *Nonlinear Problems in Applied Mathematics*, edited by T. Angell, L. P. Cook, R. Kleinman, and W. E. Olmstead, SIAM, 1996, pp. 20–29.
3. S. S. Antman and H. Koch: Self-sustained oscillations of nonlinearly viscoelastic layers, *SIAM J. Appl. Math.*, to appear.
4. S. S. Antman and J.-G. Liu: Errors in the numerical treatment of hyperbolic conservation laws caused by lack of invariance, in preparation.
5. S. S. Antman and R. Malek-Madani: Travelling waves in nonlinearly viscoelastic media and shock structure in elastic media, *Quart. Appl. Math.* **46** (1988) 77–93.
6. S. S. Antman, R. S. Marlow and C. P. Vlahacos: The complicated dynamics of heavy rigid bodies attached to light deformable rods, *Quart. Appl. Math.* **56** (1998), 431–460.
7. S. S. Antman and F. Schuricht: Incompressibility in rod and shell theories, *Math. Modelling Num. Anal.*, **33** (1999), 289–304.
8. S. S. Antman and T. I. Seidman: Quasilinear hyperbolic-parabolic equations of nonlinear viscoelasticity, *J. Diff. Eqs.* **124** (1996), 132–185.
9. S. S. Antman and T. I. Seidman: Parabolic-hyperbolic systems governing the spatial motion of nonlinearly viscoelastic rods, in preparation.
10. R. E. Caflisch and J. H. Maddocks: Nonlinear dynamical theory of the elastica, *Proc. Roy. Soc. Edin.* **99A**, 1–23.
11. H. Freistühler: Dynamic stability and vanishing viscosity: A case study of a non-strictly hyperbolic system, *Comm. Pure Appl. Math.* **45** (1992), 561–582.
12. H. Freistühler and P. Szmolyan: Existence and bifurcation of viscous profiles for all intermediate magnetohydrodynamic shock waves, SIAM J. Math. Anal. **26** (1995), 112–128.
13. S. Klainerman: The null condition and global existence to nonlinear wave equations, *Lectures in Applied Mathematics*, Vol. 23, Part 1, edited by B. Nicolaenko, D. D. Holm, and J. M. Hyman, 1986, pp. 293–326.
14. H. Koch and S. S. Antman: Stability and Hopf bifurcation for fully nonlinear parabolic-hyperbolic equations, *SIAM J. Math. Anal.*, to appear.
15. R. J. LeVeque: *Numerical Methods for Conservation Laws*, Birkhäuser, 1992.
16. R. Saxton: Existence of solutions for a finite nonlinearly elastic rod, *J. Math. Anal. Appl.* **105** (1985) 59–75.
17. D. R. Smith: *Singular Perturbation Theory*, Cambridge Univ. Press, 1985.
18. A. B. Vasil'eva, V. F. Butuzov, and L. Kalachev: *The Boundary Function Method for Singular Perturbation Problems*, SIAM, 1995.
19. J. P. Wilber: *Global Attractors for a Degenerate Partial Differential Equation from Nonlinear Viscoelasticity*, Doctoral dissertation, Univ. Maryland, 1999.
20. S.-C. Yip, S. S. Antman, and M. Wiegner: The motion of a particle on a light viscoelastic bar: Asymptotic analysis of its quasilinear parabolic-hyperbolic equation, in preparation.

Multiscale Approach to a Basic Problem of Materials Mechanics (Propagation of Phase-Transition Fronts)

Gérard A. Maugin

Laboratoire de Modélisation en Mécanique, UMR 7607
Université Pierre et Marie Curie, Case 162,
4 place Jussieu, 75252 Paris cedex 05, France
e-mail: gam@ccr.jussieu.fr

Summary. Various routes to the understanding of the propagation of phase-transition fronts in crystalline substances are examined in the light of recent works. These include lattice dynamics, meso-scopic considerations, and the fully macroscopic thermomechanical approach on the material manifold that combines an engineering interest with an invariance-theoretical viewpoint. Numerical approaches (FDM, FEM, FVM, continuous-cellular automata) corresponding to these different levels of apprehending a single phenomenon are also discussed.

1. Introduction

A full understanding of the phenomenon of the propagation of phase-transition fronts in deformable crystals – metals, alloys – is one of the essential problems of contemporary materials science and mechanics at both theoretical and application levels. This unique problem can be examined at three different scales (cf. Figure 1): (i) *a microscopic scale* (lattice dynamics) in the absence of thermodynamical irreversibility, (ii) *a mesoscopic scale* (exploitation of continuum thermomechanical equations in a structured front), and (iii) *a macroscopic scale*, that of engineering applications. The first scale inspired by the Landau-Ginzburg theory, although discrete to start with, deals with nonlinear localized waves (solitonic structures: solitary wave, soliton complexes) where nonlinearity and dispersion (discreteness) are the main ingredients. The second scale involves nonlinearity, dispersion and dissipation (viscosity). The third scale is that one at which the front is seen as an irreversibly driven singular surface and where macroscopic thermodynamics (theory of irreversible processes) and numerical methods such as finite-element and finite-volume methods are used in conjunction with a criterion of progress. The three scales are reconciled by the fact that all solutions satisfy the same *Hugoniot conditions* sufficiently far away from the front, whether structured as a solitonic or dissipative structure, or without thickness such as a singular surface. This multi-level, multi-physics approach gathers the view points of condensed-matter physicists (micro-scale), applied mathematicians (meso-scale) and engineers (macro-scale), and even that of the theoretical physicist via the inclusive notion of quasi-particles and the underlying and pervasive invariant-theoretical framework.

2. Microscopic condensed-matter-physics approach: solitonics

The first approach considered is that dealing with the *micro scale* of lattice dynamics *in a perfect lattice*, so that there is no dissipation and effects of temperature are not involved, except perhaps in the phase-transition parameter. Following works by Falk [1] Pouget [2], and Maugin and Cadet [3], this allows one to readily obtain a dynamical representation of a phase boundary (here a kink) as a *solitonic structure* for a two-degrees of freedom, but essentially one-dimensional, system. The reason for this is that, unless one wants to study the *lateral stability of this system*, the "theorem of the flea" applies: the "flea" sees only the *first-order geometrical description* of the transition layer, hence essentially the normal direction to a layer of constant thickness. Notice that the continuum model obtained in the *long-wave length limit* is that of a *nonlinear elastic body with first gradients of strains* taken into account but no dissipation. This long-wave limit is admissible because the transition layer between two phases, although thin (perhaps a few lattice spacings), is nonetheless large enough. Numerical simulations can be performed *directly* on the lattice. The elastic potential is non-convex in general (cf. [3]). To exemplify this approach, we consider a one-dimensional (x), two-degrees of freedom, lattice with transverse (main effect) and longitudinal (secondary effect) displacements from the initial position. In the so-called *long-wave limit* where the discrete dependent variables (strains) s_n and e_n vary slowly from one lattice site to the next and they can be expanded about the reference configuration (na, O), the discrete equations yield the following system of two (nondimensionalized) coupled partial – in (x,t) – differential equations (with an obvious notation for partial x and t derivatives)

$$\begin{aligned} &s_{tt} - c_T^2 s_{xx} + \left(s^3 - s^5 + 2\gamma\, es + \alpha s_{xx}\right)_{xx} = 0\,, \\ &e_{tt} - c_L^2 e_{xx} + \gamma(s^2)_{xx} = 0\,,\ \alpha \equiv \beta - \left(c_T^2/2\right)\,, \\ &s \equiv \nu_x\ e \equiv u_x\,, \end{aligned} \tag{2.1}$$

where s and e are the shear and elongation strains, γ is a coupling coefficient, and α and β are nonlocality parameters. Parameters c_T and c_L are the characteristic speeds of the linear elastic system. This corresponds to stresses and energy density given by

$$\begin{aligned} &\sigma_s = \bar{\sigma}_s - \frac{\partial}{\partial x}M = \frac{\delta}{\delta x}\psi\,,\ \sigma_e = \frac{\partial}{\partial e}\psi\,, \\ &\bar{\sigma}_s = \frac{\partial}{\partial s}\psi\,,\ M = \frac{\partial}{\partial s_x}\psi\,, \\ &\psi(s,e,s_x) = \frac{1}{2}\left(c_T^2 s^2 - \frac{1}{2}s^4 + \frac{1}{3}s^6 + c_L^2 e^2 - 2\gamma s^2 e + \alpha(s_x)^2\right)\,. \end{aligned} \tag{2.2}$$

In other words, equations $(2.1)_{1-2}$ are none other than the x-derivatives of the balance of (physical) linear momentum for a continuum made of a nonlinear, homogeneous elastic material with strain gradients – with both nonlinearity and strain gradients relating only to the *shear* deformation. Complicated as they look, equations $(2.1)_{1-2}$ still admit exact dynamical solutions of the *solitonic type*. A thorough discussion of the existence of such solutions connecting two different or equivalent minimizers (i.e., two phases) of the potential energy was given by Maugin and Cadet [3] to which we refer the reader. The remarkable fact is that such complicated solutions are shown (by computation) to satisfy the following (temperature-independent) *HUGONIOT* condition between *states at infinity*:

$$Hugo := \left[\bar{W}(s, e \text{ fixed}) - \langle \bar{\sigma}_s \rangle\, s\right] = 0\,, \tag{2.3}$$

where $\bar{\sigma}_s$, is the shear strain *without strain-gradient effect*, and $\bar{W}$ is the elastic energy with such effects similarly neglected. Obviously, gradient effects play a significant role only within the rapid transition zone that the kink solution represents, while outside the state is practically spatially uniform, although different on both sides of the localized front. Here we have used the following definitions for the jump and mean value of any quantity a:

$$[a] := a(+\infty) - a(-\infty)\,,\ \langle a \rangle := \frac{1}{2}(a(+\infty) + a(-\infty))\,. \tag{2.4}$$

Equation (2.3) 2.1) is typical of the *absence of dissipation* during the transition, in general a working hypothesis that is not realistic. Furthermore, it can in fact be rewritten as the celebrated *Maxwell's rule of equal areas.*

3. Macroscopic engineering approach: singular surface and thermodynamic criterion

In this second approach the phase boundary is considered *locally* and *macroscopically* as a discontinuity front Σ in the first-order derivatives – hence, like a *shock wave*, a *singular surface* of the first order in Hadamard's classification – of the basic field (e.g., the physical placement $\mathbf{x} = \chi(\mathbf{X}, t)$ of a "particle" $\mathbf{X}$); Σ has no thickness. The "theorem of the flea" applies again: only the first-order geometrical description of the singular surface – its normal – is involved unless one introduces some kind of surface tension. But the front itself is not necessarily flat. It may curve and even form cusps in the worst situations. The local viewpoint refers to the fact that it is assumed at each instant of time that the thermoelastic solution is known by any means – analytical, but more than often, numerical – on both sides of Σ so that one can compute a *driving force* acting on Σ. Further progress of Σ must not contradict the second law of thermodynamics. The latter, therefore, governs the local evolution of Σ which is generally *dissipative*, although no microscopic details are made explicit to justify the proposed expressions. The approach is *thermodynamical* and *incremental* (in total analogy with modern plasticity). All physical mechanisms responsible for the phase transformation are contained in the *phenomenological-macroscopic* relationship given by the *local criterion of progress* of Σ. Without entering details which can be found in recent papers (Maugin and Trimarco [4]; Maugin [5, 6]) and considering from the outset the finite-strain framework, we remind the reader that at any regular point in the body (i.e., on both sides of Σ) we have the balance of (physical) linear momentum and the future heat equation written in the Piola-Kirchhoff form for a *heat-conducting thermoelastic material* ($W(\mathbf{F},\theta)$ in general is different on both sides of Σ, and generally non-convex in its first argument and concave in the second one – the thermodynamical temperature θ). But while each phase is materially homogeneous, the presence of Σ is a patent mark of a loss of translational symmetry on the overall body, hence the consideration of a global material inhomogeneity. The field equation capturing this breaking of symmetry is the jump relation associated with the equation of momentum *on the material manifold*, i.e, what we called the balance of pseudo-momentum [5, 6]. This jump equation, together with that for entropy, governs the phase-transition phenomenon at Σ. These equations apriori read

$$\begin{aligned} &\mathbf{N}.[\mathbf{b} + \bar{\mathbf{V}} \otimes P] + \mathbf{f}_\Sigma = 0 \;, \\ &\mathbf{N}[\bar{\mathbf{V}} S - (\mathbf{Q}/\theta)] = \sigma_\Sigma \geq 0 \;, \end{aligned} \tag{3.1}$$

where the last inequality is a statement of the second law of thermodynamics at Σ, $\mathbf{N}$ is the unit normal to Σ oriented from the *minus* to the *plus* side, and we defined the jumps and mean values at Σ by (compare to (2.4)):

$$[a] := a^+ - a^- \;, \; \langle a \rangle := \frac{1}{2}(a^+ + a^-) \;, \tag{3.2}$$

where $a^\pm$ are the uniform limits of a in approaching Σ on its two faces along $\mathbf{N}$. $\bar{\mathbf{V}}$ is the material velocity of Σ, S is the entropy density, θ is the thermodynamical temperature, and the material or *pseudo* momentum and the *Eshelby stress tensor* are defined by

$$\begin{aligned} &P = -\rho_0 \mathbf{v}.\mathbf{F} \;, \; \mathbf{b} = (L\mathbf{1}_R + \mathbf{T}.\mathbf{F}) \;, \\ &L := \frac{1}{2}\rho_0 \mathbf{v}^2 - W(\mathbf{F},\theta) = \bar{L}(\mathbf{F},\theta,\mathbf{v}) \;, \end{aligned} \tag{3.3}$$

and

$$\begin{aligned} &\mathbf{v} = \left.\frac{\partial \chi}{\partial t}\right|_{\mathbf{X}} \;, \; \mathbf{F} = \left.\frac{\partial \chi}{\partial \mathbf{X}}\right|_t \equiv \nabla_R \chi \;, \; \mathbf{v} = -\mathbf{F}.\mathbf{V} \;, \\ &W = W(\mathbf{F},\theta) \;, \; \mathbf{T} = \partial W/\partial \mathbf{F} \;, \; S = -\partial W/\partial \theta \;. \end{aligned} \tag{3.4}$$

The "force" $\mathbf{f}_\Sigma$, just like σ_Σ, is an unknown and like other thermodynamical material forces it acquires a physical meaning only in the computation of the power it expands here in the material velocity $\bar{\mathbf{V}}$ of Σ. On performing this computation (e.g., [5]) in the case of a *coherent phase-transition front* Σ for which there holds the continuity conditions $[\mathbf{V}] = \mathbf{0}$ and $[\theta] = 0$ (no dislocations at Σ that is also homothermal – the transition occurs at a temperature shared by the two phases), we obtain the compatibility condition between $\mathbf{f}_\Sigma$ and σ_Σ :

$$\mathbf{f}_\Sigma.\bar{\mathbf{V}} = f_\Sigma \bar{V}_N = \theta_\Sigma \sigma_\Sigma \geq 0 \,, \tag{3.5}$$

where

$$f_\Sigma = -Hugo_{PT} \;,\; Hugo_{PT} := \mathbf{N}.[\mathbf{b}_S].\mathbf{N} = [W - \langle \mathbf{N}.\mathbf{T} \rangle.\mathbf{F}.\mathbf{N}] \,, \tag{3.6}$$

where $\mathbf{b}_S$ is the quasi-static part of $\mathbf{b}$ (although the computation is made without neglecting inertia). If this inertia is really neglected, then we have following reduction:

$$Hugo_{PT} = [W - tr(\langle \mathbf{T} \rangle.\mathbf{F})] \,. \tag{3.7}$$

In this canonical formalism the driving force $\mathbf{f}_\Sigma$ happens to be purely normal but it is contrained to satisfy, together with the propagation speed $\bar{V}_N$, the surface dissipation inequality indicated in the last of (3.5). In other words, any relationship between these two quantities must be such that the inequality (3.5) be verified. This is the basis of the formulation of a *thermodynamically admissible criterion of progress* for Σ. Indeed, we look for a relationship $\bar{V}_N = g\,(f_\Sigma; \theta_\Sigma \; fixed)$ which satisfies the last of (3.5). For illustrative purposes we may consider the *cartoonesque* case (cf. Figure 2) where the phase-transition process does not involve any characteristic time (just like rate-independent plasticity), in which case the dissipation (3.5) must be homogeneous of degree one only in $\bar{V}_N$; the threshold type of progress criterion corresponds to this. That is, $\bar{V}_N \in \partial I_f = N_C(f_\Sigma)$, where I_f is the indicator function of the closed segment $F = [-f_c, +f_c]$ – a convex set – and N_C is the "cone of outward normals" to this convex set. If we "force" the system evolution to be such that there is effective progress of the front at $\mathbf{X} \in \Sigma$ while there is *no* dissipation, then we must necessarily enforce the following condition

$$f_\Sigma = 0 \; i.e. \;,\; Hugo_{PT} \equiv [W - \langle \mathbf{N}.\mathbf{T} \rangle \mathbf{F}.\mathbf{N}] = 0 \,. \tag{3.8}$$

On account of the fact that temperature (θ_Σ) is fixed, and the thickness of the front is taken as zero, so that uniform states are reached immediately on both sides of Σ, (3.8) is none other than the condition of "Maxwell" (2.3) (2.1) in the one-dimensional pure-shear case. Thus a macroscopic approach dear to the engineer has allowed us to obtain, in general, a more realistic (in general, dissipative) progress of the front. The case of Section 2 appears then as a "*zoom*"– in the nondissipative case – on the situation described in the present section since the front acquires, through this zoom magnification (asymptotics), a definite, although small, thickness and a structure while rejecting the immediate vicinity of the zero-thickness front to infinities. The next approach allows one to introduce both a *thickness* and *dissipation*.

4. Mesoscopic applied-mathematics approach: structured front

Here the front of phase transformation is looked upon as a mixed "*viscous-dispersive*" structure at a "meso" scale. We refer to this as the *applied-mathematician approach*. This dialectical approach in which one applies macroscopic concepts at a smaller scale to obtain an improved phenomenological description is finally fruitful. Here we follow Truskinowsky [7] to whom we refer for details. We therefore consider a one-dimensional model (along the normal to the structured front – "theorem of the flea") and we envisage a *competition* between viscosity (i.e., a simple case of dissipation) and some *weak nonlocality* accounted for through a strain-gradient theory (compare Section 2). The critical nondimensional parameter which compares these two effects is defined by

$$\omega = \eta / \sqrt{\varepsilon}\,, \tag{4.1}$$

where η is the viscosity and $\varepsilon \approx L^2$ is the nonlocality parameter (size effect). Progressive-wave solutions $u = u(\xi = x - \bar{V}_N t)$ of the continuous system that relate two minimizers (uniform solutions at infinities that minimize $\bar{W}$) over a distance of the order of $\delta = \sqrt{\varepsilon}$ are discussed in terms of this parameter. The mathematical problem reduces to a *nonlinear eigenvalue problem* of which the specification of the points of the discrete spectrum constitutes the looked for *kinetic relation* $\bar{V}_N = g(f;\varepsilon)$, where $f = \bar{\sigma} - \bar{\sigma}(+\infty)$ plays the role of *driving force*. As a matter of fact the speed of propagation $\bar{V}_N$ satisfies the *Rankine-Hugoniot equation* $\bar{V}_N^2 = [\bar{\sigma}]/[s]$, where strain gradients and viscosity play no role and the jumps are taken between asymptotic values at infinity – cf. (2.4). The evolution of the kinetic law in function of ω is shown in Figure 2 where such a law compares to the schematic one suggested by the thermodynamical analysis of Section 3.

5. Theoretical-physics approach: quasi-particle and transient motion

The approach of section 3 simply accepts the value of $\bar{V}_N$, whatever its evolution, as it is computed from the full field solution at each instant of time and each material point $\mathbf{X} \in \Sigma$. In contrast, the approaches of Sections 2 and 4 provide *progressive-wave* solutions, i.e., waves that are steady in the sense that the propagation speed, although a property of the solution (and not only of the material as in linear-wave propagation) does not vary in time along the propagation path. This is a type of *inertial motion*. What about a *non-inertial motion*? To look at such a case, we envisage the problem in the following way. The localized -but with non-zero thickness- dynamical solutions of Section 2 are looked upon as global entities behaving like mass particles in motion in the appropriate *point mechanics*, i.e., as so-called *quasi-particles* [8, 9]. All perturbing effects such as dissipation, inhomogeneities, etc, will then be treated as *perturbing forces* on the inertial motion that becomes thus *non-inertial*. To understand this view point, it is sufficient to envisage the presence of a viscous (more generally, dissipative) contribution in the right-hand side of the classical balance of linear momentum. This results in the presence of an additional *material* force $\mathbf{f}_D^{inh} = -\mathbf{f}_D.\mathbf{F}$ in the right-hand side of the canonical momentum equation. The latter equation is used, after integration over the path of the wave, to treat the material force as a perturbation on the solution in the absence of $\mathbf{f}_D$. The essential problem consists then in identifying the *point mechanics* that is associated with a particular system of partial-differential equations on account of some of its exact integrals. This point mechanics – that is, a coherent system of relations between mass, momentum and energy of a point particle – can be completely new and a priori unforeseeable. For this viewpoint we refer the reader to published works [10, 11]. In particular, a perturbative approach of the canonical, quasi-particle, type was suggested by Fomethe and Maugin [12] to study the varied motion of a phase-transition front under the action of a temperature gradient.

6. Numerics: from finite-differences to continuous cellular automata

6.1 Finite-Difference Method

The *Finite-Difference Method* (FDM) is the numerical realm of nonlinear hyperbolic systems. This discretization method finds its origin in Analysis and the approximation of space and time derivatives of various orders by finite differences. The school of Courant (Lax, Wendroff, etc) in the USA and that of Godunov and Yanenko in Russia are responsible for its successes in treating problems including sharp field discontinuities such as shock waves in fluids. The accuracy of the devised FDM schemes is measured by their property to more or less conserve the global energy of the system. That is, in the present case, the numerical simulation of nonlinear-wave propagation of Section 2 is performed either directly on the set of lattice equations for a sufficiently long chain, or on the FDM discretized version of the *field* equation (2.1). For a conservative system – which is the case – , it is checked whether the global (space integral) version of the energy equation holds good for the numerically obtained solution. From that view point, checking that the global version of the here redundant pseudomomentum balance law holds good for the numerical solution obtained is also a valid criterion for the accurarcy of the scheme of the FDM scheme, except for the *vectorial nature* of that equation which may mean a check along three different axes instead of the conservation of a single scalar entity as is the case for the energy equation. This places energy and pseudomomentum on equal footing. In particular, for systems of field equations, both canonical equations and accuracy criteria concern the global physical system and not only one degree of freedom in spite of the 3D vectorial nature of pseudomomentum. Thus, following Maugin and Christov [12], we may speak of FDM schemes that more or less conserve *both* energy *and* pseudomomentum. We refer the reader to these authors for the design of accurate FDM schemes capturing the subtlety of the *stiff* problems involved in the dynamical simulation of strongly localized waves such as phase-transition fronts. In a finer analysis (Christov et al., [10]) it was shown that strongly dispersive systems such as (2.1) or (6.1) which are also simultaneously "more" nonlinear than the usual Boussinesq equation, favor the existence of more intrincate structures for the phase-transition fronts, such as so-called *Kawahara solitons*, and bound-states of four solitons, what we have referred to as *soliton complexes*. The dynamics of these "shapes" is stable, contrary to what could have been thought (cf. Bogdan et al., [13]). These "shapes" are, therefore, able candidates to represent the structure of nondissipative phase-transition fronts.

6.2 Finite-Element Method

The *Finite-Element Method* (FEM) has become the traditional numerical approach in engineering problems. The *incremental* (in time) approach mentioned in Section 3 naturally lends itself to a numerical treatment of the spatial problem by the FEM. This is being implemented by various authors, including the author and co-workers (S. Imatani) for the progress of phase-transition fronts and a rather similar problem, that of material growth. For the full thermomechanical dynamics of phases in crystalline substances the next method seems more promising.

6.3 Finite-Volume Method and cellular automaton

The *Finite-Volume Method* (FVM) considers a fixed grid and cells, rather than discrete points, as discrete elements. It directly reflects the notion of conservation laws which apply to these cells. The theory presented in Section 3 clearly emphasizes the notion of conservation laws. Therefore, the FVM seems, to the price of some adjustments, to be appropriate to the treatment of dynamical problems such as generally posed in Section 3. This is all the more true that cells thus considered may also be viewed as the elementary blocks of a thermodynamics of so-called *discrete systems* in the manner of Schottky [14]; cf. Muschik, [15]. In this thermodynamics the state in one discrete system (e.g., one of the above cells) is defined in terms of its environment which may or may not be in thermodynamical equilibrium. *Contact thermodynamical quantities* (e.g., contact temperature, contact stresses, contact velocity) are introduced to characterize the state of the discrete system (in fact defined at the boundary surface of a cell in the FVM). This idea of making a cell's state depend on that of its neighbors is tantamount to introducing a *strategy* for the propagation of the thermodynamic state. This is tantamount to introducing the notion of *cellular automaton*, although discretization here is based on continuous balance laws, so that we may refer to this method as that of *continuous cellular automata*. The above referred to strategy is essential in the case of the dynamics of a *phase-transition front*. It is along this line of thought that our most recent works develop (Berezovski [16], Berezovski and Maugin [17, 18]). In this new scheme, all thermomechanical balance laws are expressed for each cell, and the bulk quantities within each cell are related to the contact ones through the thermodynamics of discrete systems. Thereby a high-performance wave-propagation algorithm is exploited – using Lax-Wendroff and Godunov's ideas – that yields extremely good results in the simulation of the rapid progression of sharp wave fronts in 2D thermoelasticity -cf. [16, 17]. In this approach the time evolution of the Eshelby stress b_{ij} (in fact its "quasi-static" part) is given by the cell equation (volume V and boundary ∂V):

$$\begin{aligned}\frac{\partial}{\partial t}\int_V\left(b^w_{ji}+\frac{1}{2}\rho_0\mathbf{v}^2\delta_{ji}\right)dV &= -\int_{\partial V}(\sigma_{jk}V_k n_i)\,dA+\\ +\int_V\left(\left(\sigma_{pq}\frac{\partial\nu_p}{\partial x_q}-C\frac{\partial T}{\partial t}\right)\delta_{ji}-\frac{\partial\sigma_{jk}}{\partial t}\frac{\partial u_k}{\partial x_i}+\frac{\partial\sigma_{jk}}{\partial x_i}\nu_k\right)&dV\,,\end{aligned}\tag{6.1}$$

where C is the heat capacity, $\mathbf{V}$ is the contact velocity, $\mathbf{v}$ is the bulk velocity, $\mathbf{u}$ is the elastic displacement, and σ is the Cauchy stress. This evolution will govern the phase transition. Related numerical works are in progress.

7. Conclusion

In all we have presented four approaches to the problem of the propagation of phase boundaries in thermoelastic solids. The true microscopic one avoids any dissipative process and yields the notion of *solitonic structures* (Section 2). The macroscopic thermodynamical one of Section 3 duly accounts for irreversibilities and materializes in the formulation of an *engineering type of progress criterion*. Both of these coincide asymptotically. The mesoscopic approach of Section 4 introduces viscosity in a very traditional manner, but its helps refine the progress criterion. As to that of Section 5, extracted from the microscopic view but looking at the structured transition zone as a single object, it does allow for an elegant treatment of *transient motions*. All of these, accompanied by the appropriate numerical approach (Section 6), are needed for a good understanding of the phenomenon, illustrating a true *multi-physics* approach.

Acknowledgments: This work was done within the European TMR Networks 98-0229 "Phase Transitions in Crystalline Substances" and 96-0062 "Instabilities in Nonlinear Materials Science".

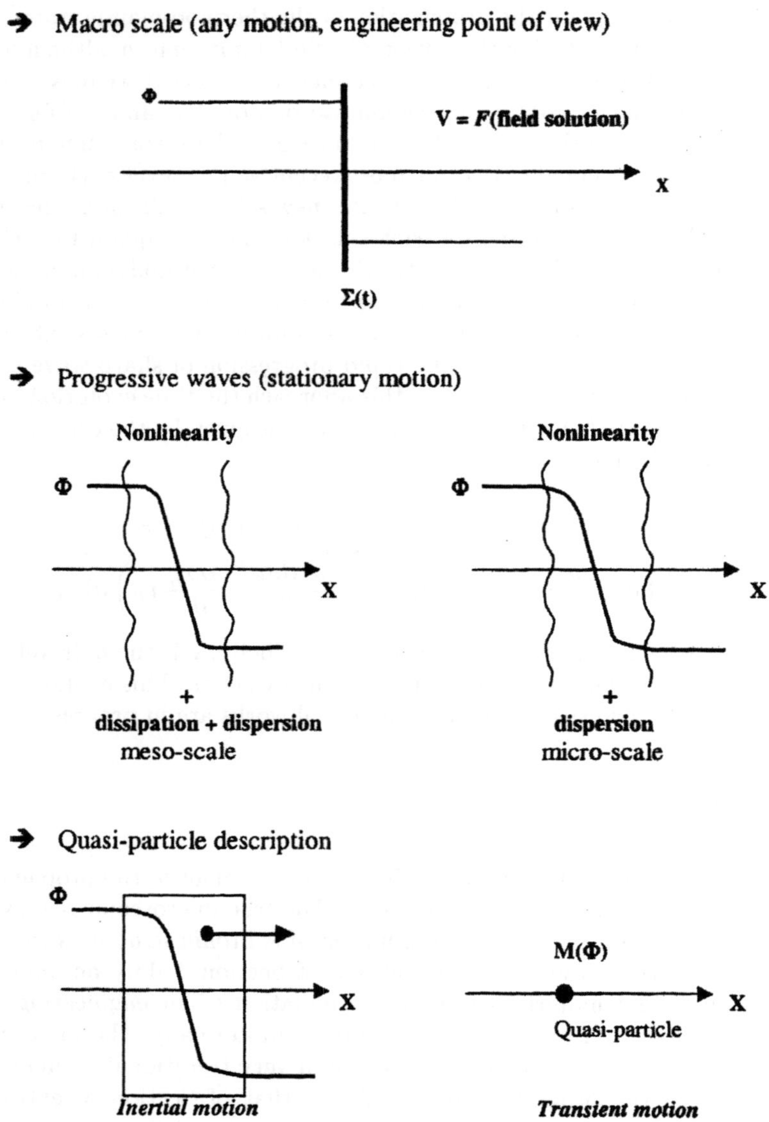

Figure 1: Four viewpoints on the propagation of phase-transition fronts

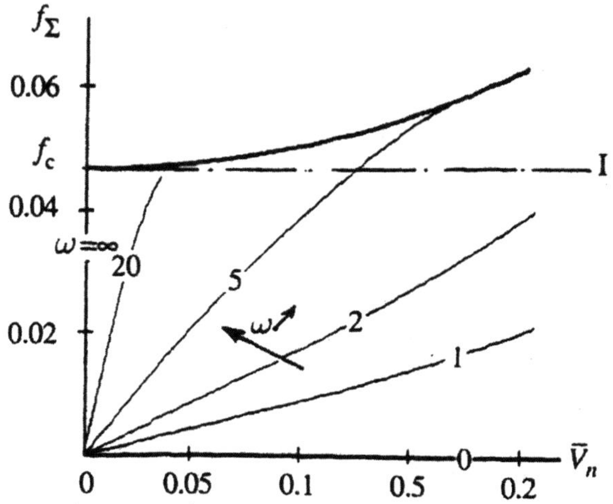

Figure 2: Kinetic curves $f_\Sigma v.s$ dimensionless $\bar{V}_N$ for different values of the parameter ω and a cubic stress-strain relation (after Truskinowsky, 1994). The plateau I corresponds to a Griffith-like criterion (threshold) similar to plastic flow. See Section 3.

References

1. Falk F. (1983) Ginzburg-Landau Theory of Static Domain Walls in Shape-memory Alloys, Zeit. Phys. C. Cond. Matter: 51, 177-185.
2. Pouget J. (1988) Nonlinear Dynamics of Lattice Models for Elastic Continua, in: NATO Summer School on Physical Properties and Thermodynamical Behavior of Minerals, Oxford, 1988, ed. K. Saljé, 359-402, Reidel, Dordrecht.
3. Maugin G.A., Cadet S. (1991) Existence of Solitary Waves in Martensitic Alloys, Int. J. Engng. Sci.: 29, 243-255.
4. Maugin G.A., Trimarco C. (1995) The Dynamics of Configurational Forces at Phase-transition Fronts, Meccanica: 30, 605-619.
5. Maugin G.A. (1997) Thermomechanics of Inhomogeneous-heterogeneous Systems: Application to the Irreversible Progress of two- and Three-dimensional Defects, ARI (Springer-Verlag): 50, 41-56.
6. Maugin G.A. (1998) On Shock Waves and Phase-transition Fronts in Continua, ARI (Springer-Verlag): 50, 141-150.
7. Truskinowsky L.M. (1994) About the Normal Growth Approximation in the Dynamical Theory of Phase Transitions, Cont. Mech. & Thermodyn.: 6, 185-208.
8. Christov C.I., Maugin G.A. (1995) An Implicit Difference Scheme for the Long time Evolution of Localized Solutions of a Generalized Bousinesq System, J. Computational Physics: 116, 39-51.
9. Maugin G.A. (1999) Nonlinear Waves in Elastic Crystals, Oxford Mathematical Monographs, Oxford University Press, UK.

10. Christov C.I., Maugin G.A., Velarde M.G. (1996) Well-posed Boussinesq Paradigm with Purely Spatial Higher-order Derivatives, Phys. Rev.: E54, 3621-3638.
11. Maugin G.A., Christov C.I. (1999) Nonlinear Waves and Conservation Laws: Nonlinear Duality Between Elastic waves and Quasi-particles. In: Nonlinear Wave Phenomena, Eds. A. Guran and J. Wegner, 40p., Birkhauser, Boston (in press).
12. Fomethe A., Maugin G.A. (1997) Propagation Criterion for Phase-transition Fronts in Thermoelastic Ferromagnets, Proc. Est. Acad. Sci. Math. Phys.: 46, 48-54.
13. Bogdan M.M., Kosevich A.M., Maugin G.A. (1999) Formation of Soliton Complexes in Dispersive Systems, Condensed Matter Physics: 2(2), 255-265.
14. Schottky W. (1961) Thermodynamik, Springer-Verlag, Berlin.
15. Muschik W. (1990) Aspects of Non-Equilibrium Thermodynamics, World Scientific, Singapore, New York.
16. Berezovski A. (1997) Continuous Cellular Automata for Simulation of Thermoelasticity, Proc. Est. Acad. Sci. Phys. Math.: 46, 5-13.
17. Berezovski A., Maugin G.A. (1999a) Application of High-resolution Wave-propagation Algorithm to 2D Elastic Propagation, J. Computational Physics (submitted, 1999).
18. Berezovski A., Maugin G.A. (1999b) Application of Wave-propagation Algorithm to 2D Thermoelastic Wave Propagation in Inhomogeneous Media, in: Proc. Intern. Symposium on Godunov Method, Oxford, UK, October 1999, Kluwer (in the press).

Section 1

Two and Multiphase Flows

Edited by H. Herrmann and S. Schwarzer

Analysing Particle Sedimentation in Fluids by Measure-Valued Stochastic Processes

Christian Hesse and Armin Dunz

Mathematics Institute A, University of Stuttgart, Germany

Summary. We present a new tool for the study of particle sedimentation in viscous fluids. It is based on measure-valued stochastic processes and is applicable for general systems. We first introduce the methodology and then, as an example, apply it to the prototypical situation of 1-d sedimentation under the influence of gravity.

1. Introduction

Particle sedimentation in viscous fluids is an important paradigm problem in the theory of two-phase flows. Consider a system where n identical, spherical particles of radius r are immersed in a fluid medium of viscosity η. Assume that sedimentation takes place in a cylindrical container from some starting configuration, the fluid being initially at rest. Particle size is assumed to be small enough for molecular diffusion to be relevant. Molecular diffusion is the motion of particles that is induced by the thermal movement of the fluid molecules. In addition to gravity, buoyancy and friction, particle motion is also influenced by the fluid flows that the particle ensemble as a whole generates in the process of settling. These influences are cumulatively known as hydrodynamic interactions. Other forces, such as those resulting from internal pressures or superposed velocity fields are not considered here. However, the method may also be extended to those more general situations. In summary, we have here restricted ourselves to systems whose evolution is governed by

gravity
buoyancy
friction
molecular diffusion
hydrodynamic interactions

Particle sedimentation is a many-body problem with complex interaction both between solids and fluid and also among the macroscopic particles themselves. Its study presents difficulties from a theoretical, numerical and from an experimental perspective. Theoretical modelling attempts may be coarsely divided into stochastic and deterministic approaches. In previous work (Hesse and Dunz (1997)) we have studied a stochastic model whose model dynamics in 6-d position-velocity phase space are governed by the following system of (Itô) stochastic differential equations:

$$\begin{aligned} dY_k^n(t) &= F\big(Y_k^n(t), c_n(Y_k^n(t),t)\big)dt + D\big(c_n(Y_k^n(t),t)\big)dB_k(t) \\ Y_k^n(0) &= \xi_k \qquad\qquad k = 1,2,\ldots,n\,. \end{aligned} \tag{1.1}$$

Here

$$Y_k^n(t) = \begin{pmatrix} X_k(t) \\ V_k(t) \end{pmatrix} \tag{1.2}$$

and $X_k(t) = \big(X_{k1}(t), X_{k2}(t), X_{k3}(t)\big)'$, $V_k(t) = \big(V_{k1}(t), V_{k2}(t), V_{k3}(t)\big)'$ denote position and translational velocity, resp., of the center of the kth particle, the first component referring to the vertical direction, the direction in which gravity acts. The stochastic processes $B_k(t)$, $k = 1,2,\ldots,n$, are independent 6-d standard Brownian motions. Also, with $y = (x_1, x_2, x_3, v_1, v_2, v_3)'$ the drift term in (1.1) is

$$F(y,c) = (v_1, v_2, v_3, -\beta v_1 + \mu(c), -\beta v_2, -\beta v_3)' \tag{1.3}$$

and the diffusion matrix $D(c)$ in (1.1) is a singular (6×6) diagonal matrix with

$$diag(D(c)) = \big(0,0,0,\sigma_1(c),\sigma_2(c),\sigma_3(c)\big)\,. \tag{1.4}$$

Finally, the concentration functional is defined by

$$c_n(Y_k(t),t) = \int K(X_k(t)-x)P_k^t(x)dx\,, \tag{1.5}$$

where $K : \mathbb{R}^3 \to \mathbb{R}$ is a kernel function integrating to one and

$$P_k^t(x) = \sum_{\substack{i=1 \\ i\neq k}}^{n} 1_{\{x\in\mathbb{R}^3:\ \|x-X_k(t)\|\leq r\}} \tag{1.6}$$

is the (kth particle) configuration process. In (1.3), $\mu(c)$ signifies a concentration-dependent drift in the vertical direction due to gravity. The friction parameter β is taken to be independent of concentration, an assumption resulting from experimental evidence, Hesse and Ramos (1994). An early and considerably simpler version of the model structure goes back to Pickard and Tory (1977). This has subsequently been extended and refined (Tory and Pickard (1982)). Even earlier, McKean (1967) has also studied the dynamics of systems in which the entire system configuration influences parametrization. The underlying heuristics of the model (1.1) are discussed in Hesse and Tory (1996). For experimental evidence supporting the model, see Tory and Hesse (1996) and Hesse and Ramos (1994).

The structure (1.1) may also be viewed as resulting from the general, two-stage, modelling procedure introduced in Hesse (1998). We explain this briefly. At stage one, the main influences on incremental particle velocity evolution are identified (here friction, buoyancy, gravity and molecular diffusion) and used to set up a system of stochastic differential equations for particle motion. It is solved by a parametric family of stochastic processes. In the present context, the solution is a class of Ornstein-Uhlenbeck processes with fixed parameters μ, β, σ. Thus results the coarse-grained model structure. Additional fine structure is supplemented at stage two. Those influences of systems configuration on particle velocity that have not yet been considered at stage one (here the hydrodynamic interactions) are being worked into the originally fixed parameters of the stochastic process. The consequence is a stochastic parametrization of the Ornstein-Uhlenbeck processes by functions of a smoothed version of the stochastic configuration process P_k^t, as defined in (1.5), (1.6).

Our aim here is to introduce a new tool into sedimentology. It is based on stochastic processes whose realizations are trajectories of probability measures over some low-dimensional measurable space rather than continuous functions taking values in high-dimensional spaces. In the present setting, these measure-valued stochastic processes afford a description of system dynamics by a trajectory of a measure over a 6-dimensional space rather than by the trajectory of a point in $6n$-dimensional space, n being the number of particles. Beyond this dimensionality reduction, the merit of the methodology lies in the close connection that measure-valued processes entertain both to martingale theory and to Itô-calculus. Martingale theory and Itô-calculus provide a hierarchy of key results for the analysis of the model dynamics.
In the next section, we first give a definition of measure-valued stochastic processes and note some special cases. We then demonstrate how to obtain a Fokker-Planck-type equation for the limiting probability density of a stochastic process that solves the system (1.1) – (1.6). In the final section, we apply the results to 1-d sedimentation.

2. System analysis by measure-valued stochastic processes

Let $\mathcal{M}(\mathbb{R}^d)$ be the space of measures on (the Borel σ-Algebra $\mathcal{B}(\mathbb{R}^d)$ of) $\mathbb{R}^d$ equipped with the weak topology and let $C_b^2(\mathbb{R}^d)$ be the space of functions from $\mathbb{R}^d$ to $\mathbb{R}$ with bounded continuous partial derivatives up to second order. We adjust the general definition of measure-valued stochastic processes to our needs.

Definition 2.1. *Let $(\Omega, \mathcal{F}, \mathcal{F}_t, P)$ be a filtered probability space. $\{S_t : t \geq 0\}$ is a measure-valued stochastic process if $S_t(\omega, \cdot)$ for $t \geq 0$, $\omega \in \Omega$ is a measure on $(\mathbb{R}^d, \mathcal{B}(\mathbb{R}^d))$.*

In the present case $d = 6$. For $\nu \in \mathcal{M}(\mathbb{R}^d)$ and $f \in C_b^2(\mathbb{R}^d)$, we also define the inner product

$$\langle \nu, f \rangle := \int f(x)\nu(dx)$$

and in the special case that ν has a density g with respect to Lebesgue measure we also write

$$\langle g, f \rangle := \int f(x)g(x)dx = \int f(x)\nu(dx) = \langle \nu, f \rangle ,$$

accepting a small abuse of notation.

A sequence of measures $(\nu_n)_{n\in\mathbb{N}}$ from $\mathcal{M}(\mathbb{R}^d)$ converges weakly to a measure ν, if for all bounded, uniformly continuous, real-valued functions h on $\mathbb{R}^d$

$$\int h(x)\nu_n(dx) \longrightarrow \int h(x)\nu(dx) \qquad \text{as } n \to \infty .$$

An important class of measure-valued processes is given by the family of atomic measure-valued processes. Its members have the representation

$$S_t(\cdot, A) = \sum_k \alpha_k(t)\delta_{\xi_k(\cdot,t)}(A)$$

with $\mathbb{R}^d$-valued random variables $\xi_k(\cdot, t)$, $\alpha_k(t) \in \mathbb{R}_+$, $A \in \mathcal{B}(\mathbb{R}^d)$ and δ is Dirac measure.
Continuous measure-valued processes make up another important class. In this case there exists some non-negative function-valued process G so that for all $A \in \mathcal{B}(\mathbb{R}^d)$

$$S_t(\cdot, A) = \int_A G_t(\cdot, x)dx .$$

To study the dynamics of an n-particle system with coordinates $Y_k^n(t)$, $k = 1, \dots, n$, in position-velocity phase space, we utilize a special atomic measure-valued process, namely

$$\overline{Y}_t^n = \frac{1}{n}\sum_{k=1}^{n} \delta_{Y_k^n(t)} . \tag{2.1}$$

It is obvious that $(\overline{Y}_t^n,\ t \geq 0)$ affords a description of system evolution which is equivalent to $Y(t) := (Y_1^n(t), \dots, Y_n^n(t),\ t \geq 0)$. Our goal is now to identify the limit dynamics of (1.1) – (1.6) as the number of particles n approaches infinity and, simultaneously, as particle radius $r = r(n)$ approaches zero in such a way that $nr^3(n)$ remains constant. Since real sedimentation systems consist of a large number of small particles whose total volume remains constant, the above limit dynamics constitutes a plausible approximation to the dynamics of finite systems.
To obtain the limiting dynamics, first derive a representation for the inner product $\frac{1}{n}\sum_{k=1}^{n} f(Y_k^n(t))$, with $f \in C_b^2(\mathbb{R}^d)$, using Itô's formula and (1.1). See also Oelschläger (1985). This results in

$$\begin{aligned}
&\langle \overline{Y}_t^n, f(\cdot)\rangle - \langle \overline{Y}_0^n, f(\cdot)\rangle - \int_0^t \langle \overline{Y}_s^n, F(\cdot, c_n(\cdot, s)) \cdot \nabla f(\cdot)\rangle ds \\
&-\frac{1}{2}\int_0^t \langle \overline{Y}_s^n, (D^2(c_n(\cdot, s))\nabla) \cdot \nabla f(\cdot)\rangle ds \\
&= \frac{1}{n}\sum_{k=1}^n \big(D(c_n(Y_k^n(s), s))\nabla f(Y_k^n(s))\big) \cdot dB_k(s) \;=:\; M(t) \,. \qquad (2.2)
\end{aligned}$$

The merit of this approach now rests on several facts which for brevity we state under the following assumptions, not all of which are required for each theorem.

Assumptions: The functions $\mu(\cdot), \sigma_i(\cdot),\ i = 1,2,3$ and $K(\cdot)$ are bounded and Lipschitz-continuous, β is a positive constant. The initial states $\xi_k,\ k = 1,2,\ldots$ are independent and identically distributed random variables.

Under these assumptions we have

Theorem 2.1. *The real-valued stochastic process $M(t)$ as defined in (2.2) is a martingale with*

$$E\Big[\sup_{0\le t\le T} \big(M(t)\big)^2\Big] \le \textit{const.}\, n^{-1}\, T\, \|\nabla f\|_\infty^2 \longrightarrow 0 \qquad \textit{as } n \to \infty \,.$$

Hence $M(t)$ is asymptotically negligible in mean square, uniformly over finite time intervals.

Theorem 2.2. *For any fixed k, $(Y_k^n(t),\ t \in [0,T], n = k, k+1, \ldots)$ is a Cauchy sequence of stochastic processes in mean square.*

Hence for any fixed k the sequence $(Y_k^n(t))$ converges to a limiting process, say $Y_k(t)$.

Theorem 2.3. *The limiting processes $Y_k(t),\ k = 1,2,\ldots$ are independent and identically distributed.*

Theorem 2.4. *The limiting process $Y_k(t),\ t \in [0,T])$, is the unique solution of the non-linear stochastic differential equation*

$$\begin{aligned}
dY_k(t) &= F\big(Y_k(t), c_\infty(Y_k(t), P_t)\big)dt + D(c_\infty(Y_k(t), P_t))dB_k(t) \\
Y_k(0) &= \xi_k \,,
\end{aligned}$$

where P_t is the probability measure that corresponds to the distribution of $Y_k(t)$ and

$$c_\infty(Y_k(t), P_t) = \int K^*(Y_k(t) - y)P_t(dy) \,,$$

with $K^*\big((x,v)'\big) = K(x)$.

Theorem 2.5. *The measure-valued stochastic process $\overline{Y}_t^n$ converges weakly to the deterministic measure P_t almost surely, uniformly over finite intervals $[0,T]$.*

To prove these results is very lengthy. The details are given in Hesse and Dunz (1997). Taken together, Theorems 2.1-2.5 imply that

$$\langle P_t(\cdot), f(\cdot)\rangle = \langle P_0(\cdot), f(\cdot)\rangle + \int_0^t \langle P_s(\cdot), F(\cdot, c_\infty(\cdot, P_s)) \cdot \nabla f(\cdot) + \frac{1}{2}(D^2(c_\infty(\cdot, P_s))\nabla) \cdot \nabla f(\cdot)\rangle ds\,, \qquad (2.3)$$

where $\langle P_t(\cdot), f(\cdot)\rangle = \int f(y) P_t(dy)$.

If P_t has a density $p(y,t)$ with respect to Lebesgue measure then (2.3) is a weak form of the PDE

$$\frac{\partial p(y,t)}{\partial t} = -\nabla \cdot \left[p(y,t) F\big(y, c_\infty(y, p(\cdot,t))\big)\right] + \frac{1}{2}\nabla \cdot \left[\nabla p(y,t) D^2\big(y, c_\infty(y, p(\cdot,t))\big)\right], \qquad (2.4)$$

where ∇ denotes the gradient with respect to position and velocity variables and

$$c_\infty(y, p(\cdot,t)) = \int K^*(y-y')p(y',t)dy'\,. \qquad (2.5)$$

The PDE (2.4) is a Fokker-Planck-type equation. It is parabolic and non-linear. Note also that both F and D^2 are functions of p through their dependence on c_∞. PDEs of this type have sometimes been called truly non-linear, see Wehner and Wolfer (1987).

3. Sedimentation in 1-d

In 1952 Kynch developed a 1-d deterministic model of sedimentation. It has been very influential subsequently and extensions thereof are still being studied today (Bustos, Paiva and Wendland (1996), Bürger (1996)). As an application of our method, we now utilize the results of Section 2 to study the 1-d Kynch (1952) context from our perspective. Specifically, we connect the as yet unspecified functions $\mu(c)$ and $\sigma(c)$ to properties of the velocity and concentration fields. The constant β is obtained by equating the Lagrangean time scale T_L and the particle relaxation time β^{-1}. Writing x, v instead of y, we obtain from (1.3), (1.4), (2.5)

$$F\big(x,v,c_\infty(x,p(\cdot,\cdot,t))\big) = \Big(v,-\beta v+\mu\big(c_\infty(x,p(\cdot,\cdot,t))\big)\Big) \tag{3.1}$$

$$D^2\big(c_\infty(x,p(\cdot,\cdot,t)\big) = \begin{pmatrix} 0 & 0 \\ 0 & \sigma^2\big(c_\infty(x,p(\cdot,\cdot,t))\big) \end{pmatrix} \tag{3.2}$$

$$c_\infty(x,p(\cdot,\cdot,t) = \iint K(x-x')p(x',v,t)dvdx' \tag{3.3}$$

and the PDE (2.4) reduces to

$$\begin{aligned}\frac{\partial p(x,v,t)}{\partial t} &= -\frac{\partial}{\partial x}\left[vp(x,v,t)\right] - \frac{\partial}{\partial v}\Big[\Big(-\beta v+\mu\big(c_\infty(x,p(\cdot,\cdot,t))\big)\Big)p(x,v,t)\Big] \\ &\quad +\frac{1}{2}\frac{\partial^2}{\partial v^2}\Big[\sigma^2\big(c_\infty(x,p(\cdot,\cdot,t))\big)p(x,v,t)\Big]\,. \end{aligned}\tag{3.4}$$

The strategy is now as follows: From the Eulerian equations for the conservation of mass and momentum, we first derive evolution equations for concentration and for velocity moments. We then obtain corresponding identities from the PDE (3.4). A comparison of the two sets of equations allows to connect the parameter functions $\mu(c)$ and $\sigma(c)$ to properties of the velocity and concentration fields.

The Eulerian conservation equations are given by

$$\begin{aligned}\frac{\partial C}{\partial t}+\frac{\partial \mathcal{U}C}{\partial x} &= 0 \\ \frac{\partial \mathcal{U}}{\partial t}+\mathcal{U}\frac{\partial \mathcal{U}}{\partial x} &= \frac{D\mathcal{U}}{Dt}\,, \end{aligned}\tag{3.5}$$

where $\mathcal{U}$ and C denote velocity and concentration, resp., and $D\mathcal{U}/Dt$ may be interpreted as the force per unit mass acting on a particle. We now introduce two averaging operations. The operator $\overline{\cdot}$ performs averaging over the entire particle ensemble. It induces a decomposition of $\mathcal{U}$ and C into mean and fluctuating parts, namely

$$\begin{aligned}\mathcal{U} &= u+u' \\ C &= c+c'\,, \end{aligned}\tag{3.6}$$

where $\overline{\mathcal{U}} = u$ and $\overline{C} = c$. Secondly, the operator $\langle\cdot\rangle$ is defined to average near specified locations in the following sense: For any function $f(x,v,t)$

$$\langle f(x,t)\rangle := \frac{\iint K(x-x')f(x',v,t)p(x',v,t)dx'dv}{c(x,t)}$$

with

$$c(x,t) = c_\infty(x,p(\cdot,\cdot,t)) = \lim_{n\to\infty}\frac{1}{n}\sum_{k=1}^{n}K(x-X_k^n(t))\,.$$

Then, for example, $v^j(x,t)$ is the average of the jth moment of the vertical velocity of the particles that are near x at time t. As in Thompson (1987), the two averaging operations are related by the identity

$$\langle f(x,t)\rangle\,\overline{C(x,t)} = \overline{\langle f(x,t)\rangle\,C(x,t)} \tag{3.7}$$

so that in particular

$$\begin{aligned} \langle v^n(x,t)\rangle\,c &= \overline{\mathcal{U}^n c'} + \overline{\mathcal{U}^n}\,c \\ \langle \mu(c)\rangle\,c &= \mu(c)\,c \\ \langle \mu(c)\,v(x,t)\rangle\,c &= \mu(c)\overline{\mathcal{U}C}\,. \end{aligned} \tag{3.8}$$

The evolution equations that can be derived from the PDE (3.4) by first multiplying by $K(x-x')$ etc. and subsequent integration are, in terms of the averaging notation:

$$\begin{aligned} \frac{\partial c(x,t)}{\partial t} &= -\frac{\partial}{\partial x}\left[\langle v(x,t)\rangle\,c(x,t)\right] \\ \frac{\partial}{\partial t}\left[\langle v(x,t)\rangle\,c(x,t)\right] &= -\frac{\partial}{\partial x}\left[\langle v^2(x,t)\rangle\,c(x,t)\right] \\ &\quad +\langle \mu(c(x,t))\rangle\,c(x,t) \\ &\quad -\beta\langle v(x,t)\rangle\,c(x,t) \\ \frac{\partial}{\partial t}\left[\langle v^2(x,t)\rangle\,c(x,t)\right] &= -\frac{\partial}{\partial x}\left[\langle v^3(x,t)\rangle\,c(x,t)\right] \\ &\quad -2\beta\langle v^2(x,t)\rangle\,c(x,t) \\ &\quad +2\langle \mu(c(x,t))v(x,t)\rangle\,c(x,t) \\ &\quad +\langle \sigma^2(c(x,t))\rangle\,c(x,t)\,. \end{aligned} \tag{3.9}$$

Furthermore, from the Euler equations (3.5) using the decomposition (3.6), one derives

$$\begin{aligned} \frac{\partial c}{\partial t} &= -\frac{\partial \overline{u'c'}}{\partial x} - \frac{\partial uc}{\partial x} \\ \frac{\partial \overline{\mathcal{U}C}}{\partial t} &= -\frac{\partial \overline{\mathcal{U}^2 C}}{\partial x} + \overline{C\,\frac{D\mathcal{U}}{Dt}} \\ \frac{\partial \overline{\mathcal{U}^2 C}}{\partial t} &= -\frac{\partial \overline{\mathcal{U}^3 C}}{\partial x} + \overline{2\,\mathcal{U}C\,\frac{D\mathcal{U}}{Dt}}\,. \end{aligned} \tag{3.10}$$

Substituting (3.8) into (3.9) and comparing the resulting identities to (3.10), we arrive at

$$\begin{aligned} \mu(c) &= c^{-1}\beta\overline{\mathcal{U}C} + c^{-1}\overline{C\,\frac{D\mathcal{U}}{Dt}} \\ \sigma^2(c) &= -2c^{-1}\mu(c)\overline{\mathcal{U}C} + 2c^{-1}\beta\overline{\mathcal{U}^2 C} + \overline{2\mathcal{U}C\frac{D\mathcal{U}}{Dt}}\,. \end{aligned} \tag{3.11}$$

Hence, the model parameter functions have been expressed as functions of the characteristics of the velocity and concentration field, resp. The analysis

of a given sedimentation system requires, prior to the averaging operations in (3.11), a determination of the specific $D\mathcal{U}/Dt$.
Finally, we also comment on the first evolution equation in (3.9). It identifies the rate of change of concentration as the spatial derivative of flux (up to a minus sign). Hence, with our interpretation of concentration, this is exactly the fundamental identity in the Kynch (1952) model. Therefore, appropriate averaging operations reduce the coupled Langevin equation approach (which is stochastic) to Kynch theory (which is deterministic).

References

1. Bürger, R. (1996): Ein Anfangs-Randwertproblem einer quasi-linearen parabolischen entarteten Gleichung in der Theorie der Sedimentation mit Kompression. Dissertation, Department of Mathematics, University of Stuttgart.
2. Bustos, M.C., Paiva, F. and Wendland, W.L. (1996): Entropy boundary conditions in the theory of sedimentation of ideal suspensions. Math. Meth. Appl. Sci. 19, 679-697.
3. Hesse, C.H. (1993): On a stochastic model for particle sedimentation in fluids. Appl. Stoch. Models and Data Anal. 9, 73-83.
4. Hesse, C.H. (1998): A general procedure for stochastic modelling of systems consisting of a large number of interacting components. J. Appl. Stoch. Models and Data Anal., 13, 249-258.
5. Hesse, C.H. and Dunz, A. (1997): Limit dynamics for a Markov model for particle sedimentation in fluids. Technical Report 97/26, SFB404, University of Stuttgart.
6. Hesse, C.H. and Ramos, E. (1994): Dynamic simulation of a stochastic model for particle sedimentation in fluids. Appl. Math. Modelling 18, 437-445.
7. Hesse, C.H. and Tory, E.M. (1996): The stochastics of sedimentation. Advances in Fluid Mechanics 7, 199-240.
8. Kynch, G.J. (1952): A theory of sedimentation. Trans. Faraday Soc. 48, 166-176.
9. McKean, H.P. (1967): Propagation of chaos for a class of non-linear parabolic equations. Lecture Series in Differential Equations 7, 41-57.
10. Oelschläger, K. (1985): A law of large numbers for moderately interacting diffusion processes. Z. Wahrscheinlichkeitstheorie verw. Gebiete 69, 279-322.
11. Pickard, D.K. and Tory, E.M. (1977): A stochastic model for sedimentation. J. Math. Anal. Appl. 60, 349-369.
12. Thompson, D.J. (1987): Criteria for the selection of stochastic models of particle trajectories in turbulent flows. J. Fluid Mech. 180, 529-556.
13. Tory, E.M. and Hesse, C.H. (1996): Theoretical and experimental evidence for a Markov model for sedimentation. Advances in Fluid Mechanics 7, 241-280.
14. Tory, E.M. and Pickard, D.K. (1982): Extensions and refinements of a Markov model for sedimentation. J. Math. Anal. Appl. 86, 442-470.
15. van Dop, H., Nieuwstadt, F.T.M. and Hunt, J.C.R. (1985): Random walk models for particle displacements in inhomogeneous unsteady turbulent flows. Phys. Fluids 28, 1639-1653.
16. Wehner, M.F. and Wolfer, W.G. (1987): Numerical evaluation of path-integral solutions of Fokker-Planck equations. III. Time and functionally dependent coefficients. Phys. Review A, 35, 1795-1801.

Phenomenological Model of Sedimentation-Consolidation Processes

R. Bürger

Institute of Mathematics A, University of Stuttgart, Germany

Summary. This paper presents an outline of the phenomenological theory of sedimentation-consolidation processes. In one space dimension, this theory leads to an initial-boundary value problem (IBVP) of a hyperbolic-parabolic degenerate partial differential equation. A brief account of recent existence and uniqueness results for this IBVP is given. The predictive power of the phenomenological theory is illustrated by two comparisons of numerical with experimental results.

1. Introduction

The phenomenological theory of sedimentation-consolidation processes describes the settling of flocculated suspensions such as the slurries found in the mineral industries. A solid-liquid suspension is considered here as a mixture of two superimposed continuous media. The starting point of the modeling are the mass and linear momentum balances of both components. Constitutive assumptions, an order-of-magnitude study and the restriction to one space dimension reduce these equations to one scalar partial differential equation (PDE) for the volumetric solids concentration plus an algebraic relationship for the excess pore pressure. The difficulty lies in the hyperbolic-parabolic type of the resulting PDE, where the type-change interface is determined by the solution, i.e. its location is unknown a priori.

Although Kynch's kinematic sedimentation theory [18] describes well a lot of phenomena especially of the settling of dilute suspensions, its major inherent shortcoming is that it predicts that concentrations always propagate along straight characteristics (which might, of course, intersect and thus produce discontinuities). This contradicts observations of the behaviour of flocculated slurries, which form compressible sediments described by curved iso-concentration lines. The phenomenological theory models correctly this behaviour, and at the same time produces a mathematical model that can be solved easily and therefore be implemented as a simulation tool.

2. Phenomenological sedimentation-consolidation model

We assume that the solid particles are small with respect to the sedimentation vessel and have the same density ϱ_s; that the constituents of the suspension are incompressible; that the suspension is completely flocculated before the sedimentation begins; and that there is no mass transfer between the solid and the fluid. Then the mixture can be described by the local solids volume fraction ϕ, the solid and fluid phase velocities $\mathbf{v}_{\mathrm{s}}$ and $\mathbf{v}_{\mathrm{f}}$ and Cauchy stress tensors $\mathbf{T}_{\mathrm{s}}$ and $\mathbf{T}_{\mathrm{f}}$, the gravity force $\mathbf{b} = -g\mathbf{k}$ where $\mathbf{k}$ is the upwards-pointing unit vector, and $\mathbf{m}$ the solid-fluid interaction force per unit volume. With the volume-average velocity $\mathbf{q} = \phi\mathbf{v}_{\mathrm{s}} + (1-\phi)\mathbf{v}_{\mathrm{f}}$, the respective local mass balances for the solid and for the mixture are

$$\phi_t + \nabla \cdot (\phi \mathbf{v}_{\mathrm{s}}) = 0, \quad \nabla \cdot \mathbf{q} = 0. \tag{2.1}$$

The respective solid and liquid component linear momentum balances are

$$\varrho_{\mathrm{s}}\phi \mathrm{D}_t^{\mathrm{s}}\mathbf{v}_{\mathrm{s}} = \nabla\cdot\mathbf{T}_{\mathrm{s}} + \varrho_{\mathrm{s}}\phi\mathbf{b} + \mathbf{m}, \quad \varrho_{\mathrm{f}}(1-\phi)\mathrm{D}_t^{\mathrm{f}}\mathbf{v}_{\mathrm{f}} = \nabla\cdot\mathbf{T}_{\mathrm{f}} + \varrho_{\mathrm{f}}(1-\phi)\mathbf{b} - \mathbf{m}. \tag{2.2}$$

We assume that the Cauchy stress tensors take the form $\mathbf{T}_{\mathrm{s}} = -p_{\mathrm{s}}\mathbf{I} + \mathbf{T}_{\mathrm{s}}^{\mathrm{E}}$ and $\mathbf{T}_{\mathrm{f}} = -p_{\mathrm{f}}\mathbf{I} + \mathbf{T}_{\mathrm{f}}^{\mathrm{E}}$, respectively, where p_{s} and p_{f} are the phase pressures, and that the viscous solid and fluid stress tensors are given by

$$\mathbf{T}_{\mathrm{s}}^{\mathrm{E}} = \mu_{\mathrm{s}}\big[\nabla\mathbf{v}_{\mathrm{s}} + (\nabla\mathbf{v}_{\mathrm{s}})^{\mathrm{T}} - \tfrac{2}{3}(\nabla\cdot\mathbf{v}_{\mathrm{s}})\mathbf{I}\big], \quad \mathbf{T}_{\mathrm{f}}^{\mathrm{E}} = \mu_{\mathrm{f}}\big[\nabla\mathbf{v}_{\mathrm{f}} + (\nabla\mathbf{v}_{\mathrm{f}})^{\mathrm{T}} - \tfrac{2}{3}(\nabla\cdot\mathbf{v}_{\mathrm{f}})\mathbf{I}\big],$$

respectively. The phase shear viscosities $\mu_{\mathrm{s}}(\phi)$ and $\mu_{\mathrm{f}}(\phi)$ are given as constitutive functions. Alternatively, a mixture-representative viscosity $\mu(\phi) = \mu_{\mathrm{s}}(\phi) + \mu_{\mathrm{f}}(\phi)$ and the ratio $h(\phi) = \mu_{\mathrm{s}}(\phi)/\mu(\phi)$ can be prescribed. For simplicity, we assume here that bulk viscosities vanish.

The theoretical variables p_{f} and p_{s} are replaced by the pore pressure $\tilde{p}$ and the effective solid stress σ_{e}, which can be measured. The assumption that the surface porosity of the sediment in a column equals its volume porosity implies the relationships $p_{\mathrm{f}} = (1-\phi)\tilde{p}$ and $p_{\mathrm{s}} = \phi\tilde{p} + \sigma_{\mathrm{e}}$. We assume that the solid flocs begin to touch each other at a critical concentration ϕ_{c}. Then the constitutive equation for σ_{e} satisfies

$$\sigma_{\mathrm{e}}'(\phi) = 0 \text{ for } \phi \le \phi_{\mathrm{c}}, \quad \sigma_{\mathrm{e}}'(\phi) > 0 \text{ for } \phi > \phi_{\mathrm{c}}. \tag{2.3}$$

Since the flow depends only on the pore pressure in excess of the hydrostatic pressure $\varrho_{\mathrm{f}}g(H-z)$, where H is the height of the vessel, the excess pore pressure $p_{\mathrm{e}} := \tilde{p} - \varrho_{\mathrm{f}}g(H-z)$ is used instead of $\tilde{p}$. The solid-fluid interaction force $\mathbf{m}$ is decomposed into a hydrostatic part $\mathbf{m}_{\mathrm{b}} = \beta\nabla\phi$ and a dynamic part $\mathbf{m}_{\mathrm{d}} = -\alpha(\phi)\mathbf{v}_{\mathrm{r}} + \gamma(\phi)\mathrm{D}_t^{\mathrm{r}}\mathbf{v}_{\mathrm{r}}$. Taking the liquid component linear momentum balance at equilibrium shows that $\beta = \tilde{p}$. The virtual mass $\gamma(\phi)$ will not be determined since we anticipate that the term $\mathrm{D}_t^{\mathrm{r}}\mathbf{v}_{\mathrm{r}}$ can be neglected.

We assume that $\alpha(\phi)$ or the Kynch batch flux density function $f(\phi) = -\Delta\varrho g\phi^2(1-\phi)^2/\alpha(\phi)$, where $\Delta\varrho = \varrho_{\mathrm{s}} - \varrho_{\mathrm{f}}$, is given by a constitutive equation. Usually f is assumed to be a Lipschitz continuous function satisfying $f(\phi) = 0$ for $\phi \le 0$ or $\phi \ge \phi_{\max}$ and $f(\phi) < 0$ for $0 < \phi < \phi_{\max}$, where $\phi_{\max} \le 1$ is the maximum concentration.

Introducing the present assumptions into the linear momentum balances yields the following relationship for the solid-fluid relative velocity $\mathbf{v}_{\mathrm{r}} = \mathbf{v}_{\mathrm{s}} - \mathbf{v}_{\mathrm{f}}$:

$$\mathbf{v}_{\mathrm{r}} = \frac{f(\phi)}{\Delta\varrho g\phi^2(1-\phi)}\Big[\nabla\sigma_{\mathrm{e}}(\phi) + \phi\Delta\varrho g\mathbf{k} + \frac{\phi}{1-\phi}\nabla\cdot\mathbf{T}_{\mathrm{f}}^{\mathrm{E}} - \nabla\cdot\mathbf{T}_{\mathrm{s}}^{\mathrm{E}}$$
$$+\phi\big(\varrho_{\mathrm{s}}\mathrm{D}_t^{\mathrm{s}}\mathbf{v}_{\mathrm{s}} - \varrho_{\mathrm{f}}\mathrm{D}_t^{\mathrm{f}}\mathbf{v}_{\mathrm{f}}\big) - \frac{1}{1-\phi}\gamma(\phi)\mathrm{D}_t^{\mathrm{r}}\mathbf{v}_{\mathrm{r}}\Big]. \qquad (2.4)$$

We assume that $10^{-4}\,[\mathrm{m/s}]$, $1\,[\mathrm{m}]$ and $10^{-6}\,[\mathrm{m^2/s}]$ are typical values of the settling velocity of a single floc in an unbounded medium, of the height of the sedimentation vessel and of the kinematic viscosity of the pure fluid, respectively. Then an order-of-magnitude analysis justifies to reduce (2.4) to

$$\mathbf{v}_{\mathrm{r}} = \big(f(\phi)/(\Delta\varrho g\phi^2(1-\phi))\big)[\nabla\sigma_{\mathrm{e}}(\phi) + \phi\Delta\varrho g\mathbf{k}]. \qquad (2.5)$$

This equation replaces one of the component linear momentum balances. The sum of these two balances yields the linear momentum balance of the mixture,

$$-\varrho\mathrm{D}_{\mathrm{t}}\mathbf{q} + \mu(\phi)\nabla^2\mathbf{q} - \nabla p_{\mathrm{e}} = \nabla\sigma_{\mathrm{e}}(\phi) + \Delta\varrho g\phi\mathbf{k} \qquad (2.6)$$
$$-\nabla\big(\mu(\phi)\big)\cdot\big[\nabla\mathbf{q} + (\nabla\mathbf{q})^{\mathrm{T}}\big] + \varrho\Big((\mathbf{q}\cdot\nabla)\big(r(\phi)\mathbf{v}_{\mathrm{r}}\big) + \big((r(\phi)\mathbf{v}_{\mathrm{r}})\cdot\nabla\big)\mathbf{q}\Big) + \boldsymbol{\Psi},$$

where $\varrho = \varrho(\phi) = \varrho_{\mathrm{s}}\phi + \varrho_{\mathrm{f}}(1-\phi)$, $r(\phi) = \Delta\varrho\phi(1-\phi)/\varrho(\phi)$ and

$$\boldsymbol{\Psi} := \nabla\cdot\Big(\mu(\phi)\big((\phi - h(\phi))(\nabla\mathbf{v}_{\mathrm{r}} + (\nabla\mathbf{v}_{\mathrm{r}})^{\mathrm{T}} - \tfrac{2}{3}(\nabla\cdot\mathbf{v}_{\mathrm{r}})\mathbf{I}) + \nabla\phi\mathbf{v}_{\mathrm{r}} + (\nabla\phi\mathbf{v}_{\mathrm{r}})^{\mathrm{T}}$$
$$-\tfrac{2}{3}(\nabla\phi\cdot\mathbf{v}_{\mathrm{r}})\mathbf{I}\big)\Big) + \varrho\Delta\varrho\mathrm{D}_{\mathrm{t}}\big([\phi(1-\phi)/\varrho]\mathbf{v}_{\mathrm{r}}\big) + \varrho_{\mathrm{s}}\varrho_{\mathrm{f}}\nabla\cdot\big([\phi(1-\phi)/\varrho]\mathbf{v}_{\mathrm{r}}\mathbf{v}_{\mathrm{r}}\big).$$

The quantity $\boldsymbol{\Psi}$ describes the interaction of the concentration field ϕ with the average flow field $\mathbf{q}$ and the excess pore pressure distribution p_{e}. This interaction is produced by viscous stress due to concentration inhomogeneities and by advective acceleration due to solid-liquid relative motion and diffusion stress. With

$$a(\phi) := -\frac{f(\phi)\sigma_{\mathrm{e}}'(\phi)}{\Delta\varrho g\phi}, \quad A(\phi) := \int_0^\phi a(s)\,ds,$$

the continuity equations can be written as

$$\phi_t + \nabla\cdot(\phi\mathbf{q} + f(\phi)\mathbf{k}) = \Delta A(\phi), \quad \nabla\cdot\mathbf{q} = 0. \qquad (2.7)$$

Combining (2.3) with the assumptions on f, we see that

$$a(\phi)\begin{cases} = 0 \text{ for } \phi \le \phi_{\mathrm{c}} \text{ and } \phi \ge \phi_{\max}: & \text{equation } (2.7)_1 \text{ is hyperbolic,} \\ > 0 \text{ for } \phi_{\mathrm{c}} < \phi < \phi_{\max}: & \text{equation } (2.7)_1 \text{ is parabolic.} \end{cases}$$

The mathematical and numerical difficulties associated with this type degeneracy of equation (2.7) are even exacerbated by the frequent assumption that σ_{e}' and thus a is discontinuous at $\phi = \phi_{\mathrm{c}}$. After inserting (2.5) into (2.6), the final set of field equations for the determination of ϕ, $\mathbf{q}$ and p_{e} is given by (2.6) and (2.7). A detailed derivation is presented in [11].

3. Analysis of the initial-boundary value problem

In the spatially one-dimensional case, we obtain $q = q(t)$ (unless singular feed sources are considered) and are left with the equations

$$\phi_t + \big(\phi q(t) + f(\phi)\big)_x = A(\phi)_{xx}, \tag{3.1}$$

$$p_{e_x} = -\big(\sigma_e(\phi)\big)_x - \Delta\varrho g\phi - \frac{\varrho q(t)}{g}\left(\frac{f(\phi)}{\varrho\phi}\big((\sigma_e(\phi))_x + \Delta\varrho g\phi\big)\right)_x + \Psi. \tag{3.2}$$

Obviously, only equation (3.1) has actually to be solved, since by equation (3.2) the excess pore pressure can always be calculated from the concentration distribution. It is therefore sufficient to consider an initial-boundary value problem (IBVP) only of equation (3.1). For simplicity, we limit the presentation here to the problem of batch sedimentation of a flocculated suspension in a closed column, i.e. we consider the IBVP of equation (3.1) with $q \equiv 0$ on $Q_T = (0, L) \times (0, T)$ and the initial and boundary conditions

$$\phi(x, 0) = \phi_0(x),\ x \in [0, L];\ \phi(L, t) = 0,\ f(\phi) - A(\phi)_x|_{x=0} = 0,\ t \in [0, T]. \tag{3.3}$$

It is well known that, due to both the type degeneracy and to the nonlinearity of the flux density function f, solutions of (3.1) are discontinuous and have to be defined as entropy solutions. We recall from [6] that a function $\phi \in L^\infty(Q_T) \cap BV(Q_T)$ is an *entropy solution* of the IBVP (3.1), (3.3) if $A(\phi)_x \in L^2(Q_T)$, if for almost all $t \in (0, T)$, $\gamma_0\big(f(\phi) - \partial_x A(\phi)\big) = 0$ where γ_0 denotes the trace of a function for $x \downarrow 0$, if for almost all $x \in [0, L]$, $\lim_{t\downarrow 0} \phi(x, t) = \phi_0(x)$, and if for all nonnegative $\varphi \in C^\infty\big((0, L] \times [0, T]\big)$ with compact support in $(0, 1] \times (0, 1)$ and for all $k \in \mathbb{R}$ the following inequality is satisfied:

$$\begin{aligned}\iint_{Q_T} \Big\{|\phi - k|\varphi_t + \operatorname{sgn}(\phi - k)\big[f(\phi) - f(k) - A(\phi)_x\big]\varphi_x\Big\}\, dtdx \\ + \int_0^T \Big\{\operatorname{sgn}(k)\big[f(\gamma_1\phi) - f(k) - \gamma_1 A(\phi)_x\big]\varphi(1, t) \\ + \big[\operatorname{sgn}(\gamma_1\phi - k) + \operatorname{sgn}(k)\big]\big[A(\gamma_1\phi) - A(k)\big]\varphi_x(1, t)\Big\}\, dt \ \geq 0.\end{aligned} \tag{3.4}$$

Entropy inequalities like (3.4) go back to Kružkov [17] and Vol'pert [19] for first-order PDEs and Vol'pert and Hudjaev [20] for second-order PDEs. By the vanishing viscosity method including a mollifier technique, it has been shown in [6] that an entropy solution of the IBVP (3.1), (3.3) exists, even if a is allowed to have jumps (i.e., A is only Lipschitz continuous). The existence proof for this case is a fairly straightforward extension of the proof presented in [10]. However, to show uniqueness of entropy solutions for a discontinuous diffusion function, new arguments have to be invoked. In fact, the uniqueness proof in [10] is based on a particular jump condition derived by Wu and Yin [21]. This jump condition leads to the inequality

$$\iint_{Q_T} \Big\{ |\phi - \psi|\varphi_t + \operatorname{sgn}(\phi - \psi)\big[f(\phi) - f(\psi) - A(\phi)_x + A(\psi)_x\big]\varphi_x \Big\} dtdx \geq 0, \tag{3.5}$$

valid for two entropy solutions ϕ and ψ and for all test functions $\varphi \in C_0^\infty(Q_T)$, from which stability and uniqueness can be obtained in a standard fashion. However, Wu and Yin's jump condition is valid only for Lipschitz continuous diffusion functions $a(\phi)$ and is limited to one space dimension. Fortunately, a very recent result by Carrillo [13] has made it possible to avoid these limitations. He utilizes the technique known as "doubling of the variables", introduced by Kružkov [17] as a tool for proving the L^1 contraction principle for entropy solutions of scalar conservation laws, in order to establish inequality (3.5). Most notably, Carrillo's approach is *not* based on jump conditions and only presupposes that $A(\phi)$ is Lipschitz continuous, i.e. $a(\phi)$ may have discontinuities. We have employed Carrillo's results [13], there formulated for a boundary value problem in which he required $A(\phi) = 0$ on the boundary, to show uniqueness of entropy solutions of the IBVP (3.1), (3.3), see [6] for details. An extension to multidimensional IBVPs is in preparation.

4. Comparison with experimental data

An overview of numerical methods suitable for the IBVP (3.1), (3.3) is presented in [7]. One can either use monotone finite difference schemes with numerical fluxes that include both the convective and the diffusive part or select the optimal existing methods for the convective and the diffusive part and solve both separately. In particular, the convective part, i.e. the first-order conservation law $\phi_t + f(\phi)_x = 0$, may be solved more efficiently by front tracking rather than by finite differencing. We refer to [14] for an excellent introduction to discretization techniques based on operator splitting. The diffusion term $A(\phi)_{xx}$ should always be discretized conservatively to ensure convergence to the right weak solution, i.e. the entropy solution of the IBVP (3.1), (3.3).

Here, we present two numerical solutions obtained by the finite difference operator splitting method described in detail in [12, Ch. 10]. The first example from [5] consists in the recalculation of a settling experiment of estuarine mud of initial concentration $\phi_0 = 0.09231$ performed by Been and Sills [1, Exp. 15]. From the published pore pressure and permeability data, we obtained the constitutive functions $\sigma_\mathrm{e}(\phi) = 0$ for $\phi \leq \phi_\mathrm{c} = 1/12$, $\sigma_\mathrm{e}(\phi) = 15.56 \exp(14.01\phi)$ [Pa] for $\phi > \phi_\mathrm{c}$ and

$$f(\phi) = \begin{cases} -1.101 \times 10^{-4}\phi(1 - 7.226\phi)^{5.15}\,[\mathrm{m/s}] & \text{for } 0 \leq \phi \leq \phi_\mathrm{c}, \\ -1.303 \times 10^{-8}\phi^2 \exp(0.6168(1-\phi)/\phi)\,[\mathrm{m/s}] & \text{for } \phi_\mathrm{c} < \phi < 0.3084, \\ 0 & \text{otherwise} \end{cases}$$

Figure 4.1 shows a comparison of the experimental concentration and excess pore pressure profiles with the corresponding numerical results.

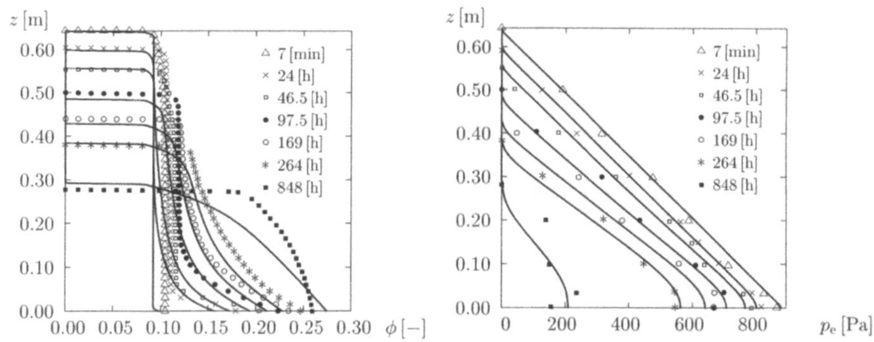

Fig. 4.1. Comparison of experimentally measured (symbols) concentration (left) and excess pore pressure profiles (right) with numerical solution of the sedimentation-consolidation model (solid lines).

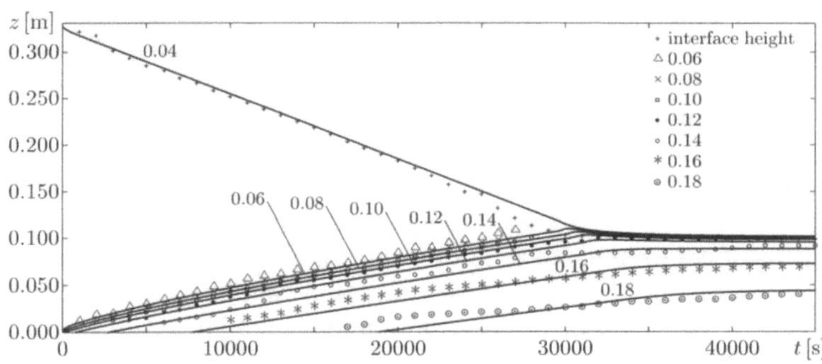

Fig. 4.2. Comparison of experimentally measured iso-concentration lines (symbols) with numerical solution of the sedimentation-consolidation model (solid lines).

In the second example, adopted from [15], we consider an experiment published recently by Holdich and Butt [16] in which a suspension of mineral talc suspended in water with $\phi_0 = 0.052$ was allowed to settle in a column. Concentration profiles were obtained by conductivity measurements. From a suggested relationship for the effective solid stress and a Kozeny type permeability expression we obtained the constitutive functions $\sigma_e(\phi) = 0$ for $\phi \leq \phi_c = 0.04$, $\sigma_e(\phi) = 2.1379 \times 10^7 \phi^{6.944}$ [Pa] for $\phi > \phi_c$ and

$$f(\phi) = \begin{cases} -4.4 \times 10^{-6}(-28.5\phi^2 + 3.0693\phi)\,[\mathrm{m/s}] & \text{for } 0 \leq \phi \leq 0.0538, \\ -4.4 \times 10^{-6}\dfrac{(1-\phi)^3}{12 - 32.5\phi}\,[\mathrm{m/s}] & \text{for } 0.0538 < \phi < 0.3692, \\ 0 & \text{otherwise} \end{cases}$$

Figure 4.2 shows the comparison of measured with numerically simulated iso-concentration lines.

Additional numerical simulations of hypothetical configurations (not directly related to experimental evidence) are presented in [2, 3, 7].

5. Conclusions and directions of future research

The existence and uniqueness results indicate that the phenomenological model is mathematically sound; and newly developed numerical methods for degenerate parabolic equations allow the efficient solution of the one-dimensional IBVP. In particular, it is not necessary to explicitly track the type-change interface $\phi = \phi_c$, which can be identified with the suspension-sediment interface. By simulation it is therefore possible to determine suitable model functions for measured concentration and excess pore pressure data.

The two numerical examples and the additional cases presented in [5, 15] show good agreement between model predictions and numerical results and underline that the theory explains most features of the settling behaviour of real materials. We therefore expect that continuing mathematical analysis of and developing numerical methods for the multidimensional equations will lead to a realistic two-dimensional thickener model.

However, still in one space dimension, a different one-dimensional thickener model will be analyzed first: it has turned out that feed and discharge mechanisms of continuous ideal thickeners can be modeled more realistically by discontinuities of the flux density function and a singular source term rather than by the concept of entropy boundary conditions [9]. In this way, the clarification zone (above the feed level) may be incorporated.

Recent and current research projects include a study of the possible behaviour of the function f in the dilute limit, where experimental evidence has shown that in certain cases the supernate-suspension interface evolves as a rarefaction wave rather than as a discontinuity [8]. Moreover, we studied a kinematical model for the settling of polydisperse suspensions [4], which leads to first-order system of N conservation laws for N particle species.

References

1. Been, K. and Sills, G.C., 'Self-weight consolidation of soft soils: an experimental and theoretical study', *Géotechnique* **31** (1981), 519–535.
2. Bürger, R., Bustos, M.C. and Concha, F., 'Settling velocities of particulate systems: 9. Phenomenological theory of sedimentation: numerical simulation of the transient behaviour of flocculated suspensions in an ideal batch or continuous thickener', *Int. J. Mineral Process.* **55** (1999), 267–282.

3. Bürger, R. and Concha, F., 'Mathematical model and numerical simulation of the settling of flocculated suspensions', *Int. J. Multiphase Flow* **24** (1998), 1005–1023.
4. Bürger, R., Concha, F., Fjelde, K.-K. and Karlsen, K.H., 'Numerical simulation of the settling of polydisperse suspensions of spheres', *Powder Technol.*, to appear.
5. Bürger, R., Concha, F. and Tiller, F.M., 'Applications of the phenomenological theory to several published experimental cases of sedimentation processes', *Separ. Purif. Technol.*, to appear in a special issue.
6. Bürger, R., Evje, S. and Karlsen, K.H., 'On strongly degenerate convection-diffusion problems modeling sedimentation-consolidation processes', Applied Mathematics Report, University of Bergen, Norway, 1999; submitted to *J. Math. Anal. Appl.*
7. Bürger, R., Evje, S., Karlsen, K.H. and Lie, K.-A., 'Numerical methods for the simulation of the settling of flocculated suspensions', *Separ. Purif. Technol.*, to appear in a special issue.
8. Bürger, R. and Tory, E.M., 'On upper rarefaction waves in batch settling', *Powder Technol.* **108** (2000), 74–87.
9. Bürger, R. and Wendland, W.L., 'Entropy boundary and jump conditions in the theory of sedimentation with compression', *Math. Meth. Appl. Sci.* **21** (1998), 865–882.
10. Bürger, R. and Wendland, W.L., 'Existence, uniqueness and stability of generalized solutions of an initial-boundary value problem for a degenerating parabolic equation', *J. Math. Anal. Appl.* **218** (1998), 207–239.
11. Bürger, R., Wendland, W.L. and Concha, F., 'Model equations for gravitational sedimentation-consolidation processes', *ZAMM* **80** (2000), 79–92.
12. Bustos, M.C., Concha, F., Bürger, R. and Tory, E.M., *Sedimentation and Thickening*, Kluwer Academic Publishers, Dordrecht, The Netherlands, 1999.
13. Carrillo, J., 'Entropy solutions for nonlinear degenerate problems', *Arch. Rat. Mech. Anal.* **147** (1999), 269–361.
14. Espedal, M.S. and Karlsen, K.H., 'Numerical solution of reservoir flow models based on large time step operator splitting algorithms', in Fasano, A. and van Duijn, H. (Eds.) *Filtration in Porous Media and Industrial Applications*, Lecture Notes in Mathematics, Springer Verlag, Berlin, to appear.
15. Garrido, P., Bürger, R., and Concha, F., 'Settling velocities of particulate systems: 11. Comparison of the phenomenological sedimentation-consolidation model with experimental results', *Int. J. Mineral Process.*, to appear.
16. Holdich, R.G. and Butt, G., 'Experimental and numerical analysis of a sedimentation forming compressible compacts', *Separ. Sci. Technol.* **32** (1997), 2149–2171.
17. Kružkov, S.N., 'First order quasilinear equations in several independent variables', *Math. USSR Sb.* **10** (1970), 217–243.
18. Kynch, G.J., 'A theory of sedimentation', *Trans. Faraday Soc.* **48** (1952), 166–176.
19. Vol'pert, A.I., 'The spaces *BV* and quasilinear equations', *Math. USSR Sb.* **2** (1967), 225–267.
20. Vol'pert, A.I. and Hudjaev, S.I., 'Cauchy's problem for degenerate second order quasilinear parabolic equations', *Math. USSR Sb.* **7** (1969), 365–387.
21. Wu, Z. and Yin, J., 'Some properties of functions in BV_x and their applications to the uniqueness of solutions for degenerate quasilinear parabolic equations', *Northeast. Math. J.* **5** (1989), 395–422.

The Structure of Bidisperse Suspensions at Low Reynolds Numbers

K. Höfler and S. Schwarzer

Institute for Computer Applications I, University of Stuttgart, Germany

Summary. We present a simulation technique for the direct simulation of the motion of spherical particles suspended in fluids at low Reynolds numbers. We use this method to simulate bidisperse suspensions at varying size ratios of the suspended spheres and compare to analytic results of Batchelor for the low concentration regime. We find that the pair correlation functions indicate that three body hydrodynamic effects are significant even at volume fractions of $\phi = 0.01$ and lead to a decrease of the settling velocity of the involved particle species. In order to describe the time evolution of the concentration profiles in a suspension that settles into a condensed sediment layer from an initially well-mixed state, we propose an advection-diffusion model. The functional forms of the required hindered settling and hydrodynamic diffusion terms are chosen to describe not only our simulation results but also match other experimental results and asymptotically known theoretical predictions.

1. Introduction

Numerous applications in chemical engineering [1, 2], fluid mechanics [3], geology [4], and biology involve systems of rigid particles immersed in a liquid or gas flow. Examples of such systems arise in the context of sedimentation processes, gas-solid or liquid-solid fluidized beds, suspension rheology describing the behavior of blood, pastes, etc., mixing processes when sediment-laden rivers enter lakes or the sea, pneumatic conveying, ticking hour glasses, flocculation in suspensions and many more. In these systems the long-ranged hydrodynamic interactions mediated by the fluid in the interstitial voids of the particle phase control the physical behavior of the particle assembly and render a rigorous treatment of systems involving many particles [5, 6] an extremely difficult task. Consequently, in order to predict suspension behavior in the semi-dilute regime, with constituent particles of complex shapes, or involving inertia effects, we almost always need to resort to appropriate numerical techniques [7, 8, 9, 10]. We will discuss one such approach to deal with a large number of suspended particles in the first part of this paper.

2. Numerical method

The continuous fluid phase of the coupled particle-fluid systems is commonly described by the Navier-Stokes equations for a incompressible fluid of constant density,

$$\frac{D}{Dt}\mathbf{v} = \frac{1}{\rho}\nabla \cdot \mathbf{T} + \mathbf{f}, \tag{2.1}$$

where $\mathbf{v}$ is the velocity field of the fluid, $\mathbf{T}$ its Cauchy stress tensor and $\mathbf{f}$ the local acceleration due to volume forces. For an incompressible Newtonian fluid, $T_{ij} = \tilde{p}\delta_{ij} + \eta(\partial_i v_j + \partial_j v_i)$, where η denotes the shear viscosity of the fluid and $\tilde{p}$ the isotropic pressure. Introducing the kinematic viscosity $\nu = \eta/\rho$ and $p = \tilde{p}/\rho$, we obtain

$$\frac{\partial}{\partial t}\mathbf{v} + (\mathbf{v} \cdot \nabla)\mathbf{v} = -\nabla p + \nu\nabla^2\mathbf{v} + \mathbf{f}. \tag{2.2}$$

The gravitational contribution to the volume acceleration term $\mathbf{f}$ cancels without dynamic consequences against an opposite equal additive hydrostatic contribution to the pressure gradient. We solve Eq. (2.2) by a combination of standard techniques on a regular mesh, which renders the numerical data structures simple and the time stepping fast [11, 12, 13].

Rigid spherical particles of radius a shall be suspended in this fluid. We need to (i) ensure that no-slip boundary conditions are met on their surfaces to the fluid and (ii) to compute the forces $\mathbf{F}^f$ exerted by the fluid on their surfaces in order to solve the equations of motion,

$$M_i\dot{\mathbf{v}}_i = \mathbf{F}_i^f + \mathbf{F}_i^p + (M_i - M_i^f)\mathbf{g} \tag{2.3}$$

for a particle i of mass M_i at position $\mathbf{x}_i$ with velocity $\mathbf{v}_i \equiv \dot{\mathbf{x}}_i$ under the influence of the fluid forces $\mathbf{F}_i^f$, forces due to direct contact with other particles $\mathbf{F}_i^p$, and to gravity $\mathbf{g}$ and the buoyancy $-M_i^f\mathbf{g}$, which is the weight of the fluid displaced by the particle. The contact forces are taken to be proportional to the virtual overlap $(a_i + a_j) - |\mathbf{x}_j - \mathbf{x}_i|$ of two touching particles; for details we refer the reader to [14].

In order to implement no-slip boundary conditions we represent the particles via a modification of the body force term $\mathbf{f}$ in (2.2), as in the immersed boundary [8] and similar to the fictitious domain method [15]. The basic idea is to consider $\mathbf{f}$ as a spatially distributed constraint force which will cause the fluid to move as a rigid body in those regions where particles shall be located. The constraint is computed by comparing the location of massless fluid tracers j at positions $\mathbf{x}_{ij}^m$ to reference positions $\mathbf{x}_{ij}^r$ fixed on "templates" T_i, entities of the same shapes as the physical particles to be modeled.

These templates also serve to include the additional inertial contributions of the particles which arise when particle and fluid densities differ. It carries the missing mass $M_i^t = M_i - M_i^f$ and is quasi rigidly coupled to the stiffened fluid region by a force $\mathbf{F}_i^c$, which will be discussed below. The template shall obey an equation of motion which comprises next to $\mathbf{F}_i^c$ all the forces of physical origin acting on the particle,

$$M_i^t\dot{\mathbf{v}}_i = \mathbf{F}_i^l + \mathbf{F}_i^p + (M_i - M_i^f)\mathbf{g} + \mathbf{F}_i^c. \tag{2.4}$$

We now consider a continuous displacement field $\mathbf{u}_i(\mathbf{x})$ on template T_i, such that $\mathbf{u}_i(\mathbf{x}_{ij}^r) = \mathbf{x}_{ij}^m - \mathbf{x}_{ij}^r$ for every tracer j associated with template i. This allows us to define a constraint fluid force density $\mathbf{f}_i^c(\mathbf{x})$ due to particle i as a penalization force that acts whenever $\mathbf{u}_i$ differs from 0,

$$\mathbf{f}_i^c(\mathbf{x} + \mathbf{u}_i(\mathbf{x})) = -k'\mathbf{u}_i(\mathbf{x}). \tag{2.5}$$

For numerical purposes, we only use the values of $\mathbf{u}_i$ at $\mathbf{x}_{ij}^r$, and obtain $\mathbf{f}_i^c(\mathbf{x})$ by interpolation from the nearest fluid grid points. The force density $\mathbf{f}_i^c$ is zero in the exterior of the fluid region $\Omega_i = \{\mathbf{x} + \mathbf{u}_i(\mathbf{x}),\ \mathbf{x} \in T_i\}$ associated with template T_i. If the external stresses are bounded, then $\mathbf{u}_i$ remains bounded and approaches zero as k' increases. We now define the coupling force $\mathbf{F}_i^c$ to be the *negative* integral over space of Eq. (2.5).

The rate of change of momentum $\dot{\mathbf{P}}_i^f$ of the fluid in the constrained region is found by integration of (2.1) over Ω_i, putting $\mathbf{f} = \mathbf{f}_i^c$,

$$\dot{\mathbf{P}}_i^f = \oint_{\partial T_i} \mathbf{T} \cdot d\mathbf{A} - \mathbf{F}_i^c. \tag{2.6}$$

Since the fluid is incompressible, the region Ω_i will always contain the same material fluid points, so that $\dot{\mathbf{P}}_i^f$ translates into a change of center of mass velocity, $\dot{\mathbf{P}} = M_i^f \dot{\mathbf{v}}_i^{CM}$. The value $1/k'$ sets the degree of accuracy up to which the fluid region resembles the template. An order of magnitude estimate shows that the center of mass velocity should deviate from that of the template at most by a term $\sim k^{-1/2}$, so that with $k \to \infty$ the velocities approach each other, $\mathbf{v}_i^{CM} = \mathbf{v}_i$. Adding the equation of motion of the template (2.4) and Eq. (2.6) we see that the internal forces $\mathbf{F}^c$ cancel, leaving only external forces

$$(M_i^t + M_i^f)\dot{\mathbf{v}}_i = M_i \dot{\mathbf{v}}_i = \oint_{\partial T_i} \mathbf{T} \cdot d\mathbf{A} + (M_i - M_i^f)\mathbf{g}. \tag{2.7}$$

To the degree that the penalization guarantees that Ω_i has the shape of the modeled particle, the surface integral equals the hydrodynamic force $\mathbf{F}_i^l$ that we expect for the fluid-particle interaction and we recover the particle equation of motion (2.3). Similarly, we can reconstruct the angular motion [13].

The simulation method above has been applied successfully to studies of monodisperse suspensions [13, 16] where, for example, the experimentally determined sedimentation velocity could be reproduced in the volume fraction range of $\phi = 0.02 \ldots 0.30$.

3. Bidisperse suspensions

We will now consider the case of bidisperse suspensions of spheres with radii a_l and a_s, where the subscripts refer to the large and small species, respectively. In addition to the volume fractions ϕ_l and ϕ_s, two other dimensionless parameters characterize the simulation, the particle radii ratio $\lambda = a_l/a_s$ and the reduced density ratio $\gamma = \Delta\rho_l/\Delta\rho_s$, where $\Delta\rho_i = \rho_i - \rho$ denotes the density difference of particle species i and the fluid.

The most important parameter remains the total volume fraction $\phi_t = \phi_l + \phi_s$, whereas the influence of λ and γ density ratio is more complex to understand and generally less important if the values do not differ significantly from unity. This is reflected in the history of models which have been proposed for the particles velocities. Only few take the influence of species volume fraction, radii ratio and density ratio into account. Perhaps the most comprehensive is the analytical study of Batchelor [5] and of Batchelor and Wen [17] who derive the average sedimentation speed and the pair correlation functions in polydisperse suspensions of spheres which are dominated by hydrodynamic pair interactions. Its validity is therefore restricted to the dilute regime, where three-body interactions can be neglected. Applying their ideas to the bidisperse case, we write for the average sedimentation velocity of a linear function of the two involved partial volume fractions ϕ_l, and ϕ_s

$$
\begin{aligned}
\langle U_s \rangle &= U_s^{(0)}(1 + S_{ss}\,\phi_s + S_{sl}\,\phi_l) \\
\langle U_l \rangle &= U_l^{(0)}(1 + S_{ls}\,\phi_s + S_{ll}\,\phi_l).
\end{aligned} \tag{3.1}
$$

Here, the $U_i^{(0)}$ denote the Stokes single particle settling velocities, $U_i^{(0)} = \frac{2}{9}a_i^2(\rho_i - \rho)g/\eta$, the $\langle U_i \rangle$ the average settling speed of species i and the S_{ij} are the sedimentation coefficients that describe the influence on the settling velocity due to the non-zero concentration of all particle species in the system. Both sedimentation coefficients S_{ss} and S_{ll} are equal to -5.6. For equal particle densities, $\gamma = 1$, the S_{ij} become functions of λ only. Their functional form as given in [17] can be fitted by a polynomial of second order

$$
\begin{aligned}
S_{sl}(\lambda) &= -3.52 - 1.04\,\lambda - 1.03\,\lambda^2 \\
S_{ls}(\lambda) &= -3.52 - 1.04\,1/\lambda - 1.03\,1/\lambda^2,
\end{aligned} \tag{3.2}
$$

where $\lambda = a_l/a_s \geq 1$ by definition. We now introduce these expressions in Eq. (3.1) to obtain for the settling velocities as a function of λ,

$$
\begin{aligned}
\langle U_s \rangle &= U_s^{(0)}(1 - 5.6\phi_s - 3.52\phi_l - 1.04\phi_l/\lambda - 1.03\phi_l/\lambda^2) \\
\langle U_l \rangle &= U_l^{(0)}(1 - 5.6\phi_l - 3.52\phi_s - 1.04\phi_s\lambda - 1.03\phi_s\lambda^2)
\end{aligned} \tag{3.3}
$$

We simulate a bidisperse suspension in a fully periodic cell where the particle volume fractions $\phi_l = \phi_s = 0.01$ are kept constant and only λ is varied by increasing the radius of the large particles. In these simulations, the Reynolds number on the scale of the small particles is $Re = a_s U_s^{(0)}/\nu = 0.1$. The periodicity also ensures that a stationary state develops in which there is no separation of the particle species on large scales.

The simulation results for the mean settling velocities U_s and U_l are compared to the theoretical predictions as functions of λ in Fig. 3.1(a). For increasing values of λ the velocity of the large particle species increases due

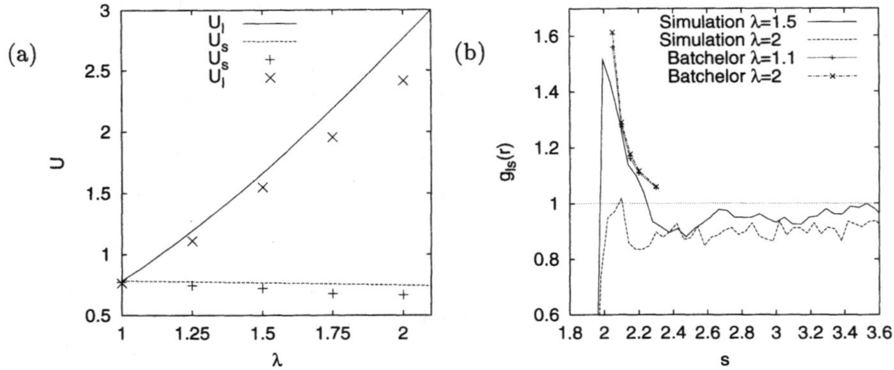

Fig. 3.1. (a) The dependence of the settling velocity of small (+) and large (×) particles on the radius ratio λ compared to (solid and dashed line) the predictions of Batchelor's theory (3.3). (b) Measured pair correlation function $g_{ls}(r)$ for $\lambda = 1.5$ (solid line) and $\lambda = 2$ (dashed line) compared to the theoretical predictions of Batchelor [17] for $\lambda = 1.1$ (dotted line) and $\lambda = 2$ (dashed-dotted line).

to the increase of the radius. In contrast, the velocity of the small particles with constant radius decreases, due to the increased back-flow caused by the faster settling large particles. The computed settling velocities are strikingly smaller than the theoretically predicted settling velocities.

To understand this effect, we need to consider the spatial structure of the suspension. To this end, we determine the pair distribution function $g_{ij}(r)$, i.e., the probability relative to an entirely homogeneous suspension to find a particle of species j at distance r from one of species i. Figure 3.1 shows the measured values for the pair distribution function for pairs of one large and one small particle for $\lambda = 1.25$ and $\lambda = 2$ as a function of the dimensionless distance $s = 2r/(a_i + a_j)$. The graph indicates that the probability for close pairs decreases as λ increases, which is a genuine three-body effect: Batchelors two-body predictions do not show a significant decrease of the pair probabilities. This decrease implies that the correlations g_{ll} and g_{ss} differ (data not shown), which likewise cannot be understood in a particle pair approximation for dilute suspensions. We find also that (not shown) $g_{ll}(r)$ shows a pronounced decrease to values significantly below 1 in the range $s = 2 \ldots 4$ for $\lambda > 1.5$. Since close pairs settle faster, a reduced probability for such configurations implies smaller settling velocities—as observed.

4. An advection-diffusion model for polydisperse suspensions

So far we have compared predicted and simulated behavior for infinite and periodic systems in which no large-scale separation of the involved particles

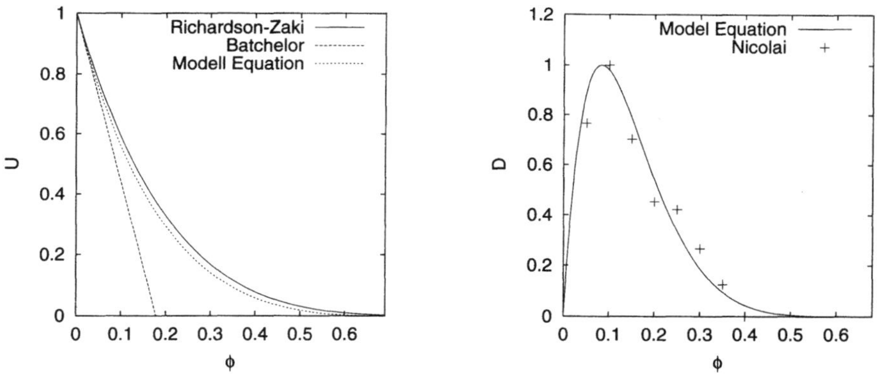

Fig. 4.1. Hindered settling function Eq. (4.2) (left side) and diffusion coefficient Eq.(4.3) (right side) used in the advection diffusion model Eq. (4.1). On the left side we also show the results of Batchelor Eq.(3.1) and Richardson and Zaki. The diffusion coefficient is compared to the normalized experimental data of Nicolai[20].

species can emerge. However, in systems, where the translation invariance in vertical direction is broken, e.g, by the presence of horizontal walls, distinct regions with different concentrations of the particle species will emerge. At the top of the vessel there is a region void of particles, because all species have moved downwards. Below, we find a region with predominantly small, slowly settling particles, then a mixed region, in which still both species are present, and, finally, at the bottom, a sediment, in which the fraction of large, quickly settling particles is increased. The transition zones display a smooth change in the particle concentrations.

In order to "model" our simulation, we define local particle volume fractions by averaging over thin horizontal slices of the system [18], in which the particle concentrations do not change appreciably as functions of the vertical coordinate. In this way, the problem is reduced to one dimension. The concentration changes due to vertical particle currents that have two components. One is due to the downward settling motion and equals the product of the local concentration and the settling velocity $\phi_i(z)U_i(\boldsymbol{\Phi})$, which is assumed to be a function of the local volume fractions $\boldsymbol{\Phi} = (\phi_l(z), \phi_s(z))$ alone. The other describes the random, diffusive component of the motion, and is proportional to the gradient of the volume fraction: $-D_i(\boldsymbol{\Phi})(\partial/\partial z)\phi_i(z)$. The continuity equations for each species then turns into an advection-diffusion equation,

$$\frac{\partial \phi_i}{\partial t} = -\frac{\partial}{\partial z}\left(\phi_i U_i(\boldsymbol{\Phi}) - D_i(\boldsymbol{\Phi})\frac{\partial \phi_i}{\partial z}\right). \tag{4.1}$$

In order to simplify the modeling, we require that the hindered settling function vanishes as we approach the maximum total volume fraction $\phi_t \equiv \sum_i \phi_i \to \phi_\infty$. We therefore extend Batchelors form (3.1) to a nonlinear equation

$$U_i = U_i^{(0)} \frac{e^{\left(1+\sum_j S_{ij}\phi_j + 2\phi_t/\phi_\infty\right)} (\phi_\infty - \phi_t)^2}{e\, \phi_\infty^2} \tag{4.2}$$

which preserves the asymptotic behavior as $\phi_i \to 0$, but vanishes at $\phi_t = \phi_\infty$. We compare the form (4.2), Batchelors hindered settling function (3.1) and the empirical relation of Richardson and Zaki [19] in Fig. 4.1.

For the diffusion coefficients $D_i(\Phi)$ we find empirically that the expression

$$D_i(\Phi) = A a_i U_i^{(0)} \phi_i \left(\frac{U_i(\Phi)}{U_i^{(0)}} \right)^2 \tag{4.3}$$

shows good agreement with the measurements of Nicolai *et al.* [20] for $A = 3.6 \times 10^3$, corresponding to a maximum of $D_i/a_i U_i^{(0)}$ of 6.4 (cf. Fig. 4.1).

In Fig. 4.2 we show our simulation results in comparison to the predictions of Eq. (4.1). We display the volume fractions at various times for the small particles (left side) and the large particles (right side). In the upper, small particle region (right part of the concentration curves) we observe an enrichment of small particles saturating at $\phi_s \approx 0.064$ in comparison to the initial volume fraction of 0.05. The width of the transition zone between small particles and clear fluid is significantly smaller than the width of the transition zone separating the mixture region of large and small particles from the zone void of large particles.

(a)

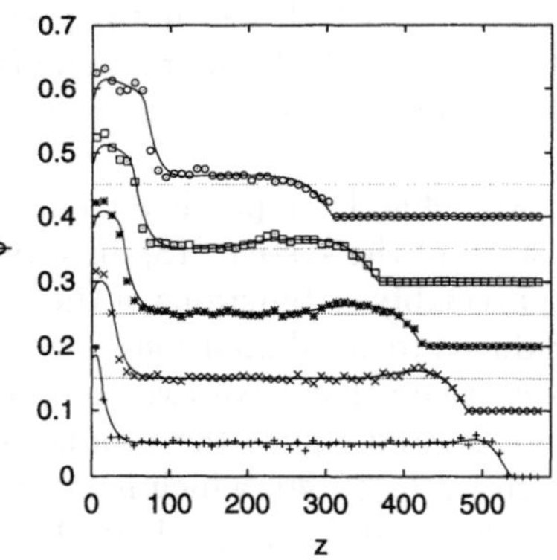

(b)

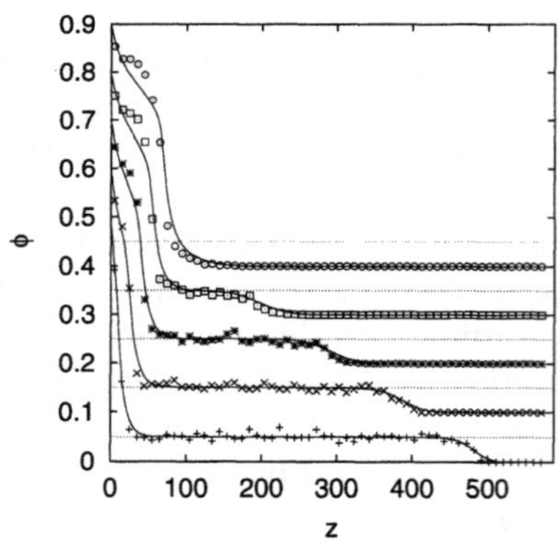

Fig. 4.2. Comparison of the predicted concentration profiles from the advection-diffusion model (solid lines) with measurements from three-dimensional simulations (data points) for (a) small particles and (b) large particles. The results are shown for the simulation times $T = 100\ (+), 200\ (\times), 300\ (*), 400\ (\square)$ and $500\ (\odot)$ in units of $a_s/U_s^{(0)}$. The data for different times are shifted by $\Delta\phi = 0.1$. The initial concentration $\phi = 0.05$ is shown as a dotted line.

The match turns out to be rather sensitive to the exact amplitude of the diffusion term whose functional dependence at small ϕ_i controls the shape of the transition profile. In simulations at low Reynolds numbers in containers of size (width$\times$ height $\times$ depth) $= 36 \times 567 \times 36$ particle radii a_s, we find the best approximation for $A = 2.3 \times 10^3$.

Acknowledgments

We would like to thank R. Glowinski, J. Goddard, E. Guazzelli, M. Müller, and B. Wachmann for discussions and helpful comments. We are grateful to the computer centers in Jülich and Stuttgart for generous allotments of computer time. K.H. acknowledges financial support by the Deutsche Forschungsgemeinschaft, SFB404.

References

1. D. Gidaspow, *Multiphase Flow and Fluidization* (Academic Press, San Diego, 1994).
2. S. L. Soo, *Particles and Continuum: Multiphase Fluid Dynamics* (Hemisphere Publishing Corporation, New York, Washington, Philadelphia, London, 1989).
3. *Mobile Particulate Systems*, edited by E. Guazzelli and L. Oger (Kluwer Academic, Dordrecht, 1995), [Proc. NATO ASI, Cargèse, July 4–15, 1994].
4. K. Pye and H. Tsoar, *Aeolian Sand and Sand Dunes* (Unwin Hyman, London, 1990).
5. G. Batchelor, J. Fluid Mech. **119**, 379 (1982).
6. E. J. Hinch, J. Fluid Mech. **83**, 695 (1977).
7. J. F. Brady and G. Bossis, Ann. Rev. Fluid Mech. **20**, 111 (1988).
8. A. L. Fogelson and C. S. Peskin, J. Comp. Phys. **79**, 50 (1988).
9. A. A. Johnson and T. E. Tezduyar, Comput. Methods Appl. Mech. Engrg. **134**, 351 (1996).
10. A. Ladd, J. Fluid Mech. **271**, 285 (1994).
11. R. Peyret and T. D. Taylor, *Computational Methods for Fluid Flow, Springer Series in Computational Physics* (Springer, New York, Berlin, Heidelberg, 1983).
12. W. Kalthoff, S. Schwarzer, and H. Herrmann, Phys. Rev. E **56**, 2234 (1997).
13. K. Höfler and S. Schwarzer, Phys. Rev. E (1999), submitted.
14. S. Schwarzer, Phys. Rev. E **52**, 6461 (1995).
15. R. Glowinski, T.-W. Pan, and J. Périaux, Comput. Methods Appl. Mech. Engrg. **112**, 133 (1994).
16. B. Wachmann and S. Schwarzer, Int. J. Mod. Phys. C **9**, 759 (1998).
17. G. Batchelor and C.-S. Wen, J. Fluid Mech. **124**, 495 (1982).
18. G. J. Kynch, Trans. Farad. Soc. **48**, 166 (1952).
19. J. F. Richardson and W. N. Zaki, Trans. Instn Chem. Engrs. **32**, 35 (1954).
20. H. Nicolai *et al.*, Phys. Fluids **7**, 12 (1995).

Lattice Boltzmann Simulations of Complex Multiphase Flows

M. Krafczyk[1], P. Lehmann[2], O. Filippova[3], D. Hänel[3] and U. Lantermann[3]

[1] LS Bauinformatik, Technische Universität München, Munich, Germany
[2] Inst. of Terrestrial Ecology, ETH Zürich, Switzerland
[3] Inst. of Combustion and Gasdynamics, University of Duisburg, Germany

Summary. This paper presents simulation results for two classes of multiphase problems obtained using Lattice-Boltzmann approaches. The first part of the paper deals with gas-particle flow investigated by the Duisburg group. A detailed numerical investigation of particle deposition on a complex geometry is presented and values of collection efficiency of dynamically obstructed filters are obtained. The second part of the paper (contributed by the first two authors) investigates air-water flow in a porous medium and, to the authors knowledge, presents for the first time the qualitative reconstruction of a hysteresis curve for a realistic soil geometry obtained by X-ray tomography.

1. Lattice-Boltzmann simulation of gas-particle flow in complex geometry

This section refers to the work of O. Filippova, D. Hänel and U. Lantermann.

1.1 Introduction

Typical for filter flows is the deposition of particles on the surfaces of filter material, resulting in irregular complex shapes, which in interaction with the flow field influences the next particle deposition. Under the assumption of small particle concentration in the stream the phases can be decoupled and computed with different numerical approaches. Lattice Boltzmann (BGK) models [1, 2] are applied for simulation of viscous incompressible flow (carrying phase) with no-slip condition on the surfaces of the filter including deposited particles. The motion of the solid (carried) phase containing large particles $D_p > 0.5\mu$m is calculated by means of a Lagrangian approach. This combined approach enables to consider the form of obstructed filter in all details, even dendrits formations can be described.

For the description of particle deposition a rather simple model is used, which assumes any particle as deposited, if it has touched the solid wall or another deposited particle. It is assumed further that the carried phase influences the carrying phase only through changing form of obstructed filter.

1.2 Method and results

For the computation of 3D viscous incompressible flow around complex geometries the 15-speed LBGK model D3Q15 [1, 2] is used. Flow through two-dimensional geometry used for validation of the code is computed with 3D code using periodical conditions in the third dimension.

Here one must emphasize that the basic LBGK scheme [1, 2] is not well suited for fine solution of such kind of problems as far as it based on the regular "underlying" molecular lattice and therefore performs macroscopically as Cartesian grid solver. Although the flows in the filters are characterized by the small Re numbers Re $\leq O(10)$ and therefore do not contain high gradients of hydrodynamic variables, the use of multi-scale LBGK schemes [3] is necessary. This is due to the fact that the filter geometry can include different length scales as, for example, the size of the wire in the screen and the distance between neighbouring screens (see Fig. 1.6) The bouncing-back conditions traditionally used in LBGK schemes for description of no-slip on the surfaces of obstacles decrease the accuracy of the scheme from 2nd to the first order. Actually they represent arbitrary curvilinear surface in "stair" manner and can strongly change the characteristics of particle deposition. This drawback can be overcome with accurate 2nd order treatment of curvilinear boundaries (boundary-fitting formulas) proposed by the authors [3]. The difference in the collection efficiency of the periodical grid of circular wires (Fig 1.1) computed with bouncing-back and boundary-fitting conditions on the surfaces of wires is shown in Fig. 1.2 for particular case of Re$= 0.5, D_w/D_p = 7.25$. Here the collection efficiency is defined as the ratio of the distance between the two limiting trajectories at inlet of periodical cell and the sum of diameters of the cylinder and the particle. Stokes numbers St$= (D_p/D_w)^2(\rho_p/\rho_g)Re/18$ are varied from 0.01 to 10. One can see from Fig. 1.2 that for St$\sim$ 0.1 the difference in collection efficiency for two different kind of boundary conditions can achieve $\sim$ 30%.

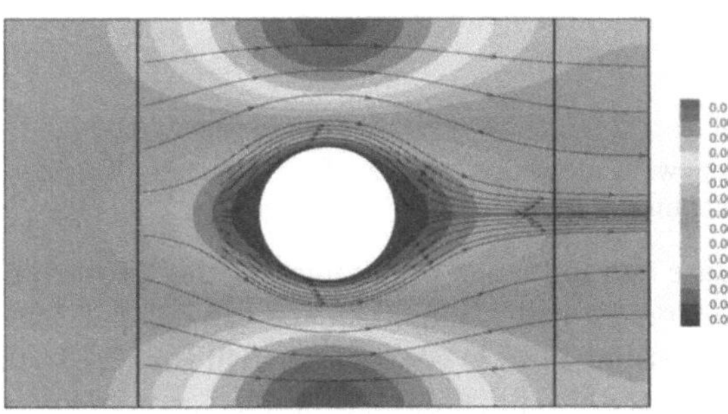

Fig. 1.1. Gas-particle flow through periodical grid of wires, Re=0.5. Embedded grid with parameter of refinement 2 around a wire. Isolines of streamwise velocity and streamlines

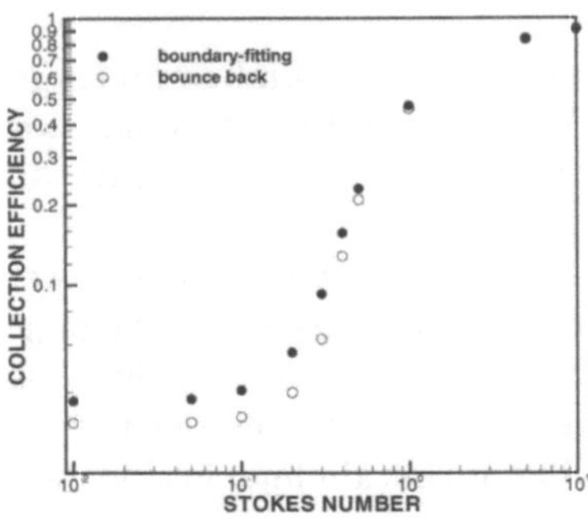

Fig. 1.2. Gas-particle flow through periodical grid of wires, Re=0.5. Collection efficiency vs Stokes number St for different kind of boundary conditions

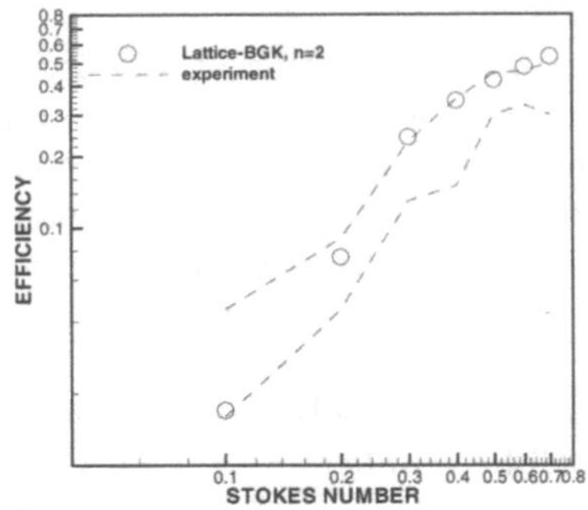

Fig. 1.3. Gas-particle flow through periodical grid of circular wires, Re=2. Collection efficiency of the wire vs Stokes number St, D_P = 0.79μm. Circles - numerical results, dashed lines - band of experimental values

To validate our 3D code the experimental data from [4] was used. Gas-particle flow through periodical grid of circular wires with the ratio between diameter of the wire and the step of periodicity 0.69 was considered. The agreement between numerical and experimental data as shown in Fig. 1.3 was found to be good.

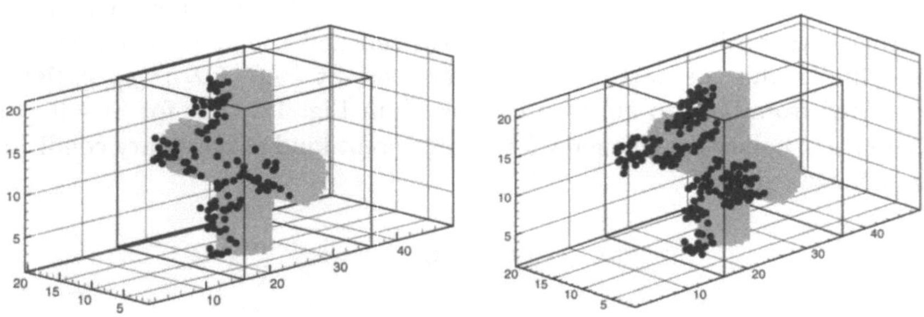

Fig. 1.4. Filter with deposited layer after passage of 1000 particles/cell, (a) changed, (b) unchanged fluid field. Re= 2, St= 0.3

To investigate the influence of changes in the hydrodynamic field in the obstructed filter on its collection efficiency 3D gas-particle flow through an infinite grid of crossing wires is considered with Re=2, St=0.3. A rectangular patch with parameter of refinement 2 is applied around the grid. In Figs. 1.4a and 1.4b the geometry of the filter with deposited layer in the periodical cell after passage of 1000 particles/cell is shown. In computations corresponding to Fig. 1.4a fluid field around new boundary is recomputed after deposition of any 10 particles, Fig. 1.4b represents the results of particle deposition in the unchanged fluid field. The collection efficiency of the filter after passage of 1000 particles/cell defined here as percentage of deposited particles is 0.117 in the case of changed fluid field and 0.208 in the case of unchanged fluid field. The dynamic change of collection efficiency after passages of smaller subparcels containing 200 particles is shown in Fig. 1.5. Figs. 1.4a,1.4b,1.5 show the necessity of recomputation of the fluid field around growing deposited layer. When the changes in fluid field are not taken into account the value of obtained collection efficiency of the obstructed filter is more than twice higher than the real one, and this discrepancy will grow further with the obstruction of the filter.

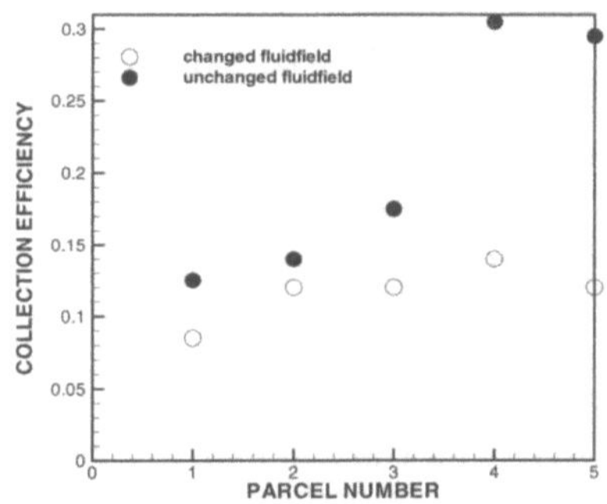

Fig. 1.5. Collection efficiency vs number of subparcel, Re=2, St=0.3

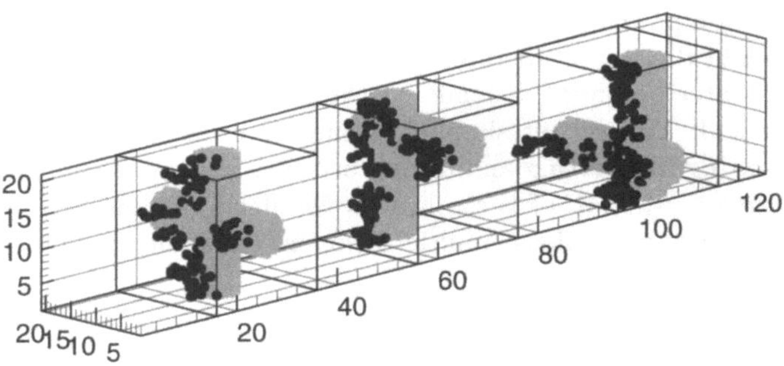

Fig. 1.6. Three infinite screens of crossing wires (model filter) with deposited layer after passage of 1000 particles. Re=2, St=0.3, n=2. Fluid field around new boundary was recomputed after deposition of any 20 particles

On Fig. 1.6 one can see the results of numerical simulations of gas-particle flow through three infinite grids of crossing wires (model filter) in which fluid field was recomputed after deposition of any 20 particles. To cluster grid points in the vicinity of obstacles three patches with parameter of refiinement 2 were applied around every screen, the collection efficiency after passage of 1000 particles/cell is equal to 0.369. Computations of gas-particle flow in the model filters described above have shown the ability of the developed code to deal with very complex irregular dynamically growing boundaries for the detailed description of particle deposition.

2. Direct calculation of hysteresis by Lattice-Boltzmann simulations

This section refers to the work of M. Krafczyk and P. Lehmann.

2.1 Introduction

Transport of contaminants from surface to groundwater is dominated by processes within the vadose soil zone. Especially the distribution of solid, gaseous and liquid phase determines the fate of dissolved components. In the vadose zone, a large amount of pore volume between solid particles may be filled with air. The percentage of water filled and air filled pore volume depends on the energy state of the soil, on topological properties of the pore space and on the wetting behaviour of solid particles.

Traditionally, the volume of soil water was calculated as a function of soil water energy potential. Based on the calculated volume fraction, water (and thus contaminant) flow is predicted with models based on empirical relationships or simplified pore models [6, 7]. This approach has shortcomings for two reasons: Water flow depends not only on its volume fraction, but on size and roughness of a pore, on the connection with other water filled pores and on the location of the water/air-interfaces. In addition, the amount of water at a specified water potential depends on the preceding wetting and drying history of the soil. This hysteresis phenomenon is caused by different pore geometries limiting the wetting and drying process. Improved understanding and quantitative prediction of these hysteresis effects is thus necessary for reliable prediction of multiphase/multicomponent transport in soils.

2.2 Hysteresis

For a single value of soil water energy potential different values of soil water volume are possible implying a hysteretic water potential - saturation relationship. If the pore space of a soil is completely water filled, the amount of water per total soil volume correponds to the porosity Φ, the volume of pore space per total soil volume. Decreasing the soil water potential Ψ will cause a replacement of water by air. In many soils it was observed, that a certain pressure must be applied before a considerable amount of water is replaced by air (air-entry value). A small volume fraction can not be replaced by air, e.g. due to being surrounded by larger pores that drain out earlier and create enclosed water volumes. This volume fraction is called residual water content. If the energy potential is increased again, air will be replaced by water, but for a given energy value, the amount of water is smaller than in the case of reverse process. If the energy potential is increased to the initial value, a part of air is enclosed and remains in the pore space. The corresponding maximum water content is called 'satiated' water content [8]. A following decrease of water content Θ to its residual value is named as main drying curve, while the wetting from residual to satiated water content is called main wetting curve.

2.3 Method and results

Because hysteresis is caused by narrow and large pores limiting drying and wetting processes respectively, the topology of the pore space must be known. An efficient way to measure pore space geometry is the application of X-ray tomography [9]. The pore structure of a sand cylinder 2.5 cm in diameter and height was measured at the Swiss Federal Laboratories for Materials Testing and Research with a resolution of 60 μm. The same sandy material was used in the hysteresis study of Lehmann et al. [5]. A subcube of 6 mm width was used as a test geometry in this work and is shown in Fig. 2.1. To predict the distribution of water and air within the measured structure, we used the D3Q15 model [1, 2] extended by multiphase extensions originating from [10]. An important advantage of the Lattice Boltzmann method is the ability to calculate multiphase flow including the evolution of fluid/fluid-interface without parametrization or explicit tracking.

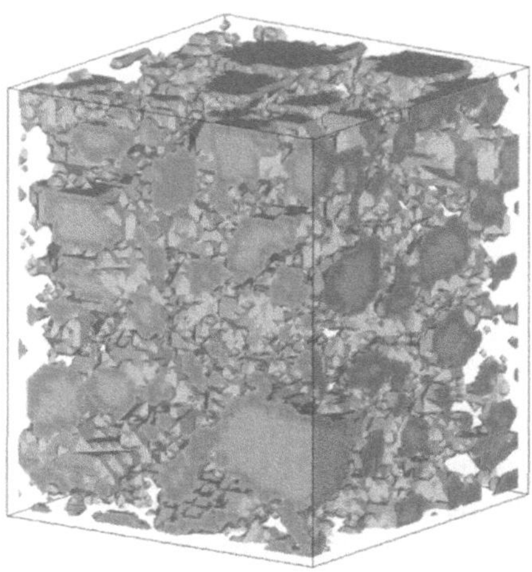

Fig. 2.1. Pore space geometry used for the simulations

For our test cube we computed the hysteresis for water and air distribution. The energy potential is equivalent to a pressure difference between a volume of water in a free water table and a water volume in soil. Thus top-bottom pressure differences were applied to replace the fluid phase within the soil. At the water/air/solid junctions capillary forces compete with the applied pressure difference. The capillary forces are caused by the attraction of the water phase to the soild matrix. A corresponding behaviour was mimicked in our simulations by tuning the interaction between water/air phases and the solid phase respectively. In a preceding numerical experiment, the spontaneous wetting of the dry wettable sand cube due to capillary forces was investigated (no pressure gradient). It was assumed that the dry soil touches a water table at the bottom boundary. The simulation verified that in this geometry a substantial volume fraction of air becomes enclosed (Fig. 2.2). Decreasing the pressure of the water phase yields a slightly decreasing of water content during the first pressure steps, but below a threshold value most of water volume flows out. Obviously even this very small sand cube exhibits a well defined air-entry value. Finally, the pressure was enhanced again and the predicted main wetting curve shows a crucially decreased water content compared to the main drying curve. The experimental and calculated wetting and drying curves are given in Fig. 2.3.

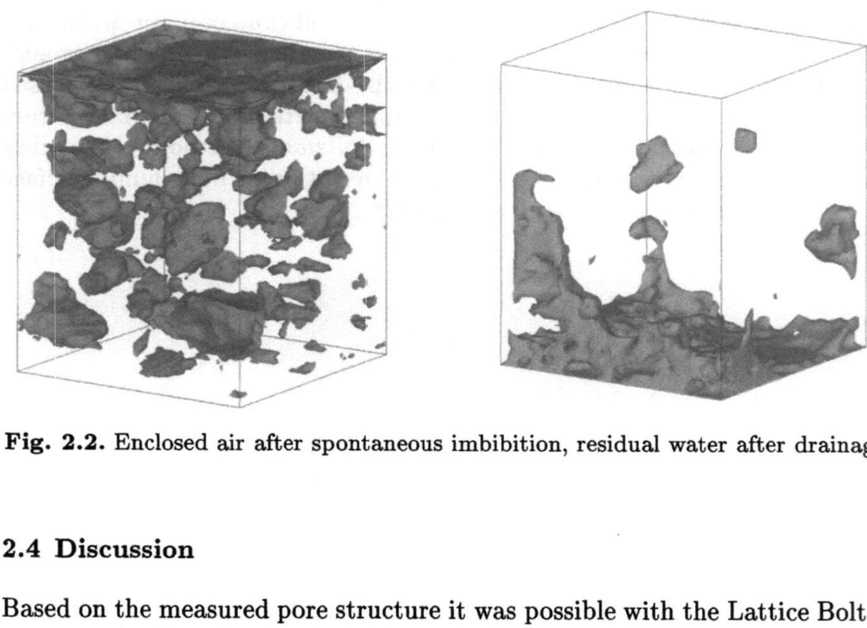

Fig. 2.2. Enclosed air after spontaneous imbibition, residual water after drainage

2.4 Discussion

Based on the measured pore structure it was possible with the Lattice Boltzmann method to predict the most important characteristics of the relationship between water pressure and volumetric water content. This seems to advocate the assumption that the small sand cube used for the calculation reasonably represents basic geometrical properties of the sand. We assume that the calculations for the whole sand cylinder will change the described effects quantitatively but the found patterns will remain. To describe the hysteresis phenomenon quantitatively problems with respect to grid dependance, contact angle hysteresis and film spreading effects have to be adressed as parts of ongoing efforts.

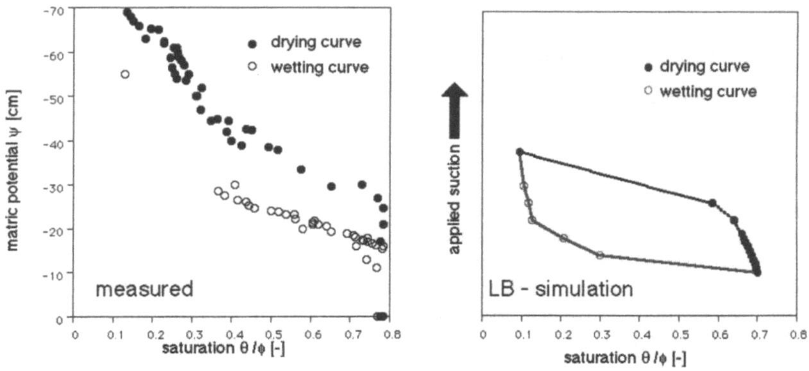

Fig. 2.3. Hysteresis curve: experiment and simulation

2.5 Acknowledgements

P. L. thanks Peter Wyss from the Swiss Federal Laboratories for Materials Testing and Research for the Micro-X-ray tomography.

References

1. Y.H. Qian, D. d'Humieres and P. Lallemand, Lattice BGK models for Navier-Stokes equation, Europh. Lett. **17**(6), 479 (1992)
2. S. Chen, Z. Wang, X. Shan and G.D. Doolen, Lattice BGK models for Navier-Stokes equation, J. Stat. Phys. **68**, 379 (1992)
3. O. Filippova and D. Hänel, Grid refinement for lattice-BGK models, J. Comp. Phys., **147**, 219 (1998)
4. C. Fan, B. Wamsley and J. W. Gentry, The effect of Stokes and Reynolds numbers on the collection efficiency of grid filters, J. of Col. Int. Sci., **65**(1), 162 (1978)
5. P. Lehmann, F. Stauffer, C. Hinz, O. Dury and H. Flühler, Effect of hysteresis on water flow in a sand column with a fluctuating capillary fringe, J. Cont. Hydr., **33**, 81-100, (1998)
6. R.H. Brooks and A.T. Corey, Properties of porous media affecting fluid flow, J. of the Irrigation and Drainage Division, ASCE, **92** (IR2), 61-88, (1966)
7. I. Fatt, The network model of porous media, I: Capillary Pressure Characteristics, Petroleum Transactions, AIME, **207**, 144 - 159, (1956)
8. Y. Mualem and E.E. Miller, A hysteresis model based on an explicit domain-dependence function, Soil Sci.Soc.Am.J., **43**, 1067-1073, (1979)
9. J.W. Hopmans, M. Cislerova and T. Vogel, X-ray tomography of soil properties, Tomography of soil-water-root processes, Special Publication 36, Soil Science Society of America, 17-28, (1994)
10. A.K. Gunstensen and D.H. Rothman, Lattice Boltzmann model of immiscible fluids, Phys. Rev. A, **V 43 N 8**, 4320-4327, (1991)

Interface Tracking in Multiphase Flows

A.-K. Tornberg and B. Engquist

Department of Numerical Analysis and Computing Science, KTH,
S-100 44 Stockholm, Sweden

Summary. Accurate numerical representation of fluid interfaces is important in high resolution simulations of immiscible multiphase flows. A new method for interface approximation, the segment projection method, is introduced. This method is coupled to a two-dimensional finite element approximation of the Navier-Stokes equations. The segment projection method is compared to level-set and front tracking methods.

1. Introduction

The dynamics of immiscible multiphase flow is strongly influenced by the location of the interfaces separating the different fluid phases. The interfaces indicate discontinuities in density and viscosity and determine the surface tension forces. In the challenging numerical simulation of immiscible multiphase flow it is therefore important to approximate the locations and effects of these interfaces in an efficient and accurate way.

It is possible to use an adaptive grid which follows the interfaces but it is often much more efficient to keep the grid for the flow variables fixed Eulerian and to represent the interfaces by some other technique.

One such method is the front tracking method [4],[6],[7]. A discrete set of points $\{\mathbf{x}^{(l)}(t)\}_{l=1}^{L}$ distributed at the interface is advected with the flow

$$\frac{d\mathbf{x}^{(l)}}{dt} = \mathbf{u}(\mathbf{x}^{(l)}, t), \tag{1.1}$$

where $\mathbf{u}$ is the flow velocity. The points are connected, for example, by spline interpolation which represents the interfaces. This method is computationally efficient and accurate but difficult to extend to flows that include merging of separated volumes of the same phase.

Another technique is the level-set method which can easily handle merging, [1]-[5]. It gives an implicit representation of the fluid interfaces. They are given as zero level sets of a continuous function $\phi(\mathbf{x}, t)$ which for two-phase flow is positive in one phase and negative in the other. The level-set function is numerically advected on an Eulerian grid as given by the differential equation

$$\frac{\partial \phi}{\partial t} + \mathbf{u} \cdot \nabla \phi = 0. \tag{1.2}$$

We introduce a new method, the segment projection method, which can be seen as a compromise between the methods mentioned above. In this method each interface in two dimensional flow is represented by a union of curve segments. The segments are chosen such that they are functions of one of the spatial coordinate variables and time, $y = f_i(x,t)$ or $x = g_j(y,t)$. Denoting the velocity field $\mathbf{u} = (u,v)$, the dynamics of an interface is given by the differential equations

$$\frac{\partial f_i}{\partial t} + u\frac{\partial f_i}{\partial x} = v, \tag{1.3}$$

$$\frac{\partial g_j}{\partial t} + v\frac{\partial g_j}{\partial y} = u. \tag{1.4}$$

The functions in the segment projection method are given on a fixed Eulerian grid which facilitates the handling of merging. One advantage of the segment projection over the level-set technique is that the segment functions $f_i(x,t)$ and $g_j(y,t)$ are only functions of one space variable. Another is better control of the merging process. The level-set method is however easier to implement.

The interface tracking methods discussed above are coupled to a finite element approximation of the Navier-Stokes equations.

2. The multiphase flow equations

The equations describing this immiscible multiphase flow are the Navier-Stokes equations for incompressible flow in two space dimensions. The contribution of the surface tension forces is in addition to the gravity forces added as a source term.

In this presentation, we assume that we have two different fluids, fluid A and fluid B. The density and viscosity at a fixed time are given by

$$(\rho(\mathbf{x}),\mu(\mathbf{x})) = \begin{cases} (\rho_A,\mu_A) & \text{for } \mathbf{x} \text{ in fluid A,} \\ (\rho_B,\mu_B) & \text{for } \mathbf{x} \text{ in fluid B.} \end{cases} \tag{2.1}$$

In general $\rho_A \neq \rho_B$ and $\mu_A \neq \mu_B$, so that the density and viscosity are discontinuous at each interface separating fluid A and B. The Navier-Stokes equations can be written as

$$\rho\left(\frac{\partial \mathbf{u}}{\partial t} + \mathbf{u}\cdot\nabla\mathbf{u}\right) = -\nabla p + \nabla\cdot\left(\mu\ \left(\nabla\mathbf{u} + \nabla\mathbf{u}^T\right)\right) + \mathbf{f} + \rho\mathbf{g} \tag{2.2}$$

$$\nabla\cdot\mathbf{u} = 0 \tag{2.3}$$

in $\Omega \subset \mathbb{R}^2$, together with boundary conditions $\mathbf{u} = v$ on $\partial\Omega$. In addition, we need to specify some appropriate initial condition for the velocity field $\mathbf{u}(\mathbf{x},t)$. In the above equations $p(\mathbf{x},t)$ denotes the pressure field. Buoyancy effects arises from the source term $\rho\mathbf{g}$, where $\mathbf{g}$ is the gravitational force. The source term $\mathbf{f}$ is the surface tension force. It is given by

$$\mathbf{f} = \sigma\kappa\hat{\mathbf{n}}\delta_\gamma, \tag{2.4}$$

where δ_γ is a measure of Dirac delta function type with support on γ. Further, $\sigma \in \mathbb{R}$ is the surface tension coefficient, $\kappa \in \mathbb{R}$ is the curvature and $\hat{\mathbf{n}} \in \mathbb{R}^2$ is a normal vector to the interface γ. At any point along an interface γ, the direction of this force is towards the local center of curvature. The interface γ is advected by the velocity field $\mathbf{u}$ and may change topology due to merging.

The singular surface tension force $\mathbf{f}$ and the discontinuous coefficients ρ and μ are better represented numerically starting from a weak form of the equations. For this formulation, we introduce the spaces

$$\mathsf{V}_\nu = \{\mathbf{v} \in H^1(\Omega)^2 : \mathbf{v} = \nu \text{ on } \partial\Omega\}, \quad \Pi = \{q \in L^2(\Omega) : \int_\Omega q \, d\mathbf{x} = 0\}.$$

Multiplying equation (2.2) by $\mathbf{v} \in \mathsf{V}_0$, (2.3) by $q \in \Pi$, and integrating over the domain using Green's formula, the following variational formulation of (2.2)-(2.3) is obtained.

Find $\mathbf{u}(\mathbf{x}, t) \in \mathsf{V}_\nu$ and $p(\mathbf{x}, t) \in \Pi$ such that $\forall t \in [0, T]$

$$m(\rho, \mathbf{u}_t, \mathbf{v}) + \tilde{a}(\mu, \mathbf{u}, \mathbf{v}) + b(\mathbf{v}, p) + c(\rho, \mathbf{u}, \mathbf{u}, \mathbf{v}) = f_\gamma(\mathbf{v}) + m(\rho, \mathbf{g}, \mathbf{v}) \tag{2.5}$$

$$b(\mathbf{u}, q) = 0 \tag{2.6}$$

hold for all $\mathbf{v}$ in V_0 and for all q in Π, respectively. The forms in the weak formulation are given in [4], i.e.

$$\tilde{a}(\mu, \mathbf{u}, \mathbf{v}) = \int_\Omega \mu \sum_{i,j=1}^{2} \left(\frac{\partial u_i}{\partial x_j} \frac{\partial v_i}{\partial x_j} + \frac{\partial u_i}{\partial x_j} \frac{\partial v_j}{\partial x_i} \right) d\mathbf{x}, \tag{2.7}$$

$$f_\gamma(\mathbf{v}) = \sigma \int_\gamma \kappa \hat{\mathbf{n}} \cdot \mathbf{v} \, d\gamma. \tag{2.8}$$

The singular terms $\nabla \cdot (\mu \, (\nabla \mathbf{u} + \nabla \mathbf{u}^T))$ and $\mathbf{f} = \sigma\kappa\hat{\mathbf{n}}\delta_\gamma$ in (2.2) are easier to approximate numerically in their weak forms (2.7) and (2.8).

2.1 Discretization

The discretization of the Navier-Stokes equations is based on a finite element approximation with piecewise quadratic polynomials in space and a second order BDF difference method in time.

The time-stepping scheme for the weak formulation of the Navier-Stokes equations is given by

$$\begin{aligned}
&\frac{1}{2\Delta t}(3\,m(\rho^{n+1}, \mathbf{u}^{n+1}, \mathbf{v}) - 2\,m(\rho^{n+1}, \mathbf{u}^n, \mathbf{v}) + m(\rho^{n+1}, \mathbf{u}^{n-1}, \mathbf{v})) + \\
&\qquad \tilde{a}(\mu^{n+1}, \mathbf{u}^{n+1}, \mathbf{v}) + b(\mathbf{v}, p^{n+1}) + c(\rho^{n+1}, \mathbf{u}^{n+1}, \mathbf{u}^{n+1}, \mathbf{v}) = \\
&\qquad\qquad 2\,f_{\gamma^n}(\mathbf{v}) - f_{\gamma^{n-1}}(\mathbf{v}) + 2\,m(\rho^n, \mathbf{g}, \mathbf{v}) - m(\rho^{n-1}, \mathbf{g}, \mathbf{v}).
\end{aligned}$$

To resolve the nonlinearities, we use a direct fixed point iteration. In each iteration, it remains to solve a Stokes problem. The divergence-free constraint is enforced using an iterated penalty method (see [4]). With this method, the problem is reformulated as an iterative procedure, where the divergence of the velocity field is removed as the pressure is obtained through iterative updates.

The elements used here are quadratic Lagrange triangles, based on a triangulation of Ω. The number of local nodes is 6. The element integrals are evaluated by using a 13-point quadrature formula (see [4]). For discontinuous functions such a quadrature formula is generically only of first order. Discontinuities will occur at γ in the forms containing ρ and μ.

By regularizing ρ and μ in a particular way, the quadrature error can be reduced and second order accuracy achieved. Introduce the modified Heaviside function

$$H_w(t) = \begin{cases} 1 & t > w \\ \nu(t/w) & |t| \leq w \\ 0 & t < -w \end{cases} \tag{2.9}$$

where $\nu(\xi)$ is a smooth transition function such that $\nu(-1) = 0$ and $\nu(1) = 1$. To define the density and viscosity fields, we introduce the approximating characteristic function $I_w(\mathbf{x})$ which is defined as

$$I_w(\mathbf{x}) = H_w(d(\mathbf{x})), \tag{2.10}$$

where $d(\mathbf{x})$ is denoting the signed distance to γ, with $d(\mathbf{x}) > 0$ in fluid A. The density $\rho(\mathbf{x})$ is given by

$$\rho(\mathbf{x}) = \rho_B - I_w(\mathbf{x})(\rho_B - \rho_A), \tag{2.11}$$

and $\mu(\mathbf{x})$ is defined analogously.

To complete this definition, the transition function $\nu(\xi)$ must be chosen. The result of the numerical evaluation of integrals with the approximating indicator function $I_w(\mathbf{x}) = H_w(d(\mathbf{x}))$ should differ as little as possible compared to the exact evaluation of the integral containing $H(d(\mathbf{x}))$. This error can be split into two parts, the analytical error made by replacing $H(d(\mathbf{x}))$ by $H_w(d(\mathbf{x}))$ and the numerical error made when evaluating the integral containing $H_w(d(\mathbf{x}))$.

The analytical error can be controlled by requiring a number of matching moments

$$\int_{-\infty}^{\infty} (H(t) - H_w(t))\, t^r dt = 0, \qquad r = 0, 1, \ldots R. \tag{2.12}$$

The numerical error decreases with increased regularity of $H_w(t)$, see [4].

3. The segment projection method

In the segment projection method, each curve is represented by a union of curve segments. The segments are chosen such that they can be given as a function of one space coordinate variable, i.e. the segments are, for each fixed time, represented by functions $f_i(x)$ and $g_j(y)$. The domain of the independent variable of such a function is the projection of the segment onto the coordinate axes.

A simple example is the description of a circle. A circle can be described by four segments, two 'x-segments', i.e. functions $f_1(x)$ and $f_2(x)$, and two 'y-segments', i.e. functions $g_1(y)$ and $g_2(y)$, see Figure 3.1.

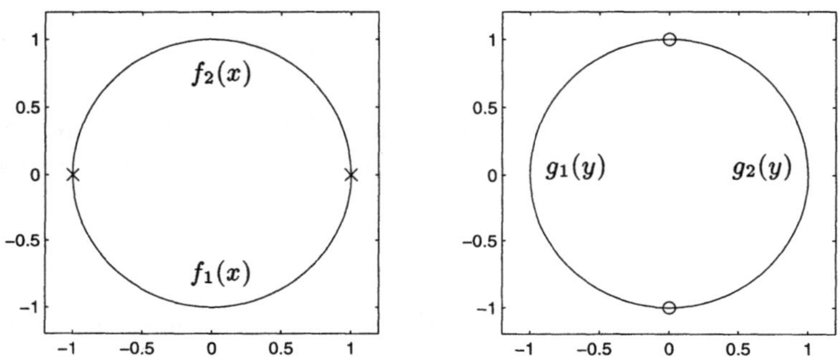

Fig. 3.1. Segments representing a circle. Extrema in x-segments yield separating break points in y-segments and vice versa.

To update the description, the segments need to be moved by the equation of motion, (1.3)-(1.4). To be able to describe the evolution of curves, we need to introduce dynamic creation and destruction of segments. After each advection step, the segment representation needs to be reinitialized and the information regarding the connectivity of segments needs to be updated.

To evaluate the interfacial force term in the multiphase flow case, given by (2.8), we need to have a representation of the interface. Since we have overlapping segments, we need to define which part(s) of each segment that should be used to define the interface. The curvature and normal vectors can easily be computed from the definition of the segments. The product $\kappa\,\hat{\mathbf{n}}$ that is included in the surface tension forces is defined by

$$\kappa(x)\,\hat{\mathbf{n}}(x) = \frac{f''(x)\,(-f'(x),1)}{(1+f'(x)^2)^2}, \tag{3.1}$$

for an x-segment. The definition for a segment in the y-direction is analogous.

In order to define the indicator function $I_w(\mathbf{x}) = H_w(d(\mathbf{x}))$, we need to define on which side of a segment there is fluid A and on which side there is fluid B. This is used to determine the sign of $d(\mathbf{x})$ as it is computed from the segment representations.

When two distinct volumes of the same fluid are close to each other, the segments may be reconnected to represent merging. We have full control of this procedure and physical conditions may be used to define if and when a merger should occur.

The segments are discretized using uniform grids. The equations for advecting the segments are approximated by the Lax-Wendroff scheme. If the spatial discretization of the segments is much finer than the resolution of the finite element grid an extra dissipative term is added. In Figure 3.2 results of a simulation including mergers are presented.

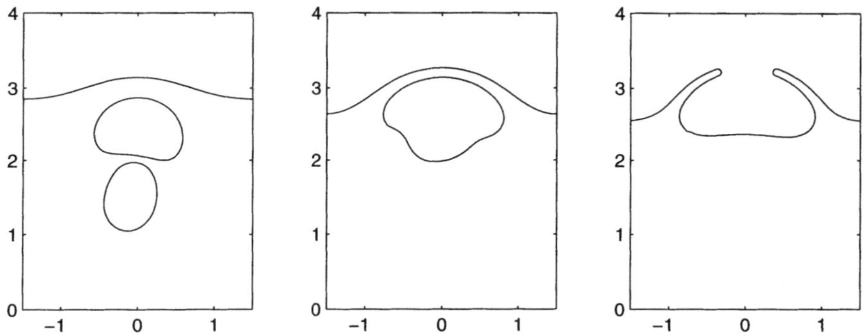

Fig. 3.2. A sequence of plots of two bubbles merging while rising due to buoyancy and breaking the surface. The simulation is based on the segment projection method.

4. Comparison of three methods for a buoyant bubble

The three techniques for representing the interface, the segment projection, the level-set and the front tracking methods were applied in a finite element simulation of a single buoyant bubble.

In the first experiment, the order of numerical accuracy was computed for the different methods after a short physical time such that the approximation is close to the asymptotic range for small time steps Δt and element sizes Δe. The results are given in Table 4.1 and indicate quite similar outcome for the three methods.

A uniform grid was used with periodic boundary conditions on the sides, and no-slip conditions at the top and bottom. The bubble was circular and at rest initially.

In the second simulation, we compared the three methods and changed the mesh to an irregular mesh defined on a taller domain to let the bubble

method	$\|\mathbf{u}_h - \mathbf{u}_{h/2}\|_1$	$\|\mathbf{u}_{h/2} - \mathbf{u}_{h/4}\|_1$	$\frac{\|\mathbf{u}_h - \mathbf{u}_{h/2}\|_1}{\|\mathbf{u}_{h/2} - \mathbf{u}_{h/4}\|_1}$	p
segment projection	$3.73 \cdot 10^{-2}$	$1.04 \cdot 10^{-2}$	3.58	1.84
level-set	$3.28 \cdot 10^{-2}$	$1.01 \cdot 10^{-2}$	3.26	1.70
front tracking	$3.38 \cdot 10^{-2}$	$1.00 \cdot 10^{-2}$	3.37	1.75

Table 4.1. Measured order of convergence p for the three different methods. Runs with $\Delta t = \Delta t_0$ for $\Delta e = h$, $\Delta t = \Delta t_0/2$ for $\Delta e = h/2$, $\Delta t = \Delta t_0/4$ for $\Delta e = h/4$. $\Delta t_0 = 1.25 \cdot 10^{-3}$, $h = 0.2$.

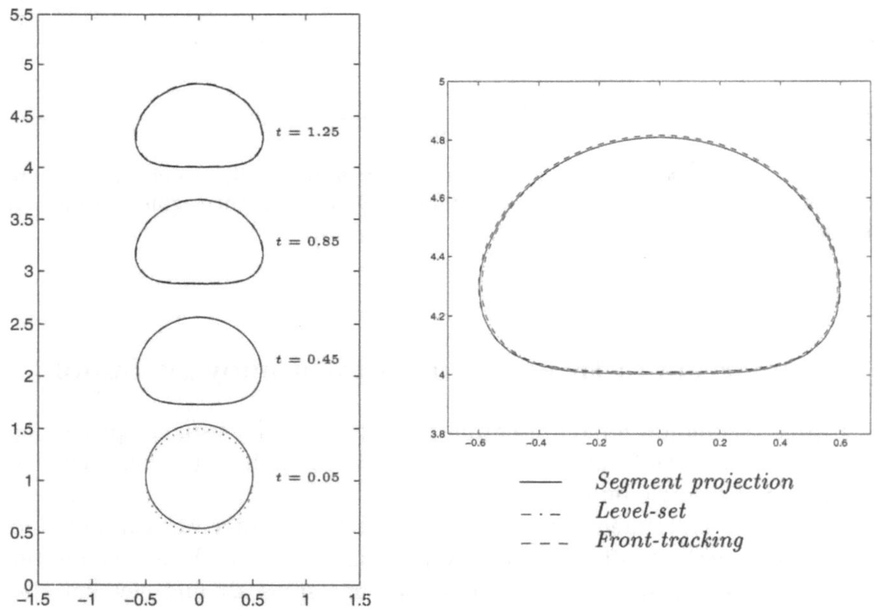

Fig. 4.1. The bubble interface plotted at different times. The dotted contour at the bottom indicates the initial position. Results for all three methods are plotted on top of each other. The plot to the right shows a close up of the bubble interface at the last time. $\rho_B/\rho_A = 100$ and $\mu_B/\mu_A = 10$.

rise further. The initial configuration and the boundary conditions were the same as above. In Figure 4.1, the bubble contours from the three different methods are plotted at a sequence of different times. The contours from the three methods are in good agreement, both when it comes to position and shape of the contours. The bubble in the level-set simulation has however lost most mass (i.e area since ρ_A is constant). At the last time plotted, the decrease compared to the initial area is 2.26%. For the segment projection method, we have a decrease of 0.76%. The front tracking method yields the best area conservation for this run, the area is increased by 0.15%.

All of the three methods give similar results. The level-set method is the least accurate with respect to mass conservation, but the simplest to implement for merging flows. The new segment projection method shows potential to be a good compromise between the front tracking and the level-set methods.

References

1. S. Osher and J.A. Sethian. Fronts Propagating with Curvature Dependent Speed: Algorithms Based on Hamilton-Jacobi Formulations. *Journal of Computational Physics*, 79:12–49, 1988.
2. M. Sussman, M. Fatemi, P. Smereka, and S. Osher. An Improved Level Method for Incompressible Two-Phase Flows. *Computers & Fluids*, 27:663–680, 1998.
3. M. Sussman, P. Smereka, and S. Osher. A Level Set Approach for Computing Solutions to Incompressible Two-Phase Flow. *Journal of Computational Physics*, 114:146–159, 1994.
4. A-K. Tornberg. A Finite Element Based Level Set Method for Multiphase Flow Simulations. Licentiate's thesis, Royal Institute of Technology, Department of Numerical Analysis and Computer Science, September 1998. ISBN 91-7170-317-9, TRITA-NA-9817.
5. A-K. Tornberg, and B. Engquist. A Finite Element Based Level Set Method for Multiphase Flow Applications. To appear in *Computing and Visualization in Science.*
6. A-K. Tornberg, R. W. Metcalfe, R. Scott, and B. Bagheri. A Fluid Particle Motion Simulation Method. In *Computational Science for the 21st Century*, pages 312–321. John Wiley and Sons, New York, 1997.
7. S.O. Unverdi and G. Tryggvason. A Front-Tracking Method for Viscous, Incompressible, Multi-Fluid Flows. *Journal of Computational Physics*, 100:25–37, 1992.

Section 2

Mechanics of Materials and Multiscaling

Edited by C. Miehe

Finite Element Computation of Macroscopic Quantities in Nonconvex Minimisation Problems and Applications in Materials Science

Sören Bartels[1], Carsten Carstensen[1] and Petr Plecháč[2]

[1] Mathematisches Seminar der Christian-Albrechts-Universität zu Kiel, Germany
[2] Mathematical Sciences Department, University of Delaware, USA

Summary. Many physical processes can be modelled by minimising a given energy functional $E(u)$ amongst a class of admissable displacements. Convexity conditions guarantee sequential lower semicontinuity of E and then accumulation points of infimising seqences are minimisers, i.e., the minimum of E is attained. For some important problems involving phase transitions in nonlinear elasticity or ferromagnetic bodies, these convexity properties are not valid and so E is not sequential lower semicontinuous; in general the minimum is not attained. Typical features of infimising sequences include their weak accumulation points regarded as a macroscopic displacement of a (generalised) solution or their oscillations in terms of Young measures to describe the phases of the strains. Those macroscopic quantities can be described within relaxed formulations. This article focuses on the numerical treatment of relaxed problems where an analytic formula of the relaxed, (quasi-) convexified energy density is known.

1. Introduction

The minimisation of nonconvex functionals in applied physics leads to highly oscillating microstructures in the solution. Two examples will be adressed in this article: the scalar double-well problem and a problem from nonlinear elasticity.

Numerical schemes for these problems have to find a global minimiser (in a discrete subspace) amongst a huge number of local ones. Classical iteration methods fail to work properly here; a few thousand starting values are needed before a scheme finds a global minimiser. To avoid these severe difficulties and since one is mainly interested in the macroscopic quantities, we consider different forms of convexifications of the related functionals. This enables efficient implementation and a priori as well as a posteriori error estimates. In all the three examples the most important microscopic quantities can be recovered from the solution of the relaxed problem.

The generic example in microstructure is due to L.C. Young [18].

Problem (P_α). Minimise

$$E(u) := \int_0^1 (u'^2 - 1)^2 + \alpha u^2 \, dx$$

amongst the space $\mathcal{A}$ of all Lipschitz-continuous functions $u : (0,1) \to \mathbb{R}$ with $u(0) = u(1) = 0$.

For $\alpha = 0$, the infimum of E over $\mathcal{A}$ equals zero and there are infinitely many minimisers. Indeed, every Lipschitz-function with slopes ± 1 almost everywhere that satisfies the boundary conditions is a minimiser. For instance, $u(x) = \min\{x, 1-x\}$ is a solution and Fig. 1.1 shows other examples.

For $\alpha = 1$, the infimum of E over $\mathcal{A}$ is also zero since there exist functions with slope ± 1 almost everywhere and with arbitrary small amplitude. The infimum is not attained because there is no admissable function satisfying $u \equiv 0$ with slopes ± 1. Some functions u_j satisfying $E(u_j) = 4^{-(j+1)}/3$ are depicted in Fig. 1.1; note that they are also solutions for (P_0). The sequence

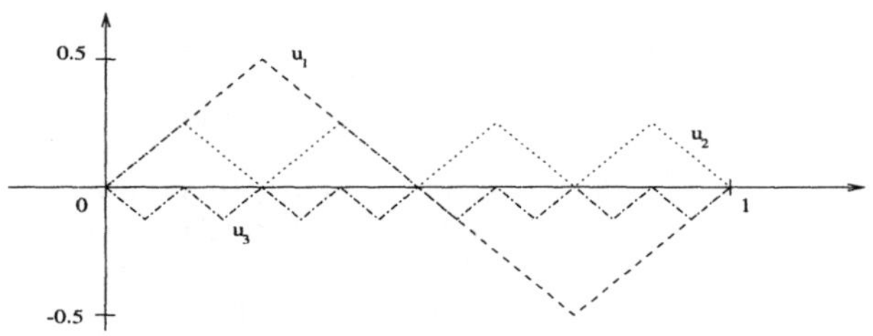

Fig. 1.1. First few members of a possible minimising sequence for (P_1)

(u_j) infimises E and converges strongly to zero in $L^\infty(0,1)$, but the sequence (u_j') only converges weakly to zero in $L^4(0,1)$. Obviously, $u \equiv 0$ is no minimiser but describes the macroscopic quantities of a (nonclassical) minimiser. The lacking strong convergence is caused by oscillations of (u_j') which can be described within a statistical notion by Young measures as follows: Let $\nu_{x,\delta}^j$ denote the probability distribution of $u_j'(y)$ where y is taken uniformly at random from the ball $B(x,\delta)$ of radius δ around x. Then, for a subsequence,

$$\nu_x := \lim_{\delta\to 0} \lim_{j\to\infty} \nu_{x,\delta}^j \quad \text{(weak* in measure)}$$

defines a probability measure, the related Young measure. It can be shown (cf. [14]) that for (P_1), any minimising sequence generates the Young measure

$$\nu_x = \tfrac{1}{2}\delta_{-1} + \tfrac{1}{2}\delta_1, \tag{1.1}$$

where $\delta_{\pm 1}$ denotes the Dirac measure with mass at $+1$ resp. -1. The representation (1.1) corresponds to the fact that we need half the slopes to be one and half of the slopes to be -1 to obtain a macroscopic slope zero. In this sense, (1.1) describes the microscopic mechanism, they do both lower the double-well energy $(u'^2 - 1)^2$ and have a macroscopic limit $u = 0$. The fact that $\nu_x \neq \delta_{u'(x)}$ causes microstructure and excludes strong convergence of infimising sequences.

A relaxation of (P_1) considers the functional E^{**} in which the nonconvex energy density $W(x) = (x^2 - 1)^2$ is replaced its convexification CW.

Problem (RP_1). Seek a minimiser $u \in \mathcal{A}$ of the functional

$$E^{**}(u) := \int_0^1 CW(u') + u^2\,dx.$$

One easily verifies that CW is given by $CW(x) = 0$ for $|x| \le 1$ and $CW(x) = W(x)$ for $|x| > 1$. Obviously, $u \equiv 0$ is the unique solution for (RP_1).

The advantage in considering problem (RP_1) instead of (P_1) is that E^{**} is a convex functional so that any classical minimisation algorithm can be used to compute an approximation of a minimiser of E^{**}. This is not the case for E and one can show that classical algorithms fail to compute a minimiser of E in a finite-dimensional space.

The remaining part of this survey article treats more-dimensional and more applied examples and is organised as follows. Section 2 starts with a scalar model problem to illustrate the most important results from relaxation in modern theory of calculus of variations and gives an error analysis for the numerical solution. We consider a problem from nonlinear elasticity in Section 3.

2. Scalar double-well problem and its numerical solution

In a variational model for phase transitions in certain crystalline alloys, fine mixtures of phases are described by oscillating infimising sequences of the stored elastic energy [2, 3]. The variational framework is based on the minimisation of an energy functional with a non-convex energy density. In the double-well problem for two distinct given vectors $F_1, F_2 \in \mathbb{R}^d$ this energy density is defined by

$$W(F) := |F - F_1|^2 |F - F_2|^2, \tag{2.1}$$

for $F \in \mathbb{R}^d$. The variational problem (P) reads as follows.

Problem (P). Find a function $u \in \mathcal{A}$ that minimises the energy functional

$$E(u) := \int_\Omega \left(W(\nabla u) + |u - f|^2\right) dx$$

in a body Ω, a bounded Lipschitz-domain in $\mathbb{R}^d$ with boundary Γ. The non-physical term $|u - f|^2$ for a given function $f \in L^2(\Omega)$ serves as regularisation term. The set of admissible functions is $\mathcal{A} := u_D + W_0^{1,p}(\Omega)$ for given $u_D \in W^{1,p}(\Omega)$ and $W_0^{1,p}(\Omega) = \{v \in W^{1,p}(\Omega) : v|_\Gamma = 0\}$.

As shown in the introduction and depending on the given data, (P) may have infinitely many or no solutions at all. The direct method of the calculus of variations does not give the existence of solutions since due to the non-convex nature of W the functional E is not weakly lower semicontinuous. Nevertheless, one is interested in information on minimising sequences.

A finite element method replaces the space of admissible functions $\mathcal{A}$ by a discrete subspace $\mathcal{A}_h \subset \mathcal{A}$ and seeks a minimiser in $\mathcal{A}_h$ which exists since $\mathcal{A}_h$ is finite-dimensional and E satisfies certain growth conditions. For an overview of approximation results in model examples we refer to [8, 9, 13]. Since local minimisers cluster around global ones, it requires a huge amount of computational work to minimise E directly over $\mathcal{A}_h$ (cf. [4]) and numerical solutions are strongly dependent on starting values.

Relaxation utilising Young measures. Minimising sequences (∇u_j) for (P) are typically weakly but *not* strongly convergent and their weak limits are in general not solutions of (P). The reason for this is their oscillatory nature. To describe this behaviour by a generalized limit, one is led to measure-valued strains: The bounded sequence (∇u_j) generates a family of probability measures $(\nu_x)_{x\in\Omega}$ on $\mathbb{R}^d$, the *gradient Young measure* $\nu \in YM(\Omega;\mathbb{R}^d)$, such that, for all $g \in C(\mathbb{R}^d)$ that vanish at infinity, there holds

$$(g(\nabla u_j)) \rightharpoonup^* \int_{\mathbb{R}^d} g(F)\, d\nu(F) \quad \text{in } L^\infty(\Omega). \tag{2.2}$$

We refer to [1, 14, 15, 16] for details.

The weak limit ∇u of a sequence of gradients (∇u_j) for a minimising sequence (u_j) for (P) is related to a Young measure by the identity

$$\nabla u(x) = \int_{\mathbb{R}^d} F\, d\nu_x(F) =: \langle \nu_x, id\rangle \quad \text{a.e. in } \Omega \tag{2.3}$$

which follows from (2.2) with $g = id$ in a neighbourhood of $\operatorname{supp}\nu$, and gives a connection between the macroscopic strain ∇u and the gradient Young measure ν. Since one is often interested only in the macroscopic quantities ∇u, $DW(\nabla u)$, and the generated Young measure, relation (2.3) motivates **Problem (*GP*).** Seek a minimiser of

$$GE(u,\nu) := \int_\Omega \langle \nu_x, W\rangle\, dx + \int_\Omega |u - f|^2\, dx \tag{2.4}$$

amongst all $(u,\nu) \in \mathcal{B} := \{(v,\mu) \in \mathcal{A} \times YM(\Omega;\mathbb{R}^d);\, \nabla v(x) =< \mu_x, id > \text{ for a.a. } x \in \Omega\}$.

Note that in contrast to (P), (GP) always has solutions and $\inf_{\mathcal{A}} E = \min_{\mathcal{B}} GE$ by construction of $\mathcal{B}$. Moreover, if (∇u_j) is a minimising sequence for (P) that generates the gradient Young measure ν and which converges weakly to ∇u, then (u,ν) is a minimiser for (GP) (cf. [15]). This situation is more involved in the vectorial case (cf. [14, 17]).

A numerical scheme for (GP) would replace the set $\mathcal{A}$ by a finite element space $\mathcal{A}_h$ and the space $YM(\Omega;\mathbb{R}^d)$ by elementwise convex-combinations of Dirac measures (cf. [7]). This approach results in a very high number of degrees of freedom.

Relaxation by convexification. Another relaxation of (P) can be obtained from (GP) by minimising the two contributions in (2.4) seperately. For fixed $F = \nabla u(x)$ one can find a probability measure ν_x such that ν_x minimises the expression $\langle \mu, W\rangle$ amongst all probability measures μ satisfying $\langle \mu, id\rangle = F$. Sometimes it is even possible to calculate an explict formula for a function QW that satisfies

$$QW(F) = \min_{\langle \mu, id \rangle = F} \langle \mu, W \rangle. \tag{2.5}$$

In the scalar case QW is given by the second Legendre transform W^{**} of W which equals the lower convex hull CW of W. Equations (2.5) and (2.4) motivate **Problem (*RP*).** Find a minimiser $u \in \mathcal{A}$ for the functional

$$E^{**}(u) = \int_\Omega W^{**}(\nabla u)\,dx + \int_\Omega |u - f|^2\,dx. \tag{2.6}$$

The map W^{**} is constructed in [5] and given by the formula

$$W^{**}(F) = \Big(\big(|F-B|^2 - |A|^2\big)_+\Big)^2 + 4\Big(|A|^2|F-B|^2 - \big(A^T \cdot (F-B)\big)^2\Big),$$

where $A := (F_2 - F_1)/2 \neq 0$, $B := (F_1 + F_2)/2$, and $(\cdot)_+ := \max\{\cdot, 0\}$.

Note that (RP) always has a solution and that $\inf_{\mathcal{A}} E = \min_{\mathcal{A}} E^{**}$. Moreover, if (∇u_j) is a minimising sequence for (P) which converges weakly to ∇u, then u is a solution for (RP) (cf. [15]).

The convexified problem (RP) plays an important role in the scalar case since the most relevant information concerning (P) such as the Young measure and the macroscopic displacement and stress field can be recovered from a minimiser u of E^{**}. Since two different energy densities might have the same lower convex envelope, one loses information on the actual problem, in general. In the vectorial case one has to consider the quasiconvex hull of W which is harder to compute than and in general different from CW.

Numerical analysis of the convexified problem allowing microstructure. In this subsection we outline the numerical analysis of the relaxation (RP) of problem (P) and seek a minimiser for E^{**} in a discrete subspace $\mathcal{A}_h \subset \mathcal{A}$ based on a *regular triangulation* $\mathcal{T}$ of Ω into closed triangles and parallelograms in the sense of [10]. We allow energy densities more general than (2.1). As mentioned before, problems like (P) may have infinitely many or no solutions at all but there usually exists a Young measure solution ν from which one can calculate the macroscopic stress field

$$\sigma := \int_{\mathbb{R}^d} DW(A)\,d\nu(A) \stackrel{!}{=} DW^{**}(\nabla u) =: \sigma^{**} \tag{2.7}$$

where a solution u of the convexified problem (RP) defines σ^{**} which actually equals σ in (2.7) and is independent of the choice of the minimiser u. Similarly, a discretisation of (RP) gives a solution $u_h \in \mathcal{A}_h$ and again, the associated stress field σ_h is independent of the choice of the minimiser.

For W given in (2.1) the gradient Young measure ν can be calculated as soon as one knows a solution of (RP): Define $A_0 := A/|A|$ and let $P = 1 - A_0 \cdot A_0^T$ denote the orthogonal projection in $\mathbb{R}^d$ onto $\operatorname{span}\{A\}^\perp$. Let

$$\begin{aligned}
\lambda(F) &:= \Big(1 + A_0^T \cdot (F-B)\big(|A|^2 - |P\cdot(F-B)|^2\big)^{-1/2}\Big)/2,\\
S_\pm(F) &:= B + P\cdot(F-B) \pm \big(|A|^2 - |P\cdot(F-B)|^2\big)^{1/2} A_0,
\end{aligned}$$

and define the map $\tilde{\mu}$, that maps a vector $F \in \mathbb{R}^d$ to a probability measure $\tilde{\mu}(F)$, by

$$\tilde{\mu}(F) := \begin{cases} \delta_F & \text{if } |A| \leq |F - B|, \\ \lambda(F)\, \delta_{S_+(F)} + (1 - \lambda(F))\, \delta_{S_-(F)} & \text{if } |F - B| < |A|. \end{cases}$$

Then, $\mu : u \mapsto \tilde{\mu}(\nabla u)$ defines a map $\mu : W^{1,p}(\Omega) \to YM(\Omega; \mathbb{R}^d)$ such that, if (∇u_j) is a minimising sequence for E in $\mathcal{A}$ which converges weakly to $u \in \mathcal{A}$ and which generates the Young measure ν, then $\nu_x = \mu(\nabla u(x))$ for a.a. $x \in \Omega$ (cf. [5]). The map also allows us to calculate discrete Young measures $\nu_x^h := \mu(\nabla u_h(x))$ for solutions u_h of (RP_h), thus recovering microscopic information from averaged quantities.

A priori error estimates. The following results hold for minimising problems related to more general energy densities than W from (2.1). We refer to [5] for a more general version of the subsequent theorem in which $\mathcal{A}_h \subset \mathcal{A}$ is a discrete subspace.

Theorem 2.1 ([5]). *There exist minimisers u and u_h of (2.6) in $\mathcal{A}$ resp. $\mathcal{A}_h$. The stress $\sigma := DW^{**}(\nabla u)$ and the discrete stress $\sigma_h := DW^{**}(\nabla u_h)$ are determined in a unique way. Furthermore, if $u, u_h \in W^{1,4}(\Omega)$*

$$\begin{aligned} \|\sigma - \sigma_h\|^2_{L^{4/3}(\Omega)} &+ \|u - u_h\|^2_{L^2(\Omega)} \\ &\leq c_1 \inf_{v_h \in \mathcal{A}_h} \left(\|u - v_h\|^2_{L^2(\Omega)} + \|\nabla u - \nabla v_h\|^2_{L^4(\Omega)} \right). \end{aligned}$$

Since no higher regularity results are known for problems of the form (RP), the error estimate is of limited use for error control in practice. Therefore, a posteriori error estimates, which allow us to estimate the error in terms of the computed approximation, are of particular interest.

A posteriori error estimates. Given a regular triangulation $\mathcal{T}$ of Ω we can define the lowest order finite element space $\mathcal{S}^1(\mathcal{T})$ which consists of all continuous functions $v : \Omega \to \mathbb{R}$ such that $v|_T$ is affine for all $T \in \mathcal{T}$. For the numerical treatment of (RP) we set $\mathcal{S}^1_0(\mathcal{T}) := \mathcal{S}^1(\mathcal{T}) \cap W^{1,4}_0(\Omega)$ and define $\mathcal{A}_h := u_D + \mathcal{S}^1_0(\mathcal{T})$. Let $\mathcal{E}$ be the set of all edges appearing in $\mathcal{T}$ and note that $\cup\mathcal{E}$, the union of all edges, is the skeleton of edges in $\mathcal{T}$. We denote by $h_T := \text{diam}(T)$, $T \in \mathcal{T}$, and $h_E = \text{diam}(E)$, $E \in \mathcal{E}$, the diameters of elements and edges in $\mathcal{T}$.

Given a solution $u_h \in \mathcal{A}_h$ of (RP_h), let $\sigma_h := DW^{**}(\nabla u_h)$. For all $T \in \mathcal{T}$ the *error indicator* η_T^R is defined by

$$\eta_T^R := h_T^{4/3} \int_T |2(f - u_h) + \operatorname{div} \sigma_h|^{4/3} dx + \sum_{E \subseteq \partial T \setminus \Gamma} h_E \int_E |[\sigma_h \cdot n_E]|^{4/3} ds. \quad (2.8)$$

The integrand $[\sigma_h \cdot n_E]$ denotes the jump of the discrete stresses $\sigma_h \cdot n_E$ along a face E of two neighbouring elements, n_E is a unit normal vector of a fixed orientation along E, and the summation in (2.8) is over all such faces of T.

Theorem 2.2 ([5]). *There exists a constant $c_2 > 0$ which is independent of $h_{\mathcal{T}}$, such that*

$$\|\sigma - \sigma_h\|^2_{L^{4/3}(\Omega)} + \|u - u_h\|^2_{L^2(\Omega)} \leq c_2 \Big(\sum_{T \in \mathcal{T}} \eta_T^R \Big)^{3/8} =: c_2\, \eta_R .$$

The proof of the theorem requires the use of the rough estimate $\|u - v_h\|_{W^{1,4}(\Omega)} \leq c_3$ for an approximation $v_h \in \mathcal{A}_h$ of u, and so loses efficiency when keeping reliability.

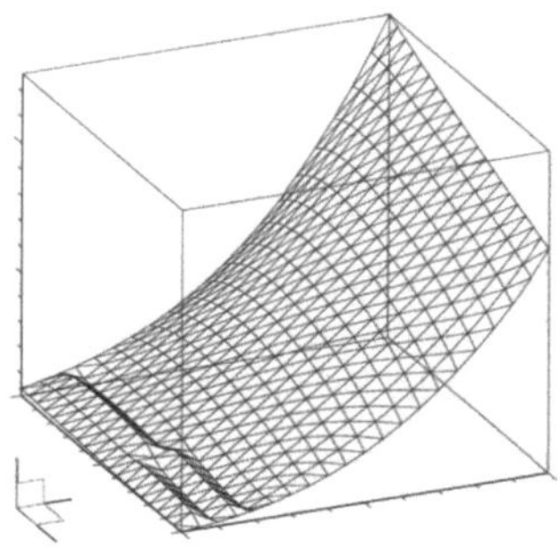

Fig. 2.1. Solution for the double-well problem. The figure shows a solution for the problem (P_h) on the unit square. Microstructure can be observed close to the left lower edge

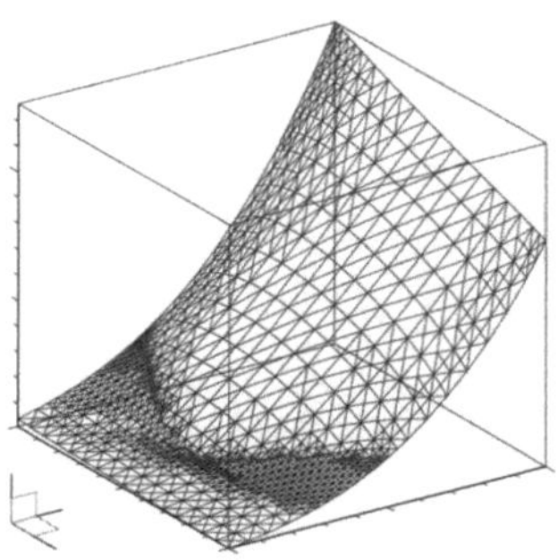

Fig. 2.2. Solution for the relaxed double-well problem using the adaptive Algorithm 1. Note that in contrast to the (discrete) solution for (P_h) no microstructure can be observed

The local error indicators η_T^R can be used in the following algorithm for adaptive mesh refinement.

Adaptive Algorithm (A). (a) Start with a coarse mesh $\mathcal{T}_0$, $k = 0$.
(b) Compute a minimiser $u_h \in \mathcal{A}_h$ of E^{**} with respect to the mesh $\mathcal{T}_k$.
(c) Compute η_T^R for each T in $\mathcal{T}$.
(d) Compute the error bound η_R and decide to terminate or to continue and then go to (e).
(e) Refine, i.e., halve the largest edge of, $T \in \mathcal{T}$ provided $\eta_T^R \geq \frac{1}{2} \max_{T' \in \mathcal{T}} \eta_{T'}^R$.

(f) Refine further elements to avoid hanging nodes and thereby define a new mesh $\mathcal{T}_{k+1}$, update k, and go to (a).

For a numerical example let $d = 2$, $\Omega = (0,1)^2$, $F_{1,2} := \mp(\cos\frac{\pi}{6}, \sin\frac{\pi}{6})$, and define $f(x,y) := f_0(x\cos\frac{\pi}{6} + y\sin\frac{\pi}{6})$ in Ω, $u_D(x,y) := f_1(x\cos\frac{\pi}{6} + y\sin\frac{\pi}{6})$ on $\partial\Omega$, for $f_0(s) := -\frac{3}{128}(s-0.5)^5 - \frac{1}{3}(s-0.5)^3$ and $f_1(s) := f_0(s)$ if $0 \le s \le 0.5$, $f_1(s) := \frac{1}{24}(s-0.5)^3 + (s-0.5)$ if $0.5 < s \le 1$.

Then, the discrete deformations for (P_h) and (RP_h) are shown Fig. 2.1 and Fig. 2.2 respectively. The microstructure seen in the solution for (P_h) vanishes in the solution for (RP_h).

3. Numerical analysis of linearised phase transitions in elastic solids

The variational model of an elastic body which occupies a bounded Lipschitz-domain $\Omega \subset \mathbb{R}^d$, $d = 2,3$, with two distinct zero-stress strain phases F_1 and F_2 in $\mathbb{R}^{d\times d}_{sym} := \{F \in \mathbb{R}^{d\times d} : F = F^T\}$ defines a double-well problem similar to the scalar one discussed above. For each well F_j, $j = 1,2$, with minimal energy $W_j^0 \in \mathbb{R}$ one has a quadratic elastic energy which is, for $F \in \mathbb{R}^{d\times d}_{sym}$, defined by (cf. [12])

$$W_j(F) := \tfrac{1}{2}\langle F - F_j, \mathbb{C}(F - F_j)\rangle + W_j^0,$$

where $\mathbb{C}$ is a fourth-order elasticity tensor and $\langle\cdot,\cdot\rangle$ denotes the scalar product in $\mathbb{R}^{d\times d}$. Energy minimisation balances the configuration of the two phases and so the strain energy density W is modelled by the minimum of W_1 and W_2 (cf. [12]), i.e.,

$$W(F) = \min\{W_1(F), W_2(F)\}.$$

The displacement $u \in \mathcal{A}$ minimises the energy functional

$$E(u) := \int_\Omega W(\epsilon(u))\,dx + \int_\Omega f\cdot u\,dx - \int_{\Gamma_N} g\cdot ds, \quad \epsilon(u) = \tfrac{1}{2}(\nabla u + \nabla u^T),$$

amongst all admissible displacements. The given functions $f \in L^2(\Omega)$ and $g \in L^2(\Gamma_N)$, $\Gamma_N \subset \partial\Omega$, describe lower order body forces. The space of admissible functions includes the Dirichlet data $u_D \in W^{1,2}(\Omega)$ and is defined by $\mathcal{A} := u_D + W_D^{1,2}(\Omega)$, where $W_D^{1,2}(\Omega) = \{v \in W^{1,2}(\Omega)^d : v|_{\Gamma_D} = 0\}$ for $\Gamma_D \subseteq \partial\Omega$ with positive surface measure.

Problem (*P*). Find a minimiser of E in $\mathcal{A}$.

As in the scalar case for (P) from Section 2, we do not have weak lower semicontinuity for E so that we cannot expect the existence of a minimiser. A relaxation allowing gradient Young measures generalising the approach in (GP) is possible but for computational reasons it is preferable to follow the ideas that led to (RP) in Section 2. The appropriate choice for the relaxed energy density is the quasiconvex envelope QW of W (cf. [11]).

Quasiconvexity is the right form of convexity to obtain weak lower semicontinuity for functionals like E to guarantee the existence of a minimiser $u \in \mathcal{A}$ of

$$QE(u) = \int_\Omega QW(\epsilon(u))\,dx + \int_\Omega f \cdot u\,dx - \int_{\Gamma_N} g \cdot u\,ds.$$

Problem (*RP*). Find a minimiser of QE in $\mathcal{A}$.

The connection between (P) and (RP) is that (RP) has solutions which are weak limits of minimising sequences for E and there holds $\inf_{\mathcal{A}} E = \min_{\mathcal{A}} QE$ (cf. [15]). Since all the important macroscopic quantities can be recovered from a solution u for (RP) it is relevant to consider this relaxed problem. For the problem at hand, the quasiconvex envelope of W is known explicitly from [12],

$$QW(F) = \begin{cases} W_2(F) & \text{if } W_2(F) + \gamma \le W_1(F), \\ \frac{1}{2}(W_2(F) + W_1(F)) - \frac{1}{4\gamma}(W_2(F) - W_1(F))^2 - \frac{\gamma}{4} & \text{if } |W_2(F) - W_1(F)| \le \gamma, \\ W_1(F) & \text{if } W_1(F) + \gamma \le W_2(F), \end{cases}$$

where γ is given by a certain projection onto the space of symmetric matrices. In case $\operatorname{rank}(F_1 - F_2) \le 1$ we have $\gamma = \frac{1}{2}\langle F_2 - F_1, \mathbb{C}(F_2 - F_1)\rangle$.

Note that in general, the quasiconvex hull of an energy density is not known explicitly.

A discretisation of (RP) consists in replacing $\mathcal{A}$ by a discrete subspace $\mathcal{A}_h$ and leads to the problem (RP_h) which has a minimiser $u_h \in \mathcal{A}_h$.

The relaxed problem and its approximation. For the numerical analysis of problem (RP) we assume that the energy wells F_1, F_2 are compatible, i.e., the difference $F_1 - F_2$ has rank one. This condition is satisfied in some applications and allows us to exploit the Galerkin-orthogonality but also implies that QW is convex.

Theorem 3.1 ([6]). *Suppose* $\operatorname{rank}(F_1 - F_2) = 1$, *and that* u *solves* (RP) *while* u_h *solves* (RP_h). *Then, for the stress fields* $\sigma := DQW(\epsilon(u))$ *and* $\sigma_h := DQW(\epsilon(u_h))$ *we have the a priori error estimate*

$$\|\mathbb{C}^{-1/2}(\sigma - \sigma_h)\|^2_{L^2(\Omega)} \le \inf_{v_h \in \mathcal{A}_h} \|\mathbb{C}^{1/2}\nabla(u - v_h)\|^2_{L^2(\Omega)}. \tag{3.1}$$

Note that u and u_h are in general not unique but (3.1) holds independent of the choice of the minimisers.

Finite element approximation and a posteriori error control. In order to derive a computable a posteriori error estimate, let $\mathcal{T}$ denote a regular triangulation of the domain Ω and suppose that $\operatorname{rank}(F_1 - F_2) = 1$.

The volume and edge residuals $R \in L^2(\Omega)^d$ and $J \in L^2(\cup\mathcal{E})^d$ are defined as residuals from the strong form of the Euler-Lagrange equations by

$$R|_T := (f + \operatorname{div} \sigma_h)|_T \quad \text{and} \quad J|_E := \begin{cases} [\sigma_h \cdot n_E] & \text{if } E \not\subseteq \Gamma, \\ g - \sigma_h \cdot n & \text{if } E \subseteq \overline{\Gamma}_N, \\ 0 & \text{if } E \subseteq \Gamma_D, \end{cases} \tag{3.2}$$

for $T \in \mathcal{T}$ and $E \in \mathcal{E}$, respectively. Moreover, we denote by $h_{\mathcal{T}}$ and $h_{\mathcal{E}}$ the functions satisfying $h_{\mathcal{T}}|_T = h_T$ and $h_{\mathcal{E}}|_E = h_E$ for all $T \in \mathcal{T}$ and $E \in \mathcal{E}$, respectively.

Theorem 3.2 ([6]). *There exists a positive constant c_4, which only depends on the shape and not on the size of the elements in $\mathcal{T}$, such that*

$$\|\mathbb{C}^{-1/2}(\sigma - \sigma_h)\|^2_{L^2(\Omega)} \leq c_4 \|u\|_{W^{1,2}(\Omega)} \big(\|h_{\mathcal{T}} R\|_{L^2(\Omega)} + \|h_{\mathcal{E}}^{1/2} J\|_{L^2(\cup\mathcal{E})}\big).$$

As in Theorem 2.2, one faces a loss of efficiency due to lacking smoothness properties of u but the estimate of the theorem still allows us to define local error indicators which can be used in Algorithm (A) for adaptive mesh refinement strategies.

Remarks on microstructures. The relaxed energy density can be seen as a convex combination of the two energies W_1 and W_2. We interpret the volume fraction θ of the area in which the material has elastic properties defined by W_1. A formula for the calculation of θ in terms of the exact strain $\epsilon(u)$ is given in [12]. A numerical approximation u_h for u thus allows us to compute an approximation θ_h of θ, although theoretical bounds are lacking.

The left plot of Fig. 3.1 shows the deformation and the modulus of the stress fields on the deformed unit square with fixed lower side and load in the vertical direction on the upper side. The material is defined through the elasticity tensor $\mathbb{C}$ and the matrices F_1, F_2 (cf. [6]). The corresponding volume fraction θ_h is shown in the right plot of Fig. 3.1.

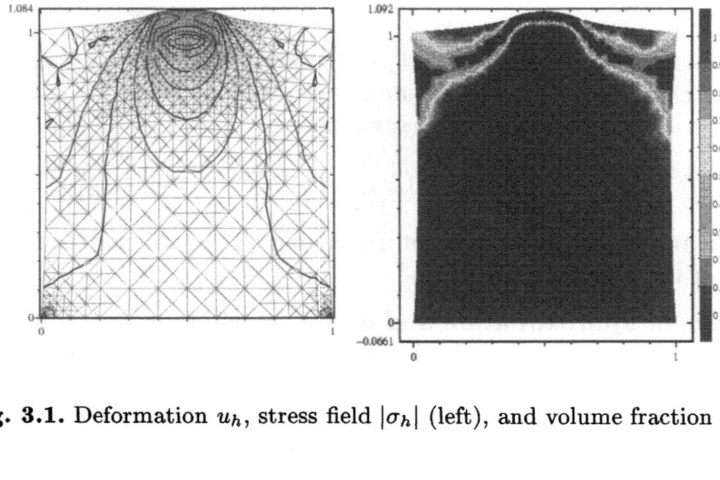

Fig. 3.1. Deformation u_h, stress field $|\sigma_h|$ (left), and volume fraction θ_h (right)

References

1. Ball, J.M. (1989), A version of the fundamental theorem for Young measures. *Partial differential equations and continuum models of phase transitions.* EDS. M RASCLE, D. SERRE, M. SLEMROD: *Lecture Notes in Physics* **344**, 207—215.
2. Ball, J.M., James, R.D. (1987), Fine phase mixtures as minimizers of energy. *Arch. Rational Mech. Anal.* **100**, 13—52.
3. Ball, J.M., James, R.D. (1992), Proposed experimental tests of the theory of fine microstructure and the two–well problem. *Phil. Trans. R. Soc. Lond. A.* **338**, 389—450.
4. Carstensen, C. (1996), Numerical Analysis of Nonconvex Minimization Problems allowing Microstructures. *Z. angew. Math. Mech.* **76** S2, 497-498.
5. Carstensen, C., Plecháč, P. (1997), Numerical solution of the scalar double-well problem allowing microstructure. *Math. Comp.* **66**, 997-1026.
6. Carstensen, C., Plecháč, P. (1999), Numerical analysis of compatible phase transitions in elastic solids. *SIAM J. Numer. Anal.*, accepted. (http://www.numerik.uni-kiel.de/reports/1998/98-16.html)
7. Carstensen, C., Roubiček, T. (1999), Numerical Approximation of Young Measures in Non-convex Variational Problems. *Numer. Math.* In press. (http://www.numerik.uni-kiel.de/reports/1998/97-18.html)
8. Chipot, M. (1991), Numerical analysis of oscillations in nonconvex problems. *Numer. Math.* **59**, 747-767.
9. Chipot, M., Müller, S. (1997), Sharp energy estimates for finite element approximations of non-convex problems. Max-Planck-Institut Leipzig, Preprint 1997-8.
10. Ciarlet, P.G. (1978), *The finite element method for elliptic problems.* North-Holland, Amsterdam.
11. Dacorogna, B. (1989), *Direct methods in the calculus of variations.* Applied Math. Sciences **78**, Springer-Verlag, Heidelberg.
12. Kohn, R.V. (1991), The relaxation of a double-well energy. *Continuum Mech. Thermodyn.* **3**, 193-236.
13. Luskin, M. (1996), On the computation of crystalline microstructure. *Acta Numerica* **5**, 191-257.
14. Müller, S. (1998), *Variational models for microstructure and phase transitions.* Max-Planck-Institut Leipzig, Lecture notes no. 2. (http://www.mis.mpg.de/cgi-bin/lecturenotes.pl)
15. Pedregal, P. (1997), *Parametrized measures and variational principles.* Birkhäuser.
16. Roubiček, T. (1997), *Relaxation in optimization theory and variational calculus.* De Gruyter Series in Nonlinear Analysis and Applications 4, New York.
17. Tartar, L. (1995), Beyond Young measures. *Meccanica* **30**, 505-526.
18. Young, L.C. (1937), Generalized curves and the existence of an attained absolute minimum in the calculus of variations. *Comptes Rendues de la Société des Sciences et des Lettres de Varsovie, classe III* **30**, 212-234.

Homogenization of an Initial-Boundary Value Problem Describing Evolving Microstructure

Hans-Dieter Alber

Fachbereich Mathematik,
Darmstadt University of Technology, Germany

Summary. In this article we formulate a mathematical model for the temporally evolving microstructure generated by phase changes in a single crystal alloy and study the homogenization of the partial differential equations in this model.

1. Introduction

In this article we formulate a mathematical model for the temporally evolving microstructure generated by phase changes in a single crystal alloy and study the homogenization of the partial differential equations in this model. The article is a summary of some parts of [2] and contains no proofs. The investigations in [2] also are partially formal, since convergence of solutions of the equations of the microstructure model to solutions of the homogenized equations is not proved.

To model the microstructure, the sharp interface approach is used. The free energy does not have a surface part. Instead, the driving traction for the interface is generated by a jump of the Eshelby tensor at the phase interfaces. This jump is caused by the misfit strain originating from the different values of the parameters of the crystal lattice in the $\gamma-$ and $\gamma'-$ phases. The mathematical model contains evolution equations for internal variables, since the different phases are assumed to be inelastic. Moreover, it contains a distributional partial differential equation for the order parameter characterizing the different phases.

At the end of this article we only provide a small number of references and refer to [2] for a large bibliography.

2. A mathematical model with sharp phase interfaces

2.1 Equations for the stress, displacement and internal variables

The model equations for the stress, displacement and internal variables coincide essentially with the equations for homogeneous inelastic materials discussed in the book [1]. The only differences are that the microstructure introduces inhomogeneity in the material and that the equations used here contain a term representing the misfit strain. We state these equations and refer the reader to [1] for a precise discussion.

Let $\Omega \subset \mathbb{R}^3$ be a bounded open set with smooth boundary $\partial\Omega$. It represents the points of a material body. By $\gamma(t)$ we denote the set of points of Ω, which at time t belong to the γ–phase, whereas $\gamma'(t)$ denotes the set of points of Ω which at time t belong to the γ'–phase. We assume that $\gamma(t)$ is closed in Ω and that

$$\gamma(t) \cap \gamma'(t) = \emptyset, \quad \gamma(t) \cup \gamma'(t) = \Omega .$$

The interface between the two phases is

$$\Gamma(t) = \gamma(t) \cap \overline{\gamma'(t)}.$$

These subsets of Ω are the cross sections at time t of the sets

$$\gamma = \{(x,t) \mid x \in \gamma(t)\}, \ \gamma' = \{(x,t) \mid x \in \gamma'(t)\} \text{ and } \Gamma = \{(x,t) \mid x \in \Gamma(t)\}.$$

To represent these sets, we introduce an order parameter $S : \Omega \times \mathbb{R}_0^+ \to \{0, 1\}$ with

$$S(x,t) = \begin{cases} 0, & x \in \gamma(t) \\ 1, & x \in \gamma'(t) . \end{cases}$$

Let $\mathcal{S}^3$ denote the set of symmetric 3×3–matrices, let $u : \Omega \times \mathbb{R}_0^+ \to \mathbb{R}^3$ denote the displacement and

$$\varepsilon\Big(\nabla_x u(x,t)\Big) = \frac{1}{2}\Big[\nabla_x u(x,t) + (\nabla_x u(x,t))^T\Big] \in \mathcal{S}^3$$

the linear strain tensor. Here $(\nabla_x\, u(x,t))^T$ denotes the transpose of the 3×3–matrix $\nabla_x\, u(x,t)$, the matrix of first order partial derivatives of u with respect to $x = (x_1, x_2, x_3)$. The function $T : \Omega \times \mathbb{R}_0^+ \to \mathcal{S}^3$ is the Cauchy stress tensor and $z : \Omega \times \mathbb{R}_0^+ \to \mathbb{R}^N$ is the vector of internal variables. Finally, $b : \Omega \times \mathbb{R}_0^+ \to \mathbb{R}^3$ is the given volume force. The strain and stress distributions are governed by the equations

$$-\operatorname{div}_x T(x,t) = b(x,t) \tag{2.1}$$

$$T(x,t) = D\big(S(x,t)\big)\Big(\varepsilon\big(\nabla_x u(x,t)\big) - \varepsilon^*\big(S(x,t)\big) - Bz(x,t)\Big) \tag{2.2}$$

$$z_t(x,t) = f\big(S(x,t),\, T(x,t),\, z(x,t)\big)\,, \tag{2.3}$$

which must be satisfied for all $t \geq 0$ and for all $x \in \Omega\backslash\Gamma(t)$. Here $D(0)$, $D(1) : \mathcal{S}^3 \to \mathcal{S}^3$ are linear, symmetric, positiv definite mappings. $D(0)$ is the elasticity tensor in the matrix phase γ, and $D(1)$ is the elasticity tensor in the γ'–phase. $\varepsilon^*(1) \in \mathcal{S}^3$ is the misfit strain in the γ'–phase. The misfit strain is equal to zero in the matrix–phase, hence $\varepsilon^*(0) = 0$.

$B : \mathbb{R}^N \to \mathcal{S}^3$ is a linear mapping, which maps the vector of internal variables to the plastic strain tensor:

$$\varepsilon_p(x,t) = Bz(x,t)\,.$$

The given function $f : \Delta(f) \subseteq \{0,1\} \times \mathcal{S}^3 \times \mathbb{R}^N \to \mathbb{R}^N$ in the evolution equation (2.3) for the vector z of internal variables determines the inelastic properties of the γ– and γ'–phases of the material. Here $\Delta(f)$ denotes the domain of definition of f. This function depends on S, since the two phases behave differently. Purely elastic behavior in the γ'–phase is obtained with $f(1,\varepsilon,z) \equiv 0$.

On the interface $\Gamma(t)$ the functions u, T and z must satisfy interface conditions. The functions $\varepsilon(\nabla_x u)$, u_t, T, and z can jump across Γ, but we assume that the displacement u and the traction vector are continuous across the interface. Thus, with a given function $g : \mathbb{R}^N \to \mathbb{R}^N$ the interface conditions for u, T and z are

$$u^+(x,t) = u^-(x,t), \tag{2.4}$$

$$T^+(x,t)n(x,t) = T^-(x,t)n(x,t), \tag{2.5}$$

$$z(x,t+) = g(z(x,t-)), \tag{2.6}$$

which must hold for all $(x,t) \in \Gamma$. Here $n(x,t) \in \mathbb{R}^3$ is the unit normal vector to $\Gamma(t)$ pointing from $\gamma'(t)$ to $\gamma(t)$. Also, $T^+(x,t)$ and $T^-(x,t)$ are the limit values of T if the argument tends to $(x,t) \in \Gamma$ from γ or from γ', respectively. Precisely, for a function w and $(x,t) \in \Gamma$ we define

$$w^+(x,t) = \lim_{\substack{\eta\to 0\\ \eta>0}} w((x,t) + \eta m(x,t))$$

$$w^-(x,t) = \lim_{\substack{\eta\to 0\\ \eta<0}} w((x,t) - \eta m(x,t))\,,$$

with the unit normal vector $m(x,t) \in \mathbb{R}^4$ to Γ pointing from γ' to γ. In the following we assume that this normal vector exists at all points of Γ and differs from the vector $(0,\pm 1) \in \mathbb{R}^3 \times \mathbb{R}$.

Finally, the boundary condition is

$$T(x,t)n(x) = 0, \quad x \in \partial\Omega,\ t \geq 0, \tag{2.7}$$

with a unit vector $n(x)$ normal to $\partial\Omega$ at x, and the initial conditions are

$$z(x,0) = z^{(0)}(x), \quad S(x,0) = S^{(0)}(x), \quad x \in \Omega. \tag{2.8}$$

2.2 Evolution equation for the phase interface, dissipation inequality

The $3+9+N$ equations (2.1)–(2.3) contain the $3+9+N+1$ unknown components of u, T, z und S. Therefore the system (2.1)–(2.3) is not closed; an evolution equation for the order parameter S is missing. The evolution of S is known if and only if the evolution of the sets $\gamma(t)$ and $\gamma'(t)$ is known, and this evolution is known, if an equation for the normal speed of the interface between the phases is known. In this subsection we first derive restrictions for the form of such an equation from the second law of thermodynamics, essentially following the standard arguments in thermodynamics. Our presentation is influenced by [4].

Let $\psi = \psi(\varepsilon, S, z)$ be the Helmholtz free energy. ψ must satisfy the basic equation

$$\rho \nabla_{\varepsilon} \psi(\varepsilon, S, z) = T\,, \tag{2.9}$$

(cf. [1]), where $\rho > 0$ denotes the constant mass density.

Second law of thermodynamics. We use the second law of thermodynamics in the following form: For every subregion R of Ω with sufficiently smooth boundary ∂R the displacement u, the stress T, the vector of internal variables z and the order parameter S must satisfy the inequality

$$\begin{aligned} \frac{d}{dt}\int_R \rho\psi(\varepsilon(\nabla_x u), S, z)\,dx \;\leq\; & \int_{\partial R} q(\varepsilon(\nabla_x u), u_t, S, z)\cdot n(x)\,d\sigma(x) \\ & + \int_R b\cdot u_t\,dx\,, \end{aligned} \tag{2.10}$$

with the negative energy flow

$$q(\varepsilon(\nabla_x u), u_t, S, z) = T(\varepsilon(\nabla_x u), S, z)u_t\,.$$

Here $n(x)$ is an unit vector normal to ∂R pointing out of R.

Theorem 2.1. (Dissipation inequalities for the internal variables and for the phase boundary) *Let $(u, T, z, S) : \Omega \times \mathbb{R}_0^+ \to \mathbb{R}^3 \times \mathcal{S}^3 \times \mathbb{R}^N \times \{0,1\}$ be a function, which is continuously differentiable on the closed set γ and on γ', and which is such that $(u, T, z, S)_{|\gamma'}$ has a continuously differentiable extension to $\overline{\gamma'}$.*

(i) *Then (2.10), the second law of thermodynamics, implies*

$$\rho\frac{\partial}{\partial t}\psi(\varepsilon, S, z) - \operatorname{div}_x q(\varepsilon, u_t, S, z) - b\cdot u_t \leq 0 \tag{2.11}$$

on $\Omega \times \mathbb{R}^+$ in the weak sense.

(ii) *Assume in addition that (2.1) is fulfilled in the weak sense in $\Omega \times \mathbb{R}^+$, that u, T, z, S satisfy (2.2), and that on the phase interface Γ*

$$u^+ = u^-.$$

Then (2.11) holds if and only if the following two inequalities are satisfied:

$$\rho\nabla_z\psi(\varepsilon(\nabla_x u(x,t)), S(x,t), z(x,t))\cdot z_t(x,t) \leq 0 \tag{2.12}$$

for almost all $(x,t) \in \Omega \times \mathbb{R}_0^+$ *, and*

$$m''(x,t)\Big(m'(x,t)\cdot\Big[C\Big(\nabla_x u(x,t), S(x,t), z(x,t)\Big)\Big]m'(x,t)\Big) \leq 0 \quad (2.13)$$

for all $(x,t) \in \Gamma$ *. Here* $m(x,t) = (m'(x,t), m''(x,t)) \in \mathbb{R}^3 \times \mathbb{R}$ *is a unit normal vector to* Γ *pointing from* γ' *to* γ *,*

$$\begin{aligned} C(\nabla_x u, S, z) &= \rho\psi(\varepsilon(\nabla_x u), S, z)I - (\nabla_x u)^T T \\ &= \rho\psi(\varepsilon(\nabla_x u), S, z)I - (\nabla_x u)^T (D(S)(\varepsilon(\nabla_x u) - \varepsilon^*(S) - Bz)) \end{aligned} \quad (2.14)$$

is the Eshelby stress tensor, and

$$[C] = C^+ - C^-$$

denotes the jump of C *along the phase boundary* Γ *.*

In the following corollary $\nu(x,t)$ denotes the normal speed of propagation of the phase interface $\Gamma(t)$ at the point x in the direction of $n(x,t)$. We extend the function ν from Γ to $\Omega \times \mathbb{R}$ by zero.

Corollary 2.1. (Constitutive equation for the normal speed of the phase interface and evolution equation for the order parameter.)
(i) *Let* $c : \mathbb{R} \to \mathbb{R}$ *be a given function with* $c(s)s \geq 0$ *for all* $s \in \mathbb{R}$ *, and let* n *be a normal vector field to the surface* $\Gamma(t)$*. If* u, T, z, S *satisfy (2.1) in the weak sense and satisfy (2.2), (2.3), if* $u^+ = u^-$ *is satisfied on the phase interface* Γ*, if this phase interface satisfies*

$$\nu(x,t) = c\Big(n(x,t)\cdot\Big[C\Big(\nabla_x u(x,t), S(x,t), z(x,t)\Big)\Big]n(x,t)\Big) \quad (2.15)$$

for all $(x,t) \in \Gamma$ *, and if the dissipation inequality*

$$\rho\nabla_z\psi(\varepsilon, S, z)\cdot f(S,T,z) \leq 0 \quad (2.16)$$

with $T = D(S)(\varepsilon - \varepsilon^*(S) - Bz)$ *is satisfied for all*

$$(\varepsilon, S, z) \in \Delta(\psi) \cap \{(\varepsilon, S, z) \mid (S,T,z) \in \Delta(f)\},$$

then the inequality (2.11) expressing the second law of thermodynamics is fulfilled.

(ii) *Equivalent to (2.15) is the evolution equation*

$$S_t(x,t) = c\Big(n(x,t)\cdot[C(\nabla_x u(x,t), S(x,t), z(x,t))]n(x,t)\Big)|\nabla_x S(x,t)|. \quad (2.17)$$

for the order parameter S*, where* $|\nabla_x S|$ *denotes the total variation measure to the measure* $\nabla_x S$.

We note that S belongs to the space $BV^{\mathrm{loc}}(\Omega \times \mathbb{R}^+)$ of functions of locally bounded variation. Therefore the total variation measure is defined. (2.17) contains normal vectors to the surface $\Gamma(t)$, which bounds the set $\{x \in \Omega \mid S(x,t) = 1\}$. It is possible to define such normal vectors if S is a general function from the space $BV^{\mathrm{loc}}(\Omega \times \mathbb{R}^+)$. However, for the derivation of the homogenized equations presented in the next section it is convenient to use a form of the evolution equation without normal vectors. We close this subsection with the presentation of this evolution equation, which is derived under the assumption that the function c is linear. Thus, in the remainder of this article c denotes a positive constant.

Lemma 2.1. *Assume that* $(u,T,z,S) : \Omega \times \mathbb{R}_0^+ \to \mathbb{R}^3 \times \mathcal{S}^3 \times \mathbb{R}^N \times \{0,1\}$ *satisfies the assumptions of Theorem 2.1 (ii). Then, in the sense of distributions,*

$$|S_t| = c|\operatorname{div}_x C(\nabla_x u, S, z) - \rho(\nabla_x z)^T \nabla_z \psi(\varepsilon, S, z) - (\nabla_x u)^T b|. \qquad (2.18)$$

3. Homogenization of the equations for materials with evolving microstructure

In this section we state the initial-boundary value problem obtained by homogenization of the initial-boundary value problem (2.1)–(2.8), (2.18). Since the derivation is formal and since we do not have a justification of the formal procedure, we call this statement a definition of the homogenized problem. We expect that the justification of the homogenization of the nonlinear distributional evolution equation (2.18) will present fundamental difficulties.

To derive the homogenized problem we consider the initial-boundary value problem (2.1)–(2.8), (2.18) to initial-data of the form

$$z(x,y,0) = z_0^{(0)}(x, \frac{x}{\eta} + y), \quad S(x,y,0) = S_0^{(0)}(x, \frac{x}{\eta} + y), \qquad (3.1)$$

where $\eta > 0$ is a positive constant and where $y \mapsto (z_0^{(0)}(x,y), S_0^{(0)}(x,y))$ is periodic in $\mathbb{R}^3$ with periodicity cell $Y \subseteq \mathbb{R}^3$. We assume that for all $0 < \eta < \eta_0$ a family of solutions $(\hat{u}_\eta, \hat{T}_\eta, \hat{z}_\eta, \hat{S}_\eta)$ of this initial-boundary value problem exists, which can be asymptotically expanded in the form

$$(\hat{u}_\eta, \hat{T}_\eta, \hat{z}_\eta, \hat{S}_\eta)(x,y,t) = (u_\eta, T_\eta, z_\eta, S_\eta)(x, \frac{x}{\eta} + y, t), \quad 0 < \eta < \eta_0, \qquad (3.2)$$

with

$$u_\eta(x,y,t) = u_0(x,t) + \eta u_1(x,y,t) + \eta^2 u_2(x,y,t,\eta) \qquad (3.3)$$

$$T_\eta(x,y,t) = T_0(x,y,t) + \eta T_1(x,y,t,\eta) \qquad (3.4)$$

$$z_\eta(x,y,t) = z_0(x,y,t) + z_1(x,y,t,\eta) \tag{3.5}$$

$$S_\eta(x,y,t) = S_0(x,y,t) + S_1(x,y,t,\eta)\,, \tag{3.6}$$

where

$$\lim_{\eta\to 0} S_1(x,y,t,\eta) = 0 \tag{3.7}$$

$$\lim_{\eta\to 0} \int_\Omega \int_Y |z_1(x,y,t,\eta)|^2 dydx = 0\,. \tag{3.8}$$

The functions

$$u_\eta,\ u_1,\ u_2(\cdot,\eta) : \Omega \times \mathbb{R}^3 \times \mathbb{R}^+ \to \mathbb{R}^3$$

$$T_\eta,\ T_0,\ T_1(\cdot,\eta) : \Omega \times \mathbb{R}^3 \times \mathbb{R}^+ \to \mathcal{S}^3$$

$$z_\eta,\ z_0,\ z_1(\cdot,\eta) : \Omega \times \mathbb{R}^3 \times \mathbb{R}^+ \to \mathbb{R}^N$$

$$S_\eta,\ S_0,\ S_1(\cdot,\eta) : \Omega \times \mathbb{R}^3 \times \mathbb{R}^+ \to \{0,\ 1\}$$

are assumed to be periodic with respect to the y–argument and have periodicity cell Y. In passing we remark that $(\hat{z}_\eta, \hat{S}_\eta)$ satisfies the initial condition (3.1) if (z_η, S_η) fulfills

$$z_\eta(x,y,0) = z_0^{(0)}(x,y), \quad S_\eta(x,y,0) = S_0^{(0)}(x,y), \quad (x,y) \in \Omega \times \mathbb{R}^3.$$

Letting tend $\eta \to 0$, we then obtain by formal considerations the homogenized initial-boundary value problem satisfied by the leading terms in (3.3)–(3.6). In the definition of this problem given below we use the *mean stress*

$$T_\infty(x,t) = M_Y T_0(x,\cdot,t) = \int_Y T_0(x,y,t)\,.dy$$

Definition 3.1. *The homogenized initial-boundary value problem associated to the initial-boundary value problem (2.1)–(2.8) is constituted by the equations*

$$-\operatorname{div}_x T_\infty(x,t) = b(x,t), \tag{3.9}$$

$$T_\infty(x,t) = \mathcal{F}_{s\le t}(\nabla_x u_0(x,s)), \tag{3.10}$$

$$T_\infty(x,t)n(x) = 0, \quad x \in \partial\Omega, \quad t \ge 0. \tag{3.11}$$

Here the history functional $\nabla_x u_0(x,\cdot) \mapsto T_\infty(x,\cdot) = \mathcal{F}_{s\le t}(\nabla_x u_0(x,s))$ *is defined by the equation*

$$T_\infty(x,t) = M_Y T_0(x,\cdot,t), \tag{3.12}$$

which couples the mean stress to the micro stress T_0*, and by an initial-boundary value problem in the* representative volume element, *which defines* T_0 *and which consists of the four equations*

$$
\begin{aligned}
\operatorname{div}_y T_0(x,y,t) &= 0, &(3.13)\\
T_0(x,y,t) &= D(S_0(x,y,t))\Big(\varepsilon(\nabla_y u_1(x,y,t)) &(3.14)\\
&\quad - \varepsilon^*(S_0(x,y,t)) - Bz_0(x,y,t) + \varepsilon(\nabla_x u_0(x,t))\Big),\\
\frac{\partial}{\partial t} z_0(x,y,t) &= f(S_0(x,y,t), T_0(x,y,t), z_0(x,y,t)), &(3.15)\\
\left|\frac{\partial}{\partial t} S_0\right| &= c\Big|\operatorname{div}_y C(\nabla_x u_0 + \nabla_y u_1, S_0, z_0) &(3.16)\\
&\quad - \rho(\nabla_y z_0)^T \nabla_z \psi(\nabla_x u_0 + \nabla_y u_1, S_0, z_0)\Big|,
\end{aligned}
$$

of the interface conditions

$$[u_0(x,y,t)] = [T_0(x,y,t)]n(x,t) = [z_0(x,y,t)] = 0, \quad (x,y,t) \in \Gamma, \qquad (3.17)$$

of the boundary condition

$$y \mapsto (u_1(x,y,t), T_0(x,y,t), z_0(x,y,t), S_0(x,y,t)) \qquad (3.18)$$

has periodicity cell Y,

and of the initial conditions

$$z_0(x,y,0) = z_0^{(0)}(x,y), \quad S_0(x,y,0) = S_0^{(0)}(x,y), \quad (x,y) \in \Omega \times \mathbb{R}^3. \qquad (3.19)$$

Remark. For every fixed $x \in \Omega$ the equations (3.13)–(3.19) form an initial-boundary value problem for the unknown function $(y,t) \mapsto (u_1, T_0, S_0, z_0)$ in the domain $Y \times \mathbb{R}^+$, which has the same form as the initial-boundary value problem (2.1)–(2.8), (2.18). Besides the micro stress T_0 also u_1 is determined by this initial-boundary value problem. The function u_1 playes the role of a micro displacement.

References

1. H.-D. Alber: Materials with memory. Springer Lecture Notes in Mathematics **1682**. Berlin: Springer 1998
2. H.-D. Alber: Evolving microstructure and homogenization. Preprint Nr. 2051, Fachbereich Mathematik, Technische Universität Darmstadt 1999
3. A. Bensoussan, J.L. Lions, G. Papanicolaou: Asymptotic analysis for periodic structures. Amsterdam: North Holland 1978
4. M.E. Gurtin: The nature of configurational forces. Arch. Rational Mech. Anal. **131**(1995), 67–100
5. G. Maugin: Material forces: Concepts and applications. Applied Mechanics Reviews **48**(1995), 213–245

Aspects of Homogenization Techniques and Multigrid Solving

C. G. Bayreuther, C. Miehe and J. Schröder

Institute for Applied Mechanics (Civil Engineering), University of Stuttgart, Germany

Summary. We discuss different concepts for the construction of transfer operators with regard to the application of multigrid methods to an efficient solving of strongly heterogeneous linear–elastic materials. The key contribution is a new physically motivated approach. In order to take into account the influence of the heterogeneity the idea is to perform a numerical homogenization within a characteristic subdomain. This subdomain is assumed to be resolved by a local finite element discretization. The efficiency of the new approach is demonstrated by means of some fundamental model problems and compared with existing methods.

1. Introduction

Multigrid solvers are efficient methods for solving large scale problems with low numerical effort. The basic multigrid idea is to damp high–frequency errors on the fine grid and to approximate low–frequency errors on coarse grids. Extensive reviews of the basic multigrid concept are provided in [1], [2], [3], [8] and [14]. Two classes of transfer operators may be differentiated. Geometric transfer operators are based on a geometric coarsening process of the involved sequence of grids, see [4] and [5]. Algebraic transfer operators are obtained by an algebraic coarsening where an artificial hierarchy of coarse grid problems is constructed based on the structure of the fine grid matrix, see [11] and [13]. A current research topic is the combination of homogenization methods and multigrid solving for an efficient solution of boundary–value–problems of heterogeneous materials, see e.g. in [6], [7] and [10]. In this paper we discuss conceptual approaches to the construction of transfer operators for micro–heterogeneous materials. The new contribution is a method based on a patch–wise numerical homogenization, which discretizes fluctuations on a characteristic micro–structure, see [9]. The obtained results are compared with existing approaches in the scope of some fundamental model problems.

2. The basic multigrid concept

Let $\mathbf{u} : \mathcal{B} \to \mathbf{R}^3$ denote the displacement field and $\boldsymbol{\varepsilon} = \mathrm{sym}[\nabla \mathbf{u}]$ the strain field of the considered continuum $\mathcal{B} \subset \mathbf{R}^3$. The boundary–value–problem of linear elastostatics is governed by the field equation $\mathrm{div}\boldsymbol{\sigma} + \bar{\mathbf{f}} = \mathbf{0}$ and by the boundary conditions $\mathbf{u} = \bar{\mathbf{u}}$ on $\partial\mathcal{B}_u$ for the displacements and $\boldsymbol{\sigma}\mathbf{n} = \bar{\mathbf{t}}$ on $\partial\mathcal{B}_t$

for the surface tractions. A finite element solution for this boundary–value–problem is based on the weak form of the equilibrium equation

$$G = \int_{\mathcal{B}} \delta\boldsymbol{\varepsilon}^T \boldsymbol{\sigma} \mathrm{dV} - \int_{\mathcal{B}} \delta\mathbf{u}^T \bar{\mathbf{f}} \mathrm{dV} - \int_{\partial\mathcal{B}} \delta\mathbf{u}^T \bar{\mathbf{t}} \mathrm{dA} = 0\,, \tag{2.1}$$

where $\delta\mathbf{u}$ represents virtual displacements and $\delta\boldsymbol{\varepsilon} = \mathrm{sym}[\nabla\delta\mathbf{u}]$. The stress–strain relation is governed by the constitutive equation $\boldsymbol{\sigma} = \mathbf{C}\boldsymbol{\varepsilon}$. The finite element discretization of the weak form (2.1) for the displacements $\mathbf{u} = \sum N_i\, \mathbf{d}_i$, $\delta\mathbf{u} = \sum N_i\, \delta\mathbf{d}_i$ and strains $\boldsymbol{\varepsilon} = \sum \mathbf{B}_i\, \mathbf{d}_i$, $\delta\boldsymbol{\varepsilon} = \sum \mathbf{B}_i\, \delta\mathbf{d}_i$ on a local element domain $\mathcal{B}^e \subset \mathcal{B}$ in terms of the ansatz functions N_i, the associated gradients $\mathbf{B}_i$ and the nodal displacements $\mathbf{d}_i$ yields a linear algebraic system

$$\mathbf{K}\,\mathbf{d} = \mathbf{f}\,, \tag{2.2}$$

where $\mathbf{K}$ and $\mathbf{f}$ represent the stiffness matrix and the load vector, respectively,

$$\mathbf{K} = \bigcup_{e=1}^{n} \int_{\mathcal{B}^e} \mathbf{B}^T \mathbf{C} \mathbf{B}\, \mathrm{dV} \quad \text{and} \quad \mathbf{f} = \bigcup_{e=1}^{n} \left\{ \int_{\mathcal{B}^e} \mathbf{N}^T \bar{\mathbf{f}}\, \mathrm{dV} + \int_{\partial\mathcal{B}^e} \mathbf{N}^T \bar{\mathbf{t}}\, \mathrm{dA} \right\},$$

where $\mathbf{N}$ and $\mathbf{B}$ are modified matrices with the ansatz functions N_i and the associated gradients $\mathbf{B}_i$ of each element. We now apply multigrid methodes to an efficient solving of (2.2) characterizing the fine grid problem $\mathbf{K}^f \mathbf{d}^f = \mathbf{f}^f$. Let $\mathcal{S}_C \subset \mathcal{S}_f$ be the coarse and fine grid ansatz space. Then, the prolongation operator $\mathbf{P}^f{}_C \in \mathbf{R}^{n_f \times n_C}$ and the restriction operator $\mathbf{R}^C{}_f \in \mathbf{R}^{n_C \times n_f}$ are defined by

$$P^f{}_C : \mathcal{S}_C \to \mathcal{S}_f \quad \text{and} \quad R^C{}_f : \mathcal{S}_f \to \mathcal{S}_C\,. \tag{2.3}$$

On each level we will smooth the high–frequency residuals within a few relaxation steps via the one–step Gauß–Seidel algorithm

$$d_i^{(k+1)} = \frac{1}{K_{ii}} \left[f_i - \sum_{j=1}^{i-1} K_{ij} d_j^{(k+1)} - \sum_{j=i+1}^{n} K_{ij} d_j^{(k)} \right]. \tag{2.4}$$

Firstly, we perform ν_1 smoothing steps of the form (2.4) to (2.2) on the finest grid. The remaining defect $\mathbf{r}^f$ will then be restricted to the coarser level by

$$\mathbf{f}^C = \mathbf{R}^C{}_f \mathbf{r}^f \quad \text{with} \quad \mathbf{r}^f = \mathbf{f}^f - \mathbf{K}^f \mathbf{d}^f\,. \tag{2.5}$$

On the coarse grid we solve the associated coarse grid problem

$$\mathbf{K}^C \mathbf{d}^C = \mathbf{f}^C \tag{2.6}$$

in terms of the coarse grid operator $\mathbf{K}^C = \mathbf{R}^C{}_f \mathbf{K}^f \mathbf{P}^f{}_C$. This Galerkin ansatz for $\mathbf{K}^C$ implies that $\mathbf{R}^C{}_f$ is the transpose of $\mathbf{P}^f{}_C$. We then prolongate the solution vector $\mathbf{d}^C$ to the fine level and update the fine grid solution vector

$$\mathbf{d}^f \;\leftarrow\; \mathbf{d}^f + \mathbf{P}^f{}_C \mathbf{d}^C\,. \tag{2.7}$$

We smooth the updated solution $\mathbf{d}^f$ by ν_2 further relaxation steps of the form (2.4). This procedure is repeated until convergence is reached within a given tolerance. We arrive at a multigrid process if this two–grid procedure is recursively applied to the solving of the coarse grid problem, respectively.

3. A numerical homogenization procedure

We now summarize basic features of a numerical homogenization technique for heterogeneous elastic materials as pointed out in [9]. Assume a micro–structure $\mathcal{B}_p \subset \mathcal{B}$ with point–wise constitutive relation $\boldsymbol{\sigma} = \mathbb{C} : \boldsymbol{\varepsilon}$, which satisfies the equlibrium equation $\operatorname{div} \boldsymbol{\sigma} = \mathbf{0}$ in $\mathcal{B}_p$. The displacement field on the micro–structure is assumed to be linked to the local strain field $\bar{\boldsymbol{\varepsilon}}$ of $\mathcal{B}$ via the ansatz

$$\mathbf{u} = \bar{\boldsymbol{\varepsilon}}\,\mathbf{x} + \tilde{\mathbf{w}} \quad \text{in } \mathcal{B}_p \tag{3.1}$$

in terms of a linear part $\bar{\boldsymbol{\varepsilon}}\mathbf{x}$ and a superimposed fluctuation field $\tilde{\mathbf{w}}$. This implies a decomposition of the strains into $\boldsymbol{\varepsilon} = \bar{\boldsymbol{\varepsilon}} + \tilde{\boldsymbol{\varepsilon}}$ with $\tilde{\boldsymbol{\varepsilon}} = \operatorname{sym}[\nabla \tilde{\mathbf{w}}]$. As depicted in Fig. 3.1, we propose the boundary condition $\tilde{\mathbf{w}}^+ = \tilde{\mathbf{w}}^-$ on

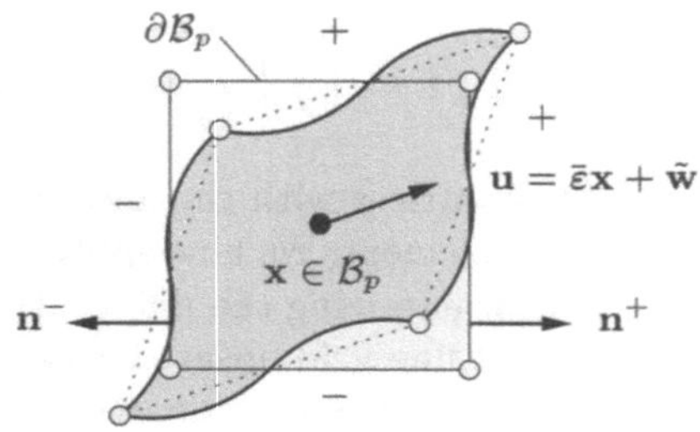

Fig. 3.1. Deformed micro–structure with periodic boundary condition

associated points $\mathbf{x}^+ \in \partial\mathcal{B}_p^+$ and $\mathbf{x}^- \in \partial\mathcal{B}_p^-$ with $\partial\mathcal{B}_p = \partial\mathcal{B}_p^+ \cup \partial\mathcal{B}_p^-$. A finite element solution of the equilibrium equation requires the weak form

$$G := \int_{\mathcal{B}_p} \delta\tilde{\boldsymbol{\varepsilon}} : \mathbb{C} : (\bar{\boldsymbol{\varepsilon}} + \tilde{\boldsymbol{\varepsilon}})\,\mathrm{dV} = 0\,. \tag{3.2}$$

The discretization of the weak form (3.2) for the fluctuation strains $\tilde{\boldsymbol{\varepsilon}}^e = \mathbf{B}\tilde{\mathbf{d}}^e$ on a local element domain $\mathcal{B}^e \subset \mathcal{B}_p$ yields the discrete fluctuation parameters

$$\tilde{\mathbf{d}} = -\mathbf{K}^{-1}\mathbf{L}\bar{\boldsymbol{\varepsilon}} \tag{3.3}$$

in terms of the two fluctuation stiffness matrices

$$\mathbf{K} = \bigcup_{e=1}^{n} \int_{\mathcal{B}^e} \mathbf{B}^T \mathbb{C} \mathbf{B}\,\mathrm{dV} \quad \text{and} \quad \mathbf{L} = \bigcup_{e=1}^{n} \int_{\mathcal{B}^e} \mathbf{B}^T \mathbb{C}\,\mathrm{dV}\,.$$

The definition of macroscopic stresses and macroscopic moduli in the form

$$\bar{\boldsymbol{\sigma}} = \frac{1}{\mathrm{V}} \int_{\mathcal{B}_p} \boldsymbol{\sigma}\,\mathrm{dV} \quad , \quad \bar{\mathbb{C}} = \partial_{\bar{\boldsymbol{\varepsilon}}}\,[\bar{\boldsymbol{\sigma}}] = \partial_{\bar{\boldsymbol{\varepsilon}}}\,\frac{1}{\mathrm{V}} \int_{\mathcal{B}_p} \mathbb{C} : (\bar{\boldsymbol{\varepsilon}} + \tilde{\boldsymbol{\varepsilon}})\,\mathrm{dV}\,, \tag{3.4}$$

respectively, finally yields the closed form presentation of the overall moduli

$$\bar{\mathbb{C}} = \frac{1}{\mathrm{V}} \int_{\mathcal{B}_p} \mathbb{C}\,\mathrm{dV} - \frac{1}{\mathrm{V}} \mathbf{L}^T \mathbf{K}^{-1} \mathbf{L}\,. \tag{3.5}$$

4. Transfer operators for heterogeneous structures

In this section we discuss different procedures for the construction of transfer operators in the case of micro–heterogeneous materials. The micro–structure is assumed to be resolved sufficiently well. For the specification of the following grid transfer strategies we investigate only two grids with the accompanying conforming hierarchical ansatz spaces $\mathcal{S}_C \subset \mathcal{S}_f$. Let us first consider the

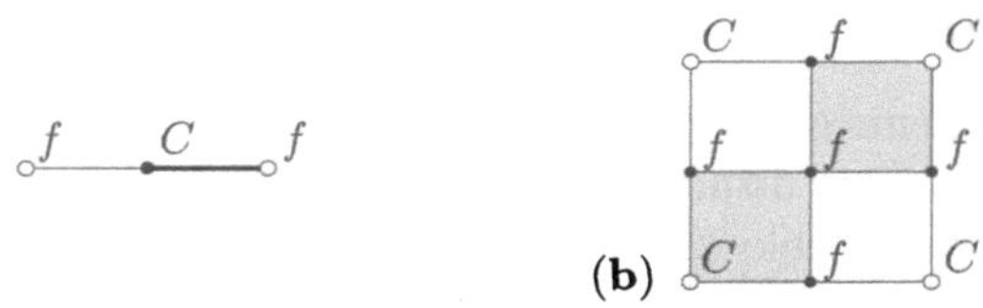

(a) (b)

Fig. 4.1. (**a**) 1–D and (**b**) 2–D heterogeneous patch structure proposed for a patch-wise grid transfer, f denoting fine grid nodes, C denoting coarse grid nodes.

simple case of the construction of the so–called geometric grid transfer operators. The finite element discretizations of the displacement fields on the fine and coarse grid are given by $u_a^f = \sum_j N_j^f(\boldsymbol{\xi}^f) d_{ja}^f$ and $u_A^C = \sum_I N_I^C(\boldsymbol{\xi}^C) d_{IA}^C$ with $a = 1, 2$ and $A = 1, 2$ indicating the dimensions. The restriction and prolongation matrices can be obtained from the relations

$$N_I^C(\boldsymbol{\xi}^C) = \sum_j R^I{}_j N_j^f(\boldsymbol{\xi}^f) \quad \text{and} \quad P^f{}_C = (R^C{}_f)^T \tag{4.1}$$

by the evaluation of the ansatz functions at fine grid nodes. In the following we focus on grid transfer operators which perform on an a priori defined patch structure as depicted in Fig. 4.1.

4.1 Smoothing adapted transfer operators

Based on the smoothing behaviour on the different grids so–called smoothing adapted transfer operators can be designed as discussed in [12]. We express the resulting fine grid error $\Delta \mathbf{d}^f \in \mathbf{R}^{n_f}$ in terms of the n_f eigenvectors $\boldsymbol{\phi}_j$ of the matrix $\mathbf{K}^f \in \mathbf{R}^{n_f \times n_f}$ in the form

$$\Delta \mathbf{d}^f = \sum_{j=1}^{n_f} \boldsymbol{\phi}_j \alpha_j \,. \tag{4.2}$$

We assume that after several relaxation steps on the fine grid the high–frequency errors associated to the $(n_f - n_C)$ largest eigenvalues of $\mathbf{K}^f$ are smoothed out. Here, n_C denotes the dimension of the coarse system matrix $\mathbf{K}^C \in \mathbf{R}^{n_C \times n_C}$. A decomposition of the remaining error in the displacement field into coarse and fine grid components leads to the expression

$$\Delta \mathbf{d}^f = \begin{bmatrix} \Delta \mathbf{d}^{C,f} \\ \Delta \mathbf{d}^{f,f} \end{bmatrix} = \begin{bmatrix} \boldsymbol{\Phi}_{CC} \\ \boldsymbol{\Phi}_{fC} \end{bmatrix} \boldsymbol{\alpha} \,. \tag{4.3}$$

From (4.3) we obtain $\boldsymbol{\alpha} = \boldsymbol{\Phi}_{CC}^{-1} \Delta \mathbf{d}^{C,f}$, which leads with the transfer condition $\Delta \mathbf{d}^{C,f} = \Delta \mathbf{d}^{C}$ to the prolongation and restriction matrix in the form

$$\Delta \mathbf{d}^f = \mathbf{P}^f{}_C \Delta \mathbf{d}^C \quad \text{with} \quad \mathbf{P}^f{}_C = \{\mathbf{R}^C{}_f\}^T := \begin{bmatrix} \mathbf{1} \\ \boldsymbol{\Phi}_{fC} \boldsymbol{\Phi}_{CC}^{-1} \end{bmatrix} . \tag{4.4}$$

In order to obtain a numerical efficient procedure, we compute the transfer operators only on a priori defined patches of the whole system.

4.2 Transfer operators associated with unit displacements

Here, the components of the transfer operators represent displacements at fine grid nodes due to unit displacements $\hat{\mathbf{d}}^{C,f}$ applied to coarse grid nodes, see [12]. This leads to the block decomposition of $\mathbf{K}^f \in \mathbf{R}^{n_f \times n_f}$ and $\hat{\mathbf{d}}^f \in \mathbf{R}^{n_f}$ into fine and coarse grid components in the form

$$\begin{bmatrix} \mathbf{K}_{CC} & \mathbf{K}_{Cf} \\ \mathbf{K}_{fC} & \mathbf{K}_{ff} \end{bmatrix} \begin{bmatrix} \hat{\mathbf{d}}^{C,f} \\ \hat{\mathbf{d}}^{f,f} \end{bmatrix} = \mathbf{0} . \tag{4.5}$$

From (4.5) we extract the fine grid displacements by $\hat{\mathbf{d}}^{f,f} = -\mathbf{K}_{ff}^{-1} \mathbf{K}_{fC} \hat{\mathbf{d}}^{C,f}$. Based on this result and the transfer condition $\hat{\mathbf{d}}^C = \hat{\mathbf{d}}^{C,f}$ we finally obtain the prolongation matrix $\mathbf{P}^f{}_C$ in the form

$$\hat{\mathbf{d}}^f = \mathbf{P}^f{}_C \hat{\mathbf{d}}^C \quad \text{with} \quad \mathbf{P}^f{}_C = \{\mathbf{R}^C{}_f\}^T := \begin{bmatrix} \mathbf{1} \\ -\mathbf{K}_{ff}^{-1} \mathbf{K}_{fC} \end{bmatrix} . \tag{4.6}$$

The coarse grid operator $\mathbf{K}^C$ obtained by the transfer matrices in (4.6) is identical to the Schur–complement of $\mathbf{K}^f$ with respect to $\mathbf{K}_{ff}$. We evaluate the transfer operators on a priori defined patches of the whole system.

4.3 Algebraical motivated transfer operators

Algebraical motivated grid transfer operators are performed completely independent of any geometry informations of the individual grids. There exist different heuristics for the sampling of coarse and fine grid nodes on the fine level, see e.g. in [11] and [13]. If a fine grid node i coincides with a coarse grid node I we constitute $P^i{}_I = R^I{}_i = 1$. Otherwise the associated component of the transfer matrices is determined based on the definition of a so–called algebraic smooth error $\Delta \mathbf{d}^f$ which induces approximately $\sum_j K_{ij}^f \Delta d_j^f = 0$. Within this relation we approach the error $\Delta d_i^{f,f}$ at the considered node i by linear interpolation of the error $\Delta d_I^{C,f}$ of neighbour coarse grid nodes I. These assumptions finally lead to the required transfer operators in the form

$$\Delta d_i^{f,f} = P^i{}_I \Delta d_I^C \quad \text{with} \quad P^i{}_I = R^I{}_i := -\frac{K_{iI}^f}{\tilde{K}_{ii}^f} . \tag{4.7}$$

Assuming now the whole system applied to a rigid body displacement $\mathbf{1}^f$ we extract $\tilde{K}_{ii}^f = -\sum_J K_{iJ}^f$ from $\sum_j K_{ij}^f \Delta d_j^f = 0$ by substituting $\mathbf{1}^f$ for $\Delta \mathbf{d}^f$ where J subscripts the coarse grid nodes in the neighbourhood of i.

4.4 Homogenization based transfer operators

We propose a new physical motivated alternative for the construction of transfer operators. The key idea is the introduction of the results of the homogenization theory in order to take into account the influence of the heterogeneity of the micro–structure. We consider a patch of finite elements on the fine grid characterizing a part of the heterogeneous micro–structure. Based on the treatment of the proposed homogenization procedure with an additive split of the displacement field in linear and fluctuation terms, i.e. $\mathbf{u}^f = \bar{\mathbf{u}}^f + \tilde{\mathbf{w}}^f$, we introduce a split of the grid transfer operator $\mathbf{P}^f{}_C$ into $\mathbf{P} = \bar{\mathbf{P}} + \tilde{\mathbf{P}}$. This split of the prolongation operator implies the relations $\mathbf{u}^f = \mathbf{P}^f{}_C \mathbf{u}^C$, $\bar{\mathbf{u}}^f = \bar{\mathbf{P}}^f{}_C \mathbf{u}^C$ and $\tilde{\mathbf{w}}^f = \tilde{\mathbf{P}}^f{}_C \mathbf{u}^C$. The interpolation of discrete nodal displacements associated to the linear displacement field is given by

$$\bar{d}^f_{ia} = \sum_I \sum_A N_I^C(\boldsymbol{\xi}_i)\, \delta_{aA}\, d^C_{IA} \,. \tag{4.8}$$

The discrete nodal displacements associated to the fluctuation field based on homogenization analysis, see (3.3), yields

$$\tilde{d}^f_{ia} = \sum_j \sum_K -\{K_p^{-1}\}_{(ia)j}\, L^f_{jK}\, \bar{\varepsilon}_K =: \sum_K H_{(ia)K}\, \bar{\varepsilon}_K \,, \tag{4.9}$$

with the patch stiffness matrix $\mathbf{K}_p$ on the fine grid and the micro–structural influence matrix $\mathbf{H}$. We relate the macroscopic strains $\bar{\varepsilon}$ to the coarse grid displacements in terms of the patch volume averages of the B–matrices. These approximations lead to the required transfer relation $d^f_{ia} = P^{ia}{}_{IA} d^C_{iA}$ within a patch which yields the explicit expression for the transfer operators

$$\boxed{P^{ia}{}_{IA} = \{R^{IA}{}_{ia}\}^T := \delta_{aA}\, N_I^C(\boldsymbol{\xi}_i) + \textstyle\sum_K H_{(ia)K}\, \bar{B}^C_{I(KA)}} \tag{4.10}$$

In general, the restricted fine grid matrix $\mathbf{R}^C{}_f \mathbf{K} \mathbf{P}^f{}_C$ is not equal to the coarse grid matrix $\mathbf{K}^C(\bar{\mathbb{C}})$ obtained with homogenized moduli $\bar{\mathbb{C}}$, see also [7].

5. Comparative study of model problems

We apply now the discussed techniques to three model problems. First, we consider a heterogeneous beam which is applied to a horizontal load $F = 1$, see Fig. 5.1a. We choose the material parameters $E_1 = 1$, $E_2 = 10$ and a

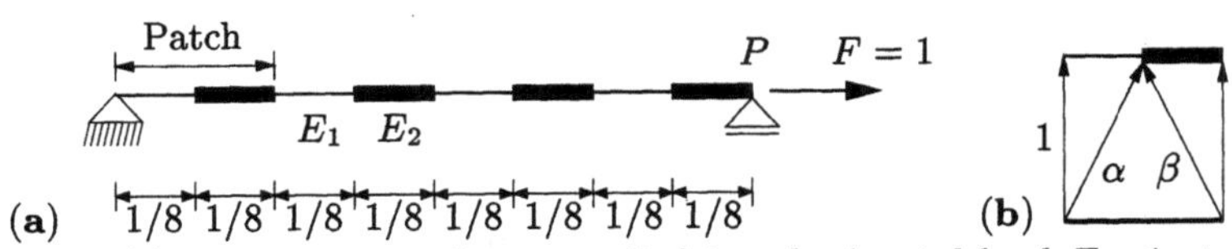

Fig. 5.1. (a) Heterogeneous beam applied to a horizontal load $F = 1$ at node P. (b) Prolongation within a patch via weighting factors 1, α and β

cross–section of the beam with $A = 1$. For the construction of the required transfer operators we focus on a patch, as depicted in Fig. 5.1b. The con-

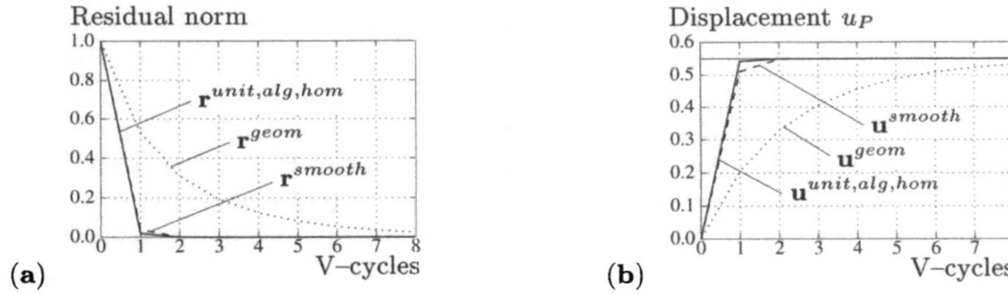

Fig. 5.2. Convergence study. (**a**) Residual norm $||\mathbf{r}||_2$ versus V–cycles. (**b**) Displacement u_P at node P versus V–cycles, $u_{exact} = 0.55$

vergence behaviour of the different approaches is depicted in Fig. 5.2 in the scope of a four–grid V–cycle–algorithm. For the transfer between the coarse levels $l = 1$ and $l = 3$ we perform the standard grid transfer based on (4.1). The solid line depicting the unit displacement, the algebraic and the homog-

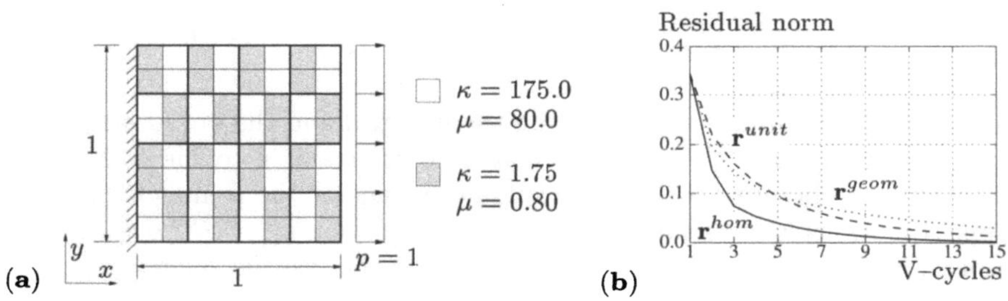

Fig. 5.3. Convergence study involving two grids. (**a**) Tension of a specimen with periodic microstructure. (**b**) Residual norm $||\mathbf{r}||_2$ versus V–cycles

enization based strategy shows a strongly decreasing residual norm and a good asymptotical behaviour of the nodal displacement u_p. The dashed line representing the smoothing adapted transfer is rather close to the solid line. The geometric transfer alternative is worse in contrast to the previous ones. Furthermore, we consider a tension test as depicted in Fig. 5.3a. We perform a two–grid procedure. The coarse grid is depicted by thick solid lines.

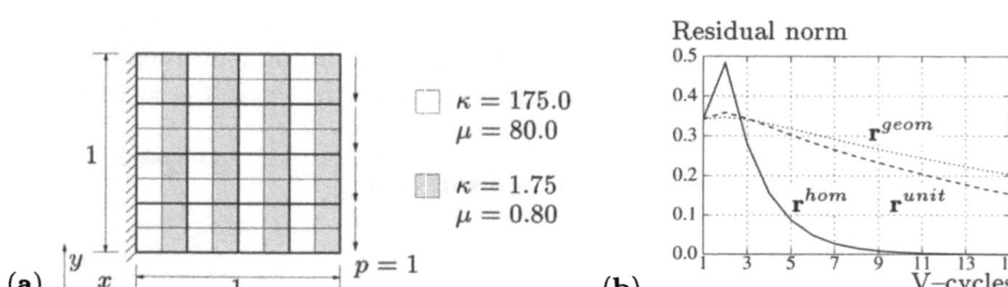

Fig. 5.4. Convergence study involving two grids. (**a**) Bending of a specimen with periodic microstructure. (**b**) Residual norm $||\mathbf{r}||_2$ versus V–cycles

In contrast to the unit displacement and the geometric based variant the proposed new homogenization approach yields a distinct decreasing of the residual norm, see Fig. 5.3b. A similar tendency due to the homogenization transfer technique is depicted in Fig. 5.4 which shows the bending of a heterogeneous specimen and the individual convergence rates of the considered transfer strategies.

6. Conclusion

In the scope of linear elastic materials with heterogeneous micro–structures different concepts for the construction of transfer operators have been discussed. The key contribution was the suggestion of a new homogenization based transfer operator. First numerical investigations indicate an excellent convergence behaviour of the proposed method when compared with other approaches.

References

1. Bastian, P., Hackbusch, W., Wittum, G. (1998) Additive and Multiplicative Multi–Grid - a Comparison, Computing **60**, 345–364
2. Brandt, A. (1982) Guide to multigrid development. In: Multigrid Methods, W. Hackbusch and U. Trottenberg, eds., Lecture Notes in Mathematics **960**, Springer–Verlag
3. Briggs, W. L. (1987) A multigrid tutorial. SIAM, Philadelphia
4. Dendy, J. E. (1982) Black box multi–grid. J. Comput. Physics **48**, 366–386
5. de Zeeuw, P. M. (1990) Matrix–dependent prolongations and restrictions in a black box multigrid solver. J. Comput. Appl. Math. **33**, 1–27
6. Fish, J., Belsky, V. (1995) Multigrid method for periodic heterogeneous media. Part 1: Convergence studies for one–dimensional case. Comput. Methods Appl. Mech. Engrg. **126**, 1–16
7. Fish, J., Belsky, V. (1995) Multigrid method for periodic heterogeneous media. Part 2: Multiscale modeling and quality control in multidimensional case. Comput. Methods Appl. Mech. Engrg. **126**, 17–38
8. Hackbusch, W. (1985) Multigrid Methods and Applications. Springer–Verlag
9. Miehe, C., Schröder, J., Bayreuther, C. G. (1998) Homogenization Analysis of Linear Composite Materials on Discretized Fluctuations on the Micro–Structure. submitted to Acta Mechanica
10. Neuss, N. (1995) Homogenisierung und Mehrgitter. Ph. D. thesis, University of Heidelberg
11. Ruge, J. W., Stüben, K. (1987) Algebraic multigrid. In: Multigrid Methods, S. F. McCormick, ed., SIAM, Philadelphia
12. Rust, W. (1991) Mehrgitterverfahren und Netzadaption für lineare und nichtlineare statische Finite–Elemente–Berechnungen von Flächentragwerken. dissertation, University of Hannover
13. Vaněk, P., Mandel, J., Brezina, M. (1996) Algebraic multigrid on unstructured meshes. Technical report, Center for Computational Mathematics, University of Colorado at Denver
14. Wesseling, P. (1992) An introduction to multigrid methods. Wiley, Chichester

Estimates on the Mixture Function for Multiphase Problems in Elasticity

Alexander Mielke

Mathematisches Institut A, Universität Stuttgart, Germany

1. Introduction

In the theory of shape memory alloys and other crystallographic problems it is the basic assumption, that each mesoscopic part of the crystal can choose to be in one of the n allowed phases. We distinguish these phases by their stored–energy densities

$$W(E, e_i) = \tfrac{1}{2}\langle\!\langle \mathbf{C}_i[E-A_i], E-A_i \rangle\!\rangle + \alpha_i, \qquad i = 1, \ldots, n,$$

where $e_i \in \mathbb{R}^n$ is the ith unit vector and $E = \nabla_{\text{sym}} u = \frac{1}{2}(\nabla u + \nabla u^T) \in \mathbb{R}^{d\times d}_{\text{sym}}$ is the linearized strain tensor.

On a macroscopic level it is often desirable to describe the evolution of the phase portions $c \in P^n = \text{conv}(P^n_{\text{pure}})$ with $P^n_{\text{pure}} = \{e_1, \ldots, e_n\}$ using a mixture theory, see [KMS99, KuM00, Lev98, MTL98, HaG99]. Here we do not consider the evolution problem of c for time-dependent processes, but only the mixture function $W(E, c)$ which is defined in (2.1). This function is derived mathematically solely from the functions $W(\cdot, e_i)$, $i = 1, \ldots, n$, without using any further physical modeling. Hence, $W(E, c)$ is given as the minimal energy in a representative volume element where the minimization is done over all possible mesoscopic arrangements of the phases with given volume fraction $c \in P^n$ and all mesoscopic deformations with macroscopic strain E, see [AvM89, SmW98].

For identical elastic tensors (i.e., $\mathbf{C}_i = \mathbf{C}$) the mixture function reads

$$W(E, c) = \sum_{i=1}^{n} c_i W(E, e_i) + w_{\text{mix}}(c), \quad c = (c_1, \ldots, c_n)^T,$$

where $w_{\text{mix}} : P^n \to (-\infty, 0]$ is convex and satisfies $w_{\text{mix}}(e_i) = 0$. While $w_{\text{mix}}(\theta e_i + (1-\theta) e_j)$ can be given explicitly for each $\theta \in [0, 1]$ and all $i \neq j$ (see [Koh91]), it is not known how to compute $w_{\text{mix}}(c)$ in the general case. Here we provide upper and lower estimates for $w_{\text{mix}}(c)$ in terms of the function $\psi : \mathbb{R}^n_* = \{\, \eta \in \mathbb{R}^n : \langle \eta, e_* \rangle = 0 \,\} \to \mathbb{R}$ given by

$$\psi(\eta) = \min\{ -\tfrac{1}{2}\langle T^{-1}(\omega)\mathbf{C}[A^\eta]\omega, \mathbf{C}[A^\eta]\omega \rangle : \omega \in \mathbb{S}^{d-1} \} \leq 0,$$

where $T(\omega)$ is the acoustic tensor and $A^\eta = \sum_{j=1}^{n} \eta_j A_j$. With $\mathbb{M}(c) = \text{diag}(c) - c \otimes c$ and $m = (n^2 - n)/2$ we obtain the lower bound

$$w^{\mathrm{low}}(c) = \min\{ \textstyle\sum_{j=1}^{m} \psi(\eta^{(j)}) : \eta^{(j)} \in \mathbb{R}^n_*,\ \sum_{j=1}^{m} \eta^{(j)} \otimes \eta^{(j)} = \mathbb{M}(c) \}. \quad (1.1)$$

An upper bound is any function $w^{\mathrm{upp}} : P^n \to \mathbb{R}$ with $w^{\mathrm{upp}}(e_i) \geq 0$ and

$$w^{\mathrm{upp}}(\theta c^{(1)} + (1-\theta) c^{(2)}) \geq \theta w^{\mathrm{upp}}(c^{(1)}) + (1-\theta) w^{\mathrm{upp}}(c^{(2)}) + \theta(1-\theta)\psi(c^{(1)} - c^{(2)}).$$

The theory uses $\mathbb{H}$–measures for minimizing sequences of optimal arrangements of the phase mixture. We follow the lines in [Koh91, SmW98]. The lower estimate is obtained by relaxing the set of $\mathbb{H}$–measures whereas the upper estimate relies on the mixing formula for laminated $\mathbb{H}$–measures.

In particular, we show that the function $w_{\mathrm{mix}} : P^n \to \mathbb{R}$ takes the form

$$w_{\mathrm{mix}}(c) = \langle\!\langle B, \mathbb{M}(c) \rangle\!\rangle \quad \text{with } B \in \mathbb{R}^{n\times n}_{\mathrm{sym}},$$

if the transformation strains satisfy $A_j = A_0 + a_j A$.

2. Definition of the mixture function

We have given the n stored–energy densities $W(F, e_j)$, $j = 1, \ldots, n$, which are assumed to be quasiconvex. The **mixture function** for given volume fraction $c \in P$ is defined via

$$\begin{aligned} W(F,c) = \inf\Big\{ \textstyle\int_{Q^d} W(F + \nabla\varphi(y), \chi(y))\,\mathrm{d}y : \varphi \in W^{1,p}_{\mathrm{per}}(Q^d), \\ \chi(y) \in P^n_{\mathrm{pure}}, \textstyle\int_{Q^d} \chi(y)\,\mathrm{d}y = c \Big\}, \end{aligned} \quad (2.1)$$

where $Q^d = (0,1)^d \subset \mathbb{R}^d$. It can be shown (cf. [Mie00]) that for each $c \in P^n$ the function $F \mapsto W(F, c)$ is quasiconvex and for each F the function $c \mapsto W(F, c)$ is convex.

Here we are interested in the special case of linearized elasticity where additionally each phase has the same elastic tensor $\mathbf{C}$:

$$W(F, e_j) = \tfrac{1}{2} \langle\!\langle \mathbf{C}[E - A_j], E - A_j \rangle\!\rangle + \alpha_j. \quad (2.2)$$

Here $E = \frac{1}{2}(F + F^T) - I$ is the linearized strain, A_j is the transformation strain, and α_j is the height of the well (which usually depends on the temperature T assumed to be constant here).

Proposition 2.1. *If all $W(F, e_j)$, $j = 1, \ldots, n$, have the form (2.2) then the mixture function takes the form $W(F, c) = \sum_{j=1}^{n} c_j W(F, e_j) + w_{\mathrm{mix}}(c)$, where the mixture term $w_{\mathrm{mix}} : P^n \to \mathbb{R}$ is convex and given by*

$$\begin{aligned} w_{\mathrm{mix}}(c) = \inf\Big\{ \textstyle\int_{Q^d} \langle\!\langle \mathbf{C}[\nabla_{\mathrm{sym}}\varphi], \tfrac{1}{2}\nabla_{\mathrm{sym}}\varphi - \sum_{j=1}^{n} \chi_j(y) A_j \rangle\!\rangle\,\mathrm{d}y : \\ \textstyle\int_{Q^d} \chi(y)\,\mathrm{d}y = c, \chi(y) \in P^n_{\mathrm{pure}}, \varphi \in W^{1,2}_{\mathrm{per}}(Q^d) \Big\}. \end{aligned} \quad (2.3)$$

This is a consequence of $W(F,e_j)$ being quadratic in (E,A_j) and the identities $\int \nabla_{\rm sym}\varphi \,{\rm d}y = 0$, $\int \chi(y)\,{\rm d}y = c$ and $\sum \chi_j(y) \equiv 1$.

Since φ appears quadratically in $w_{\min}(c)$ we may eliminate it by first minimizing with respect to φ and keeping χ fixed. This is done most easily using Fourier series

$$\varphi(y) = \textstyle\sum_{\xi\in\Gamma_*} \varphi_\xi {\rm e}^{{\rm i}\langle\xi,y\rangle}, \quad \chi(y) = \sum_{\xi\in\Gamma} \chi_\xi {\rm e}^{{\rm i}\langle\xi,y\rangle}, \quad \chi_0 = c,$$

where $\Gamma = (2\pi\mathbb{Z})^d$ and $\Gamma_* = \Gamma\setminus\{0\}$. Inserting this form into $w_{\rm mix}(c)$ we find a decoupling, and $\varphi_\xi = \overline{\varphi}_{-\xi} \in \mathbb{C}^d$ can be found by minimizing

$$\tfrac12 \langle\!\langle \mathbf{C}[{\rm i}\xi\otimes_{\rm sym}\varphi_\xi], {\rm i}\xi\otimes_{\rm sym}\varphi_\xi\rangle\!\rangle - {\rm Re}\langle\!\langle \mathbf{C}[{\rm i}\xi\otimes_{\rm sym}\phi_\xi], \textstyle\sum_j (\chi_\xi)_j A_j\rangle\!\rangle. \tag{2.4}$$

After some elementary calculations we arrive at

$$w_{\min}(c) = \inf\{\, I(\chi) : \textstyle\int_{Q^d} \chi(y)\,{\rm d}y = c, \chi(y) \in P^n_{\rm pure}\}$$

with $I(\chi) = -\frac12 \sum_{\xi\in\Gamma_*} \langle T(\xi)^{-1}\mathbf{C}[\sum_{j=1}^n (\chi_\xi)_j A_j]\xi, \mathbf{C}[\sum_{j=1}^n (\chi_\xi)_j A_j]\xi\rangle$. Here the acoustic tensor $T(\xi)$ is defined via $\langle T(\xi)u,u\rangle = \langle\!\langle \mathbf{C}[\xi\otimes_{\rm sym}u], \xi\otimes_{\rm sym}u\rangle\!\rangle$ and is homogeneous of degree 2 in $\xi\in\mathbb{R}^d$. Introducing $\widehat{G}, G(\xi) \in \mathbb{R}^{n\times n}_{\rm sym}$ via

$$G(\xi)_{jk} = \tfrac12 \langle T(\xi)^{-1}\mathbf{C}[A_j]\xi, \mathbf{C}[A_k]\xi\rangle \quad \text{and} \quad \widehat{G}_{jk} = \tfrac12 \langle\!\langle \mathbf{C}[A_j], A_k\rangle\!\rangle$$

we obtain the final formula

$$I(\chi) = -\textstyle\sum_{\xi\in\Gamma_*} \langle\!\langle G(\xi), \overline{\chi}_\xi\otimes\chi_\xi\rangle\!\rangle \text{ where } \chi(y) = \sum_{\xi\in\Gamma} \chi_\xi {\rm e}^{{\rm i}\langle\xi,y\rangle}.$$

The term in (2.4) is $\geq -\frac12 \langle\!\langle \mathbf{C}[\sum_{j=1}^n (\chi_\xi)_j A_j], \sum_{k=1}^n (\overline{\chi}_\xi)_k A_k\rangle\!\rangle$ (minimize over $E = \overline{E}^T \in \mathbb{C}^{d\times d}$ instead of ${\rm i}\xi\otimes_{\rm sym}\varphi_\xi$). Adding up these terms we find $I(\chi) \geq -\langle\!\langle \widehat{G}, \sum_{\xi\in\Gamma_*} \chi_\xi\otimes\overline{\chi}_\xi\rangle\!\rangle = -\langle\!\langle \widehat{G}, \mathbb{M}(c)\rangle\!\rangle$ (use that $\chi(y) \in P^n_{\rm pure}$ gives $\int_{Q^d} \chi(y)\otimes\chi(y)\,{\rm d}y = \sum_{\xi\in\Gamma} \chi_\xi\otimes\chi_{-\xi} = {\rm diag}(c)$). Thus, we found the rather simple bounds

$$0 \geq w_{\rm mix}(c) \geq -\langle\!\langle \widehat{G}, \mathbb{M}(c)\rangle\!\rangle. \tag{2.5}$$

Another form of the mixture function using the averaged transformation strain $A^c = \sum_{j=1}^n c_j A_j$ is sometimes prefered in engineering (cf. [KMS99, KuM00, Lev98]):

$$W(E,c) = \tfrac12 \langle\!\langle \mathbf{C}[E-A^c], E-A^c\rangle\!\rangle + \textstyle\sum_{j=1}^n c_j\alpha_j + \widehat{w}_{\rm mix}(c).$$

A simple calculation shows that $\widehat{w}_{\rm mix}$ is related to $w_{\rm mix}$ by $w_{\rm mix}(c) = \widehat{w}_{\rm mix}(c) - \langle\!\langle \widehat{G}, \mathbb{M}(c)\rangle\!\rangle$ and satisfies the simple bounds $0 \leq \widehat{w}_{\rm mix}(c) \leq \langle\!\langle \widehat{G}, \mathbb{M}(c)\rangle\!\rangle$.

3. The $\mathbb{H}$–measure associated with χ

The function $\chi : Q^d \to P^n_{\text{pure}}$ defines a microstructure of pure phases having the volume average $c = \int_{Q^d} \chi(y)\,dy$. We associate to each function χ a matrix–valued measure $\mu = \widetilde{\mu}(\chi)$ on the unit sphere $\mathbb{S}^{d-1}$. The values of μ will lie in the set $\mathbb{H}^n_{\geq 0}$ of positive semidefinite Hermitian matrices. For any Borel measurable $\Sigma \subset \mathbb{S}^{d-1}$ we let $\mu(\Sigma) = \sum_{\xi\in\Gamma_*,\,\xi/|\xi|\in\Sigma} \chi_\xi \otimes \overline{\chi}_\xi \in \mathbb{H}^n_{\geq 0}$. Thus, μ measures the correlation between the oscillations in direction $\omega \in \mathbb{S}^{d-1}$ of the different components. The measure is a homogeneous $\mathbb{H}$–measure in the sense of [Tar90], when the generating sequence $u^{(n)}(x) = \chi(nx) - c$ is considered.

Introducing the linear functional

$$J : \text{meas}(\mathbb{S}^{d-1}, \mathbb{H}^n_{\geq 0}) \to \mathbb{R};\ \mu \mapsto J(\mu) = -\int_{\omega\in\mathbb{S}^{d-1}} \langle\!\langle G(\omega), \mu(\mathrm{d}\omega)\rangle\!\rangle \tag{3.1}$$

we obtain the relation $I(\chi) = J(\widehat{\mu}(\chi))$. Moreover, the symmetry $G(-\omega) = G(\omega)$ tells us that J depends only on the real part of μ. We consider the map

$$\widehat{\mu} : \mathrm{L}^\infty(Q^d, P^n_{\text{pure}}) \to \text{meas}(\mathbb{S}^{d-1}, \mathbb{P}^n);\ \chi \mapsto \widehat{\mu}(\chi) = \mathrm{Re}\widetilde{\mu}(\chi),$$

where $\mathbb{P}^n = \{\, B \in \mathbb{R}^{n\times n} : B = B^T,\ B \text{ positive semidefinite}\,\}$. Then each $\mu = \widehat{\mu}(\chi)$ satisfies (with $e_* = (1, \ldots, 1)^T$)

$$\begin{array}{ll} \text{(i)} & \mu(\Sigma)e_* = 0 \text{ for all } \Sigma \subset \mathbb{S}^{d-1} \\ \text{(ii)} & \mu(\mathbb{S}^{d-1}) = \mathbb{M}(c) \text{ where } c = \int \chi(y)\,\mathrm{d}y \\ \text{(iii)} & \mu(-\Sigma) = \mu(\Sigma) \in \mathbb{P}^n \text{ for all } \Sigma \subset \mathbb{S}^{d-1}. \end{array} \tag{3.2}$$

For each $c \in P^n$ we define the subsets of matrix–valued measures

$$\begin{aligned} \mathcal{H}(c) &= \text{closure}\{\, \widehat{\mu}(\chi) : \textstyle\int_{Q^d} \chi(y)\,\mathrm{d}y = c\,\}, \\ \overline{\mathcal{H}}(c) &= \{\, \mu \in \text{meas}(\mathbb{S}^{d-1}, \mathbb{P}^n) : (3.2) \text{ holds}\,\}. \end{aligned}$$

Clearly, $\mathcal{H}(c) \subset \overline{\mathcal{H}}(c)$ and $\overline{\mathcal{H}}(c)$ is convex. The function w_{mix} can now be expressed as

$$w_{\text{mix}}(c) = \inf\{\, J(\mu) : \mu \in \mathcal{H}(c)\,\}. \tag{3.3}$$

Constructing $\mathcal{H}^{\text{inn}}(c)$ and $\mathcal{H}^{\text{out}}$ with $\mathcal{H}^{\text{inn}}(c) \subset \mathcal{H}(c) \subset \mathcal{H}^{\text{out}}(c)$ we obtain upper and lower bounds for w_{mix} if $\mathcal{H}(c)$ in the infimum is replaced by $\mathcal{H}^{\text{inn}}(c)$ and $\mathcal{H}^{\text{out}}$, respectively.

The difficulty in calculating $w_{\text{mix}}(c)$ is reminiscent to calculating $\mathcal{H}(c)$. Until now there are only very few methods to generate suitably large subclasses of $\mathcal{H}(c)$. In section (4.) we will use $\mathcal{H}^{\text{out}} = \widehat{\mathcal{H}}(c)$ to obtain a lower bound. In section (5.) the lamination mixture formula for $\mathbb{H}$–measures (see theorem (5.1)) is employed to find upper bounds.

4. A lower estimate

To improve the trivial lower bound in (2.5) we let

$$w^{\text{low}}(c) = \inf\{ - \textstyle\int_{\mathbb{S}^{d-1}} \langle\!\langle G(w), \mu(\mathrm{d}w)\rangle\!\rangle : \mu \in \overline{\mathcal{H}}(c)\}$$

which provides a lower bound for $w_{\text{mix}}(c)$. Since $\overline{\mathcal{H}}(c)$ is characterized solely by algebraic conditions it is easier to calculate $w^{\text{low}}(c)$ than $w_{\text{mix}}(c)$. In particular, following [SmW98] we will show that the calculation of $w^{\text{low}}(c)$ can be reduced to a finite dimensional minimization problem. (It is an open problem to find more algebraic restriction on $\mu \in \mathcal{H}(c)$ in order to make $\overline{\mathcal{H}}(c)$ smaller and to increase $w^{\text{low}}(c)$.)

The mapping $J : \text{meas}(\mathbb{S}^{d-1}, \mathbb{H}^n_{\geq 0}) \to \mathbb{R}$ given in (3.1) is linear and $\overline{\mathcal{H}}(c)$ is convex, closed and bounded. Hence we can apply the Krein–Milman theorem which states that J attains its minimum on the extremal points of $\overline{\mathcal{H}}(c)$:

$$\text{ex}(\overline{\mathcal{H}}(c)) = \{\mu \in \overline{\mathcal{H}}(c) : \overline{\mathcal{H}} \setminus \{\mu\} \text{ is convex}\}.$$

Recalling $m = (n^2-n)/2$ and using the Dirac measure $\delta_\omega \in \text{meas}(\mathbb{S}^{d-1}, \mathbb{R})$ and $\widehat{\delta}_\omega = \frac{1}{2}(\delta_\omega + \delta_{-\omega})$, this set can be characterized as follows.

Proposition 4.1. *We have $\mu \in \text{ex}(\overline{\mathcal{H}}(c))$ if and only if there exist $\omega_1, \ldots, \omega_m \in \mathbb{S}^{d-1}$ and $\eta_1, \ldots, \eta_m \in \mathbb{R}^n_*$ such that $\mu = \sum_{j=1}^m \widehat{\delta}_{\xi_j}\, \omega_j \otimes \omega_j$.*

Proof. Assume that $\mathbb{S}^{d-1}$ contains $m+1$ pairwise disjoint symmetric Σ_j (i.e. $\Sigma_j = -\Sigma_j$), such that $\mu(\Sigma_j) \neq 0$. Since all $\mu(\Sigma_j)$ lie in the m–dimensional linear space $\{M \in \mathbb{R}^{n\times n}_{\text{sym}} : Me_* = 0\}$ the equation $\sum_{j=1}^{m+1} \alpha_j \mu(\Sigma_j) = 0$ has a nontrivial solution $(\alpha_1, \ldots, \alpha_{m+1}) \in \mathbb{R}^{m+1} \setminus \{0\}$. We may assume $|\alpha_j| \leq 1$ and define the two measures $\mu^\pm \in \overline{\mathcal{H}}(c)$ via $\mu^\pm(\Sigma) = (1 \pm \alpha_j)\mu(\Sigma)$ if $\Sigma \subset \Sigma_j$ and $\mu^\pm(\Sigma) = \mu(\Sigma)$ if $\Sigma \cap \bigcup_{j=1}^m \Sigma_j = \emptyset$. Clearly, $\mu^+ \neq \mu^-$ and $\mu = \frac{1}{2}(\mu^+ + \mu^-)$. Thus, μ cannot be extremal. We conclude that extremal measures take the form $\mu = \sum_{j=1}^m \widehat{\delta}_{\omega_j} M_j$ with $M_j \in \mathbb{P}^n_* = \{M \in \mathbb{P}^n : Me_* = 0\}$.

Now assume $\sum_{j=1}^m \operatorname{rank} M_j > m$, then as above $\sum_{j=1}^m \alpha_j M_j = 0$ has a nontrivial solution with $|\alpha_j| \leq 1$. The measures $\mu^\pm = \sum_{j=1}^m (1+\alpha_j)\widehat{\delta}_{\xi_j} M_j$ are different, lie in $\overline{\mathcal{H}}(c)$ and satisfy $\mu = \frac{1}{2}(\mu^+ + \mu^-)$. Thus, μ was no extremal measure. We conclude $\sum_{j=1}^m \operatorname{rank} M_j \leq m$ and together with $M_j = \sum_{i=1}^{m_j} a_{j,i} \otimes a_{j,i}$ where $m_j = \operatorname{rank} M_j$ the result follows.

To formulate the main result of this section we define $\psi : \mathbb{R}^n \to \mathbb{R}$ via

$$\psi(\eta) = \inf\{ -\langle\!\langle G(\omega), \eta \otimes \eta \rangle\!\rangle : \omega \in \mathbb{S}^{d-1}\}.$$

Moreover, let $\mathbb{E}^n_* = \{\eta \otimes \eta \in \mathbb{P}^n_* : \eta \in \mathbb{S}^{n-1}, \langle \eta, e_* \rangle = 0\}$ and $\mathbb{K}^n_* = \{M \in \mathbb{P}^n_* : \operatorname{tr} M = 1\}$. Then $\mathbb{K}^n_* = \text{conv}(\mathbb{E}^n_*)$ and $\mathbb{E}^n_* = \text{ex}(\mathbb{K}^n_*)$. The set $\mathbb{E}^n_*$ is a smooth manifold of dimension $(n-2)$. The set $\mathbb{K}^n_*$ lies in an affine subspace of $\mathbb{R}^{n\times n}$ of dimension $(n^2-n)/2 - 1$ and has nonempty interior with respect to this subspace. Finally we define $\Psi : \mathbb{K}_*^n \to \mathbb{R}$ to be the largest convex function which satisfies $\Psi(\eta \otimes \eta) = \psi(\eta)$ for $\eta \otimes \eta \in \mathbb{E}_*^n$, namely

$$\Psi(M) = \min\Big\{ \textstyle\sum_{j=1}^m \alpha_j \psi(\eta_j) \ : \alpha_j \geq 0, \sum_{j=1}^m \alpha_j = 1, \\ \eta_j \otimes \eta_j \in \mathbb{E}_*^n, \sum_{j=1}^m \alpha_j \eta_j \otimes \eta_j = M \Big\}. \tag{4.1}$$

Here again $m = (n^2 - n)/2$ by the standard theory of convexity and the fact that $\mathbb{K}_*^n$ lies in a $(m-1)$–dimensional affine subspace of $\mathbb{R}^{n \times n}$.

In fact, we may extend ψ to $\psi : \mathbb{R}_*^n \to \mathbb{R}$ naturally such that $\psi(\alpha\eta) = \alpha^2 \psi(\eta)$, and similarly we may extend Ψ into $\Psi : \mathbb{P}_*^n \to \mathbb{R}$ such that $\Psi(\alpha M) = \alpha \Psi(M)$ for $\alpha \geq 0$. Combining the Krein–Milman theorem and the definition of Ψ leads to

Theorem 4.1. *We have the formula* $w^{\mathrm{low}}(c) = \Psi(\mathbb{M}(c))$.

Note that we do not need the function Ψ on the whole set $\mathbb{K}_*^n$. Since

$$\mathbb{M}(c) = \mathrm{diag}(c) - c \otimes c = \tfrac{1}{2} \textstyle\sum_{i,j=1}^n c_i c_j (e_i - e_j) \otimes (e_i - e_j), \tag{4.2}$$

we have $[\mathrm{tr}\, \mathbb{M}(c)]^{-1} \mathbb{M}(c) \in \mathrm{conv}\{ \frac{1}{2}(e_i - e_j) \otimes (e_i - e_j) \in \mathbb{E}_*^n : i \neq j \} \subset \mathbb{K}_*^n$.

In the case $n = 3$ the dimension $m - 1$ equals 2 and $\mathbb{K}_*^n$ can be identified with a circle, see [SmW98]. Formula (1.1) is obtained by extending (4.1) homogeneously.

5. Upper bounds

To improve the trivial upper bound $w_{\mathrm{mix}}(c) \leq 0$ we need to find a suitable inner approximation of $\mathcal{H}(c)$. In fact, in light of the Krein–Milman theorem it would be sufficient to characterize (or approximate) its extremal points.

There is no general theory to describe $\mathcal{H}(c)$ suitably, however the **lamination mixture formula**, which is due to [AvM89, Tar90, Koh91], allows us to construct new $\mathbb{H}$–measures as mixture of two given ones.

Theorem 5.1. *For $\mu^{(1)} \in \mathcal{H}(c^{(1)}), \mu^{(2)} \in \mathcal{H}(c^{(2)}), \theta \in (0,1)$ and $\omega \in \mathbb{S}^{d-1}$, we have (with $d = c^{(2)} - c^{(1)}$)*

$$\mu = \theta \mu^{(1)} + (1-\theta) \mu^{(2)} + \theta(1-\theta)\, d \otimes d\, \widehat{\delta}_\omega \in \mathcal{H}(c^{(1)} + \theta d). \tag{5.1}$$

This formula implies convexity of $\mathcal{H}(c)$, since mixing with $c^{(1)} = c^{(2)} = c$ does not generate an additional quadratic term. More important, we have

$$w_{\mathrm{mix}}(c^{(1)} + \theta d) \leq \theta w_{\mathrm{mix}}(c^{(1)}) + (1-\theta) w_{\mathrm{mix}}(c^{(2)}) + \theta(1-\theta)\psi(d). \tag{5.2}$$

To see this choose $\mu^{(j)} \in \mathcal{H}(c^{(j)})$ with $w_{\mathrm{mix}}(c^{(j)}) = J(\mu^{(j)}))$ for $j = 1$ and 2. Further choose $\omega \in \mathbb{S}^{d-1}$ such that $\psi(d) = -\langle\!\langle G(\omega), d \otimes d \rangle\!\rangle$. Then, μ defined in

((5.1)) lies in $\mathcal{H}(c^{(1)}+\theta d)$ and ((3.3)) gives ((5.2)). As $\psi(\eta)<0$ in most cases, this implies better bounds than simple convexity. Starting with $0\in\mathcal{H}(e_j)$ we may inductively apply this formula to obtain nontrivial bounds.

It was first observed by R.V. Kohn [Koh91] that this approach gives $w_{\rm mix}$ exactly in the case of two phases only, namely

$$w_{\rm mix}(\theta e_j+(1-\theta)e_k)=\theta(1-\theta)\psi(e_j-e_k). \tag{5.3}$$

This result follows immediately since the lower and upper bounds for $w_{\rm mix}(c)$ coincide. Our theorem (6.1) generalizes this fact to the case of n phases, if all transformation strains lie on one straight line.

By applying (5.2) with $c^{(1)}=\theta e_j+(1-\theta)e_k$ and $c^{(2)}=e_l$ we obtain a bound for all points in $\mathrm{conv}\{e_j,e_k,e_l\}$. Proceeding like this we obtain

Proposition 5.1. *For $c\in P^n$ let $c^{(k)}=(c_1+\ldots+c_k)^{-1}(c_1e_1+\ldots+c_ke_k)$, then*

$$w_{\rm mix}(c)\le\sum_{j=2}^n\tfrac{(c_1+\ldots+c_{j-1})c_j}{c_1+\ldots+c_j}\,\psi(c^{(j-1)}-c^{(j)}). \tag{5.4}$$

By permutation of the points $e_1,\ldots,e_n$ we can generate other upper bounds.

Of course, the upper bound in the right–hand side of (5.4) may be further reduced by applying (5.2) again. In fact, in typical cases it needs infinitely many applications of the lamination formula to reach the lowest level. This corresponds to infinite sequential lamination, see [SmW98].

In principle, $w^{\rm upp}(c)=\max\{\,\widetilde w(c):\widetilde w\text{ satisfies (5.2)},\widetilde w(e_j)\le0,\,j=1,\ldots,n\,\}$ defines the best upper bound for $w(c)$ which can be obtained from lamination. For practical purposes it is better to write $w^{\rm upp}$ as a minimum, since then each candidate $\overline w$ provides a true upper bound

$$w^{\rm upp}(c)=\min\{\,\overline w(c):\overline w\text{ satisfies (5.5)},\ \overline w(e_j)\ge0\text{ for }j=1,\ldots,n\,\},$$

where now the condition reads (with $d=c^{(2)}-c^{(1)}$)

$$\overline w(c^{(1)}+\theta d)\ge\theta\overline w(c^{(1)})+(1-\theta)\overline w(c^{(2)})+\theta(1-\theta)\psi(d). \tag{5.5}$$

6. A special case

We illustrate the effectiveness of the proved bounds by application to a special case. A major task is to calculate $\psi:\mathbb{R}^n_*\to\mathbb{R}$. We first note

Proposition 6.1. *If $\psi(\eta)=\langle\!\langle B,\eta\otimes\eta\rangle\!\rangle$, then $w_{\rm mix}(c)=\langle\!\langle B,\mathbb{M}(c)\rangle\!\rangle$.*

To proof this we first note that $w^{\rm low}(c)=\langle\!\langle B,\mathbb{M}(c)\rangle\!\rangle$ since in (1.1) every linear combination gives the same result by linearity. Similarly the upper bound (5.4) gives the same result. In fact, the assumption can be weakened severely by assuming $\psi(\eta)\ge\langle\!\langle B,\eta\otimes\eta\rangle\!\rangle$ with equality for those η which are needed in (5.4).

Now consider the case that all transformation strains lie on one straight line in $\mathbb{R}^{n\times n}_{\text{sym}}$, i.e., $A_j = A_0 + a_j A$ with $a_j \in \mathbb{R}$. The case $A_1 = -A_3$ and $A_2 = 0$ is treated in [PAM98]. For $A^\eta = \sum_j \eta_j A_j$ with $\eta \in \mathbb{E}^n_*$ we find $A^\eta = \langle \eta, a\rangle A$ and

$$\psi(\eta) = -\tfrac{\gamma}{2}\langle a,\eta\rangle^2 = -\tfrac{\gamma}{2}\langle\!\langle a\otimes a, \eta\otimes\eta\rangle\!\rangle,$$
$$\text{with } \gamma = \max\{\, \langle T(\omega)^{-1}\mathbf{C}[A]\omega, \mathbf{C}[A]\omega\rangle : \omega \in \mathbb{S}^{d-1}\,\}.$$

With proposition (6.1) we conclude the following exact formula.

Theorem 6.1. *If all $W(E, e_j)$ have the form (2.2) with $A_j = A_0 + a_j A$, then*

$$w_{\text{mix}}(c) = -\tfrac{\gamma}{2}\langle\!\langle a\otimes a, \mathbb{M}(c)\rangle\!\rangle = -\tfrac{\gamma}{2}\Big[\sum_{j=1}^n a_j^2 c_j - \langle a, c\rangle^2\Big].$$

References

[AvM89] M. Avellaneda, G.W. Milton. Bounds on the effective elasticity tensor of composites based on two-point correlations. In "*Proc. ASME Energy–Technology Conf. and Exposition*, D. Hui, T, Koszic (eds). New York: ASME, 1989.

[HaG99] G.J. Hall, S. Govindjee. A model and numerical framework for the simulation of solid–solid phase transformation. Preprint UCB/SEMM–1999/11, University of California at Berkeley, 1999.

[KMS99] M.S. Kuczma, A. Mielke, E. Stein. Modelling of hysteresis in two–phase systems. *Arch. Mech.* 51 (1999) 693–715.

[Koh91] R.V. Kohn. The relaxation of a double–well energy. *Continuum Mech. Thermodyn.* 3 (1991) 193–236.

[KuM00] M.S. Kuczma, A. Mielke. Influence of hardening and inhomogeneity on internal loops in pseudo–elasticity. *Z. angew. Math. Mech. (ZAMM)* 80 (2000) 291–306.

[Lev98] V.I. Levitas. Thermomechanical theory of martensitic phase transformations in inelastic materials. *Int. J. Solids Structures* 35 (1998) 889–940.

[Mie00] A. Mielke. Existence results for incremental problems in phase transitions with friction. In preparation, 2000.

[MTL98] A. Mielke, F. Theil, V.I. Levitas. Mathematical formulation of rate–independent phase transformations with friction using an extremum principle. Preprint IfAM, Univ. Hannover, September 1998.

[PAM98] S. Pagano, P. Alert, O. Maisonneuve. Solid-solid phase transition modelling. Local and global minimizations of non-convex and relaxed potentials. Isothermal case for shape memory alloys. *Int. J. Eng. Science* 36 (1998) 1143–1172.

[SmW98] V.P. Smyshlyaev, J.R. Willis. On the relaxation of a three–well energy. *Proc. Royal Soc. London* A 455 (1998) 779–814.

[Tar90] L. Tartar. H–measures, a new approach for studying homogenization, oscillation and concentration effects in partial differential equations. *Proc. Royal Soc. Edinburgh* A 115 (1990) 193–230.

A Two–Scale Micro–Macro–Approach to Anisotropic Finite Plasticity of Polycrystals

C. Miehe and J. Schotte

Institute of Applied Mechanics (Civil Engineering), University of Stuttgart, Germany

Summary. We discuss elements of a two–scale micro–macro–approach to the modelling of the elastic–plastic material response of polycrystals at finite strains. The underlying key idea is a direct analysis of a homogenized macro–continuum with a locally attached micro–structure, which represents a representative aggregate of single crystal grains. The first part of the paper focuses on the formulation and numerical implementation of an average–type micro–macro–transition and points out a discinct closed–form representation of the overall elastic–plastic tangent moduli. The second part adresses a computational model of finite crystal plasticity for the modelling of the grains in the microstructure and proposes a characteristic stress update algorithm based on an incremental minimizing problem.

1. Introduction

The macroscopic description of heterogeneous materials such as polycrystals can be based on the introduction of at least two different phenomenological scales. In this context, the *macro–scale* is usually associated with a homogenized continuous medium. The *micro–scale* is often characterized by a representative volume which contains micro–structural constituents. For an idealized polycrystal, these constituents are defined by the single crystal grains as indicated in Fig. 1.1. With regard to the prediction of texture developments in polycrystalline materials, a direct constitutive modelling on the macro–scale turns out to be very complicated. Effects of anisotropy due to the crystallographic texture evolution, i.e. the reorientation of the crystal lattices of the grains during the deformation, are very hard to describe within the direct macroscopic approach. In order to circumvent these difficulties we consider here a two–scale micro–macro–approach where the overall constitutive response of a polycrystal is described on the basis of two ingredients, the structural arrangements of the single crystal grains and a constititive model of finite crystal plasticty for the description of the grain behaviour. Thus the overall anisotropic stress response of the polycrystal is governed by taking into account a distinct micro–structure of randomly orientated grains and considering the volume average of the micro–stress variables, as indicated in Fig. 2.1. The paper adresses key aspects of the formulation and numerical implementation of (i) average–type micro–to–macro transitions for the analysis of the structural effects based on the works [6], [7] and (ii) the constitutive modelling of finite single crystal plasticity for the description of the grain response, following [4], [5], [6], [8].

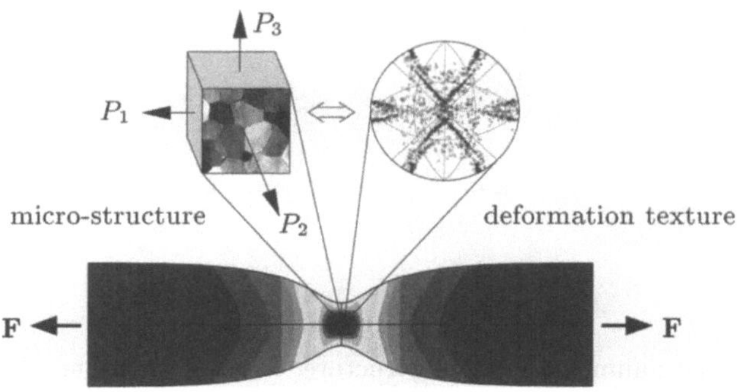

Fig. 1.1. Necking of a polycrystalline specimen under uniaxial tension. Zoomed micro–structure with single crystal grains and developed grain–orientation–texture.

2. A class of mirco-macro-transitions at large strains

Let $\bar{\boldsymbol{x}} : \bar{\mathcal{B}} \to \bar{\mathcal{S}}$ denote the nonlinear deformation map of the macro–structure $\bar{\mathcal{B}} \subset \mathcal{R}^3$ depicted in Fig. 2.1 at time $t \in \mathcal{R}_+$ and $\bar{\boldsymbol{F}} = \nabla\bar{\boldsymbol{x}}$ the associated deformation gradient with Jacobian $\bar{J} = \det[\bar{\boldsymbol{F}}] > 0$. $\bar{\boldsymbol{x}}$ maps points $\bar{\boldsymbol{X}} \in \bar{\mathcal{B}}$ onto points $\bar{\boldsymbol{x}}(\bar{\boldsymbol{X}}) \in \bar{\mathcal{S}}$ of the current configuration $\bar{\mathcal{S}} \subset \mathcal{R}^3$. The macroscopic boundary value problem of quasistatic finite elastoplasticity is governed by the equilibrium equation

$$\mathrm{Div}\bar{\boldsymbol{P}} + \bar{\gamma} = \boldsymbol{0} \quad \text{in } \bar{\mathcal{B}} \,. \tag{2.1}$$

The boundary conditions for the deformation and the surface tractions are

$$\bar{\boldsymbol{x}} = \bar{\boldsymbol{x}}_b \quad \text{on } \partial\bar{\mathcal{B}}_{\bar{x}} \qquad \text{and} \qquad \bar{\boldsymbol{t}} := \bar{\boldsymbol{P}}\bar{\boldsymbol{N}} = \bar{\boldsymbol{t}}_b \quad \text{on } \partial\bar{\mathcal{B}}_{\bar{t}} \tag{2.2}$$

with outward normal $\bar{\boldsymbol{N}}$ on $\bar{\mathcal{B}}$, where we consider a decomposition of the surface of the reference configuration $\partial\bar{\mathcal{B}} = \partial\bar{\mathcal{B}}_{\bar{x}} \cup \partial\bar{\mathcal{B}}_{\bar{t}}$ and $\partial\bar{\mathcal{B}}_{\bar{x}} \cap \partial\bar{\mathcal{B}}_{\bar{t}} = \emptyset$. A formal initial condition

$$\text{macro-state}(\bar{\boldsymbol{X}}, t = 0) = \text{macro-state}_0(\bar{\boldsymbol{X}}) \tag{2.3}$$

sets start values of all internal variables at $\bar{\boldsymbol{X}} \in \bar{\mathcal{B}}$ used for the description of the macroscopic inelastic stress reponse.

In (2.1), $\bar{\boldsymbol{P}}$ with symmetry property $\bar{\boldsymbol{P}}^T\bar{\boldsymbol{F}}^{-T} = \bar{\boldsymbol{F}}^{-1}\bar{\boldsymbol{P}}$ denotes the first Piola or nominal stress tensor and $\bar{\gamma}$ a given body force with respect to the initial volume of the macro–structure. Key ascpect of the subsequent two–scale constitutive modelling is to assume the macroscopic stresses $\bar{\boldsymbol{P}}$ be determined by the *volume average* $(2.4)_1$ of the associated variable $\boldsymbol{P}$ defined at $\boldsymbol{X} \in \mathcal{B}$ on a micro–structure $\mathcal{B}$ attached to the point $\bar{\boldsymbol{X}} \in \bar{\mathcal{B}}$ of the macro–continuum $\bar{\mathcal{B}}$. The two–scale structural character, visualized in Fig. 2.1, offers a possibility to model the overall macroscopic response in terms of both a *local material response* and an *assumed heterogeneous*

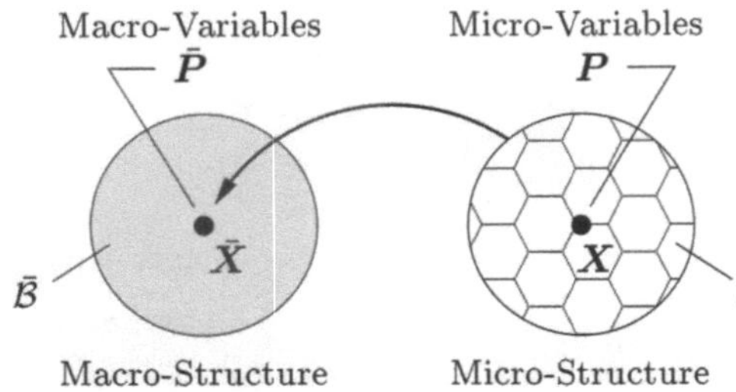

Fig. 2.1. Continuum with micro–structure. A micro-structure $\mathcal{B}$ (micro-stress-variable $\boldsymbol{P}$) attached at each point of the macro-structure $\bar{\mathcal{B}}$ (macro-variable $\bar{\boldsymbol{P}}$)

structure of the micro–structure. In this paper we focus on micro–structures consisting of a representative assembly of coherent single crystal grains, which undergo local elastic–plastic deformations at large strains. Then the stresses $\boldsymbol{P}$ at points $\boldsymbol{X} \in \mathcal{B}$ are governed by a constitutive model of finite crystal plasticity, which we comment on in Section 3 below. In this context, the initial condition (2.3) is understood to set the *field of start values* for the internal local history variables on the micro–structure, see e.g. $(3.3)_2$ below.

The deformation of the micro–structure is driven by the macroscopic deformation gradient $\bar{\boldsymbol{F}}$ as follows. Let $\boldsymbol{x} : \mathcal{B} \to \mathcal{S}$ denote the nonlinear deformation map of the micro–structure $\mathcal{B} \in \mathcal{R}^3$ at time $t \in \mathcal{R}_+$ and $\boldsymbol{F} = \nabla \boldsymbol{x}$ the associated deformation gradient with Jacobian $J = \det[\boldsymbol{F}] > 0$. $\boldsymbol{x}$ maps points $\boldsymbol{X} \in \mathcal{B}$ onto points $\boldsymbol{x}(\boldsymbol{X}) \in \mathcal{S}$ of the current configuration $\mathcal{S} \subset \mathcal{R}^3$. The macroscopic and microscopic deformation gradients $\bar{\boldsymbol{F}}$ and $\boldsymbol{F}$ and associated work–conjugate stress measures $\bar{\boldsymbol{P}}$ and $\boldsymbol{P}$ are assumed to be related via the *volume averages*

$$\bar{\boldsymbol{P}} = \frac{1}{V} \int_{\mathcal{B}} \boldsymbol{P} \, dV \quad \text{and} \quad \bar{\boldsymbol{F}} = \frac{1}{V} \int_{\mathcal{B}} \boldsymbol{F} \, dV \tag{2.4}$$

where $V := vol(\mathcal{B})$ is the volume of the micro–structure $\mathcal{B}$. Assuming a static equlibrium state of the micro–structure governed by the field equation

$$\mathrm{Div}\boldsymbol{P} = \boldsymbol{0} \quad \text{in } \mathcal{B} \,, \tag{2.5}$$

a reformulation of (2.4) using Gauss–type theorems yields $\bar{\boldsymbol{P}} = \frac{1}{V} \int_{\partial\mathcal{B}} \boldsymbol{t} \otimes \boldsymbol{X} dA$ and $\bar{\boldsymbol{F}} = \frac{1}{V} \int_{\partial\mathcal{B}} \boldsymbol{x} \otimes \boldsymbol{N} dA$. Thus, the volume averages $\bar{\boldsymbol{F}}$ and $\bar{\boldsymbol{P}}$ can be exclusively expressed in terms of the deformation $\boldsymbol{x}$ and the traction $\boldsymbol{t} = \boldsymbol{P}\boldsymbol{N}$ at points $\boldsymbol{X} \in \partial\mathcal{B}$ on the boundary $\partial\mathcal{B}$ of the micro–structure with outward normal $\boldsymbol{N}$. The link of the deformation of the micro–structure to the local deformation of the macro–continuum at $\bar{\boldsymbol{X}} \in \bar{\mathcal{B}}$ is assumed to be given by

$$\boldsymbol{x} = \bar{\boldsymbol{F}}\boldsymbol{X} + \tilde{\boldsymbol{w}} \quad \text{in } \mathcal{B} \,. \tag{2.6}$$

Thus the deformation of the micro–structure consists of a linear part $\bar{\boldsymbol{F}}\boldsymbol{X}$ and a nonlinear superimposed field $\tilde{\boldsymbol{w}}(\boldsymbol{X})$, usually referred to as the *fluctuation field.* As a consequence, we have the relationship

$$\boldsymbol{F} = \bar{\boldsymbol{F}} + \tilde{\boldsymbol{F}} \qquad \text{with} \qquad \tilde{\boldsymbol{F}} = \nabla \tilde{\boldsymbol{w}} \tag{2.7}$$

between the microscopic and macroscopic deformation gradients. For given macroscopic deformation $\bar{\boldsymbol{F}}$, we compute the fluctuations from the equilibrium condition (2.5) by taking into account alternatively the constraints

$$(i)\ \ \tilde{\boldsymbol{w}} = \boldsymbol{0} \text{ in } \mathcal{B}\ ; \quad (ii)\ \ \tilde{\boldsymbol{w}} = \boldsymbol{0} \text{ on } \partial\mathcal{B}\ ; \quad (iii)\ \ \tilde{\boldsymbol{w}}^{+} = \tilde{\boldsymbol{w}}^{-} \text{ on } \partial\mathcal{B}\,. \tag{2.8}$$

The first trivial condition enforces a homogeneous deformation of the entire domain of the microstructure and yields the so–caleed Voigt– or Taylor–bound of the overall stiffness. The second condition demands linear deformations on the boundary of the microstructure, the third ansatz states the periodicity of the superimposed deformation field $\tilde{\boldsymbol{w}}$ on $\partial\mathcal{B}$. All three conditions satisfy the so–called micro–heterogenity theorem, see [3], [11], [6], which demands that the average of the work (or stress power) decomposes into the product of work–averages, i.e. $\bar{\boldsymbol{P}}:\bar{\boldsymbol{F}} = \frac{1}{V}\int_{\mathcal{B}} \boldsymbol{P}:\boldsymbol{F}\, dV$.

The stresses $\boldsymbol{P}$ at $\boldsymbol{X} \in \mathcal{B}$ and current time t are goverened by a constitutive model of finite single crystal plasticity, see (3.5) and the algorithmic counterpart (3.14) below. Observe that for a given macroscopic deformation map $\bar{\boldsymbol{F}}$ the microscopic boundary value problem is governed by the equilibrium condition (2.5), the evaluation of the constitutive stress update (3.14) for the microscopic stresses and the boundary conditions (2.8) for the fluctuations $\tilde{\boldsymbol{w}}$. The solution of this quasistatic boundary value problem in a typical time interval $[t_n, t_{n+1}]$ of interest can be obtained based on a spatial discretization of the fluctuation gradient $\nabla\tilde{\boldsymbol{w}}_{n+1}$ on the microstructure

$$\tilde{\boldsymbol{F}}_{n+1} = \boldsymbol{B}\tilde{\boldsymbol{d}}_{n+1} \tag{2.9}$$

in terms of the matrix $\boldsymbol{B}$ of *assumed fluctuation shapes* and the matrix $\tilde{\boldsymbol{d}}_{n+1}$ of *discrete fluctuation variables.* In a typical time interval of interest, one then computes the current discrete fluctuations from a Newton iteration based on the linearization of the weak form of (2.5), yielding the typical update formula

$$\tilde{\boldsymbol{d}}_{n+1} \leftarrow \tilde{\boldsymbol{d}}_{n+1} - \boldsymbol{K}^{-1}\boldsymbol{r} \quad \text{for} \quad ||\boldsymbol{r}|| > tol \tag{2.10}$$

in terms of the *structural matrices*

$$\boldsymbol{r} := \textstyle\int_{\mathcal{B}} \boldsymbol{B}^T \boldsymbol{P}_{n+1} dV\ ; \quad \boldsymbol{K} := \int_{\mathcal{B}} \boldsymbol{B}^T \mathbb{C}_{n+1} \boldsymbol{B} dV\ ; \quad \boldsymbol{L} := \int_{\mathcal{B}} \boldsymbol{B}^T \mathbb{C}_{n+1} dV \tag{2.11}$$

of the discretized micro–structure. These structural matrices govern also the sensitivity of the macroscopic stresses

$$\Delta\bar{\boldsymbol{P}}_{n+1} = \bar{\mathbb{C}}_{n+1} : \Delta\bar{\boldsymbol{F}}_{n+1} \tag{2.12}$$

with respect to a change in macroscopic deformation, which are needed for the iterative solution and the stability analysis of the macro–structure. As shown in [7], a closed–form representation of the *overall macroscopic nominal tangent moduli* has the form

$$\bar{\mathbb{C}}_{n+1} = \frac{1}{V}\int_{\mathcal{B}} \mathbb{C}_{n+1}\, dV - \frac{1}{V}\boldsymbol{L}^T \boldsymbol{K}^{-1} \boldsymbol{L}\,. \tag{2.13}$$

The stress updates $\boldsymbol{P}_{n+1}$ and associated algorithmic tangent moduli $\mathbb{C}_{n+1}$ associated with a model of finite crystal plasticity are given in (3.14) and (3.16) below.

3. A computational model of finite crystal plasticity

In what follows we summarize a computational model of finite crystal plasticity in a multisurface setting for ideal elastic–plastic Schmid–type flow response, see [2], [1], [6] , [8] for extended formulations which include hardening effects. The *anisotropic elastic response* of a single crystal grain at $\boldsymbol{X} \in \mathcal{B}$ is governed by the lattice deformations. We consider a stored free energy function

$$\psi = \hat{\psi}(\boldsymbol{F}^e, \boldsymbol{X}) \quad \text{with} \quad \boldsymbol{F}^e \in GL(3)_+\,, \tag{3.1}$$

which is assumed to satisfy the objectivity demand $\hat{\psi}(\boldsymbol{Q}\boldsymbol{F}^e, \boldsymbol{X}) = \hat{\psi}(\boldsymbol{Q}\boldsymbol{F}^e, \boldsymbol{X})$ $\forall\, \boldsymbol{Q} \in SO(3)$. Here, $\boldsymbol{F}^e$ is assumed to describe the deformation of the crystal lattice with respect to a given reference configuration. This lattice deformation map is *defined* by the composition

$$\boldsymbol{F}^e := \boldsymbol{F}\boldsymbol{F}^{p-1} \quad \text{with} \quad \boldsymbol{F} \in GL(3)_+ \quad \text{and} \quad \boldsymbol{F}^p \in SL(3)_+ \tag{3.2}$$

in terms of the local deformation gradient $\boldsymbol{F}$ of the micro–structure introduced above and an unimodular plastic deformation map $\boldsymbol{F}^p$, which describes the multiple shearing on crystallographic slip planes. We consider $\boldsymbol{F}^p$ as a formal internal variable *determined* by a flow rule

$$\dot{\boldsymbol{F}}^p = \textstyle\sum_{\alpha=1}^{m} \dot{\gamma}^\alpha \boldsymbol{M}^\alpha \boldsymbol{F}^p \quad \text{with} \quad \boldsymbol{F}^p(\boldsymbol{X}, t=0) = \boldsymbol{1} \tag{3.3}$$

in terms of constant crystallographic (deviatoric) structural tensors $\boldsymbol{M}^\alpha = \hat{\boldsymbol{M}}^\alpha(\boldsymbol{X})$, which govern the *anisotropic plastic response* of a grain at $\boldsymbol{X} \in \mathcal{B}$ associated with $\alpha = 1...m$ crystallographic slip systems. The local stress power $\mathcal{P} := \boldsymbol{P} : \dot{\boldsymbol{F}}$ per volume of the reference configuration of the micro–structure can be decomposed into an partial reversible part and a partial dissipative part

$$\mathcal{P} := \dot{\psi} + \mathcal{D} \quad \text{with} \quad \mathcal{D} \geq 0\,. \tag{3.4}$$

Exploiting standard arguments of continuum–thermodynamics, the insertion of (3.1) into (3.4) identifies the constitutive equations

$$\boldsymbol{P} = \partial_{\boldsymbol{F}^e}\hat{\psi}\boldsymbol{F}^{p-T} \tag{3.5}$$

for the nominal stress tensor and internal dissipation

$$\mathcal{D} = \sum_{\alpha=1}^{m} \tau^\alpha \dot{\gamma}^\alpha \geq 0 \quad \text{with} \quad \tau^\alpha := \boldsymbol{F}^{eT}\partial_{\boldsymbol{F}^e}\hat{\psi} : \boldsymbol{M}^\alpha \ . \tag{3.6}$$

Here, τ^α is the thermodynamic definition of the resolved shear stress or so–called Schmid stress, i.e. thermodynamic variable conjugate to the plastic shearing γ^α on the slip system α. For the rate–independent response, the evolution of the plastic slip is determined by the plastic loading conditions

$$\dot{\gamma}^\alpha \geq 0 \ , \ (\tau^\alpha - \tau_c) \leq 0 \ , \ \dot{\gamma}^\alpha(\tau^\alpha - \tau_c) = 0 \tag{3.7}$$

written in Kuhn–Tucker form. Here, τ_c is the critical resolved shear stresses, which is considered in the context of ideal plasticity as a *constant* material property. Considering the currently active slip systems $\mathcal{A} := \{\alpha = 1...m | \ \dot{\gamma}^\alpha > 0\}$ we can recast the dissipation into the form

$$\mathcal{D} = \sum_{\alpha\in\mathcal{A}} \tau_c \dot{\gamma}^\alpha \geq 0 \ . \tag{3.8}$$

The numerical impementation of the model outlined above is based on a time integration algorithm applied to (3.3) in time interval $[t_n, t_{n+1}]$ under consideration. Assuming all variables at time t_n as known, we consider the algorithm

$$\boldsymbol{F}^p_{n+1}[\gamma^\alpha_{n+1}] = \exp[\sum_{\alpha\in\mathcal{A}}(\gamma^\alpha_{n+1} - \gamma^\alpha_n)\boldsymbol{M}^\alpha]\boldsymbol{F}^p_n \tag{3.9}$$

which preserves algorithmically exact the geometric structure of the unimodular deformation map $\boldsymbol{F}^p \in SL(3)$. Following conceptually ideas of [9],[10], we base the stress update in the time interval of interest on an *incremental variational formulation.* To this end, we consider the increment of total work with respect to the reference volume

$$\Psi := \int_{t_n}^{t_{n+1}} \mathcal{P} dt \tag{3.10}$$

as a potential function in the time interval $[t_n, t_{n+1}]$. Insertion of (3.4) and integration of (3.8) yields the closed–form expression

$$\hat{\Psi}(\boldsymbol{F}_{n+1}, \gamma^\alpha_{n+1}) = \hat{\psi}(\boldsymbol{F}_{n+1}\boldsymbol{F}^{p-1}_{n+1}[\gamma^\alpha_{n+1}]) - \psi_n + \sum_{\alpha\in\mathcal{A}} \tau_c(\gamma^\alpha_{n+1} - \gamma^\alpha_n) \tag{3.11}$$

at a frozen active set $\mathcal{A}$. A minimizing path for the total deformation is obtained when (3.11) is minimized with respect to the plastic parameters $\gamma^\alpha_{n+1}, \gamma^\alpha_{n+1} \in \mathcal{A}$. Then

$$\partial_{\gamma^\alpha_{n+1}}\hat{\Psi} = \partial_{\gamma^\alpha_{n+1}}\hat{\psi}(\boldsymbol{F}_{n+1}\boldsymbol{F}^{p-1}_{n+1}[\gamma^\alpha_{n+1}]) + \tau_c = 0 \tag{3.12}$$

can be considered as an *energetic consistency condition* for the determination of the current plastic slip $\gamma^\alpha_{n+1} \in \mathcal{A}$. The variation of (3.11)

$$\delta\hat{\Psi} = \partial_{\boldsymbol{F}_{n+1}}\hat{\Psi} : \delta\boldsymbol{F}_{n+1} + \sum_{\alpha\in\mathcal{A}} \partial_{\gamma^\alpha_{n+1}}\hat{\Psi} : \delta\gamma^\alpha_{n+1} = 0 \tag{3.13}$$

indicates that, in case of plastic parameters evaluated by (3.12), $\hat{\Psi}(\boldsymbol{F}_{n+1};\gamma^{\alpha}_{n+1})$ serves as a potential for the nominal stresses $\boldsymbol{P}_{n+1}$ at time t_{n+1}. Thus the stress update algorithm of finite plasticity in the time interval under consdieration has the formal structure of an pseudoelastic constitutive equation

$$\boldsymbol{P}_{n+1} = \partial_{\boldsymbol{F}_{n+1}} \hat{\Psi} \; . \tag{3.14}$$

The sensitivity of this algorithmic expression with respect of a change of the deformation gradient

$$\Delta \boldsymbol{P}_{n+1} = \mathbb{C}_{n+1} : \Delta \boldsymbol{F}_{n+1} \tag{3.15}$$

defines the consistent tangent moduli

$$\mathbb{C}_{n+1} = \partial^2_{\boldsymbol{F}_{n+1}\boldsymbol{F}_{n+1}} \hat{\Psi} - \textstyle\sum_{\alpha\in\mathcal{A}}\sum_{\beta\in\mathcal{A}} G^{\alpha\beta-1} \partial^2_{\boldsymbol{F}_{n+1}\gamma^{\alpha}_{n+1}} \hat{\Psi} \otimes \partial^2_{\gamma^{\beta}_{n+1}\boldsymbol{F}_{n+1}} \hat{\Psi} \tag{3.16}$$

with the definition $G^{\alpha\beta} := \partial^2\hat{\Psi}/\partial\gamma^{\alpha}_{n+1}\partial\gamma^{\beta}_{n+1}$. Note that the update procedure provides *symmetric algorithmic moduli* $\mathbb{C}_{n+1}$ which is an important ingredient in view to an efficient implementation of the model of finite strain plasticity. For further details concerning the treatment of the exponential map, its derivatives, alternative projection algorithms, active set search algorithms and uniqueness considerations we refer to the recent papers [6], [8].

4. Texture evolution in homogeneous compression test

In a numerical example we analyse the response of a polycrystalline micro–strucure which undergoes an isochoric compression mode governed by the macroscopic deformation gradient $\bar{\boldsymbol{F}} = e^{\lambda}\boldsymbol{e}_1 \otimes \boldsymbol{e}_1 + e^{\lambda}\boldsymbol{e}_2 \otimes \boldsymbol{e}_2 + e^{-2\lambda}\boldsymbol{e}_3 \otimes \boldsymbol{e}_3$ as depicted in Figure 4.1. The polycrystalline micro–structure considered is a copper unit cube, which has been discretized using $7^3 = 343$ mixed finite elements. A random crystal orientation model [6] assignes to each element a distinct initial orientation of a single crystal grain. The overall response of the polycrytstal is simulated on the basis of the micro–macro–transition considered above where alternatively the Taylor–type and the periodic conditions in (2.8) are investigated.

During the simulation the specimen is compressed down to 37% of its initial height, where Figure 4.1 shows the initial mesh and final deformation of the unit cube reached after 1000 equal load steps. The pole figures of the micro-structure at the final state are then compared to experimental data given in [1] and to the predictions of a corresponding Taylor-type calculation. Figure 4.2 depicts the $\langle 111\rangle$-pole figures. The results of both simulations are in good agreement with the experimental textures; however, the Taylor-type calculation yields sharper pole figures because of the additional constraints.

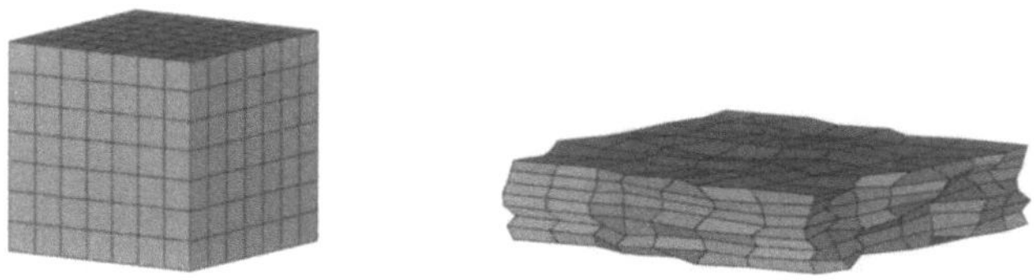

Fig. 4.1. Simple compression test. (**a**) Initial mesh and geometry, unit cube built of 343 elements. (**b**) Deformed structure with periodic boundary conditions after uniaxial compression (reduction of 63 %)

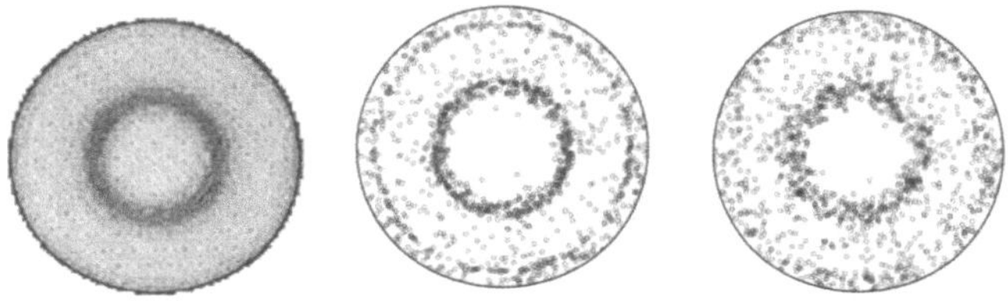

Fig. 4.2. Simple compression test. ⟨111⟩-pole figures of deformed structure after reduction of 63 %. (**a**) Experimental result. (**b**) Taylor-type calculation using 343 grains. (**c**) 3d-calculation with periodic boundary conditions

References

1. Bronkhorst, C.A., Kalidindi, S.R., Anand, L. (1992) Polycrystalline plasticity and the evolution of crystallographic texture in fcc metals, Phil. Trans. R. Soc. London, Ser. A, **341**, 443–447.
2. Cuitino, A. M., Ortiz, A. (1992) Computational modelling of single crystals. Mod. Simul. Mater. Sci. Eng. **1**, 225–263.
3. Hill, R. (1972) On constitutive macro–variables for heterogeneous solids at finite strain, Proc. R. Soc. London, Ser. A, **326**, 131–147.
4. Miehe, C. (1994) On the Representation of Prandtl-Reuss Tensors within the Framework of Multiplicative Elasto-Plasticity, Int. J. Plasticity, **10**, 609–621.
5. Miehe, C. (1996) Multisurface thermoplasticity for single crystals at large strains in terms of Eulerian vector updates. Int. J. Solids Structures **33**, 3103–3130
6. Miehe, C., Schröder, J., Schotte, J. (1999). Computational homogenization analysis in finite plasticity. Simulation of texture development in polycrystalline materials. Comp. Meth. Appl. Mech. Eng. **171**, 387–418.
7. Miehe, C., Schotte, J., Schröder, J. (1999) Computational micro–macro–transitions and overall moduli in the analysis of polycrystals at large strains. Comp. Materials Science **16**, 372–382.
8. Miehe, C. (2000) Computational plasticity of crystals at large strains for rate–independent and rate–dependent flow response, Int. J. Plasticity, to appear.
9. Ortiz, M., Repetto, E.A. (1999) Nonconvex energy minimization and dislocation structures in ductile single crystals, J. Mech. Phys. Solids **47**, 397–462.
10. Ortiz, M., Stainier, L. (1999) The variational formulation of viscoplastic constitutive updates, Comp. Meth. Appl. Mech. Eng. **171**, 419–444.
11. Suquet, P.M. (1986) Elements of homogenization for inelastic solid mechanics, in: Homogenization techniques for composite materials, Lecture notes in physics 272, edited by E. Sanchez–Palenzia, A. Zaoui, Springer–Verlag, 193–278.

Section 3

Solid–Fluid–Interaction

Edited by W. Ehlers

Boundary Conditions at the Viscous Sliding Interface of Incompressible Porous Deformable Media

G.A. Ateshian and X. Wang

Department of Mechanical Engineering, Columbia University, USA

Summary. The boundary conditions at the interface between sliding porous deformable media are derived, taking into account the viscosity of the interstitial fluid. These boundary conditions are shown to reduce to all the known limiting conditions, including those of Beavers and Joseph (1967), and Hou et al. (1989). These equations can be used to investigate the frictional interactions at the sliding interface of soft hydrated biological tissues such as articular cartilage.

1. Introduction

An important topic of research in biotribology is the sliding contact of articular cartilage, as occurs in diarthrodial joints, e.g., the knee, hip, or shoulder. Cartilage may be suitably modeled as a biphasic mixture of a solid phase, representing the collagen matrix and its enmeshed proteoglycan molecules, and a fluid phase representing the interstitial water which comprises between 65% and 85% of the tissue by weight [17]. However, few solutions exist for the problem of sliding contact of such porous deformable materials. In a previous study [1], we analyzed the problem of rolling and sliding contact of cylindrical biphasic cartilage layers bonded to a rigid bony substrate, for the case when the fluid phase of cartilage was assumed inviscid (i.e., when deviatoric stresses in the fluid are neglected and the effect of the drag between the solid and fluid phases is modeled with a Darcy-like relation). This assumption made it possible to consider frictionless conditions at the contact interface. However, despite the fact that cartilage friction is generally low (e.g., [11]), such ideal conditions do not prevail in actual joints as friction may result from solid-to-solid, solid-to-fluid, and fluid-to-fluid interactions at the contact interface of biphasic media. In a more recent study [2, 3], we proposed a boundary friction model for biphasic materials whereby interfacial solid-to-solid frictional interactions are assumed to dominate the frictional response, while the remaining solid-to-fluid and fluid-to-fluid interactions are taken to be negligible. In these analyses, it was similarly convenient to assume an inviscid fluid phase.

If the viscosity of the interstitial fluid is no longer neglected, a boundary layer develops at the surface of a biphasic material subjected to a shear traction [4, 15, 16]. For rigid porous materials, inclusion of viscous terms marks the distinction between Darcy's law and Brinkman's law [10]. This boundary layer is due to the competing influence of the viscous drag imposed by the surface shear traction (the Couette-like flow predicted from the Navier-Stokes terms of the governing equations) and the resistive drag resulting from the momentum exchange between the fluid and solid phases (the Darcy-like flow typical of porous media). The thickness of the boundary layer depends on the properties of the biphasic material, and it may be very thin in the case of articular cartilage [15]. It is possible to anticipate that the sliding contact of biphasic media whose fluid phase is viscous might similarly induce boundary layers near the surface, and consequently viscous shear tractions at the contact interface. Though fluid-film lubrication problems involving bearing surfaces comprised of biphasic materials have been solved previously [5, 13, 14, 16, 19], and while a general solution for sliding tractions on a porous deformable biphasic layer with viscous interstitial fluid has been described [4], the sliding boundary contact of biphasic media with viscous interstitial fluid has not been presented previously.

The primary obstacle for the solution of such sliding contact problems is the absence of adequate boundary conditions to describe the continuity requirements for the kinematic and kinetic biphasic variables at the contact interface. Boundary conditions for related problems have been proposed by Beavers and Joseph [6] for the interface between a viscous fluid and a rigid-porous solid, by Biot [7] and Deresiewicz and Skalak [12] for the interface between poroelastic media with inviscid interstitial fluid, and by Hou et al. [15] for the adhesive interface between biphasic mixture media with viscous fluid phase, or the interface between a viscous fluid and a biphasic mixture medium. Therefore, the objective of this study is to extend the work of Hou et al. [15] by developing the boundary conditions at the contact interface between sliding porous deformable media with intrinsically incompressible solid and fluid phases, where the viscosity of the fluid phase is no longer neglected. A biphasic (binary) incompressible mixture model is adopted to describe the porous deformable material, based on the work of Mow et al. [17], and Bowen [9]. The primary contribution is the introduction of a pseudo-noslip kinematic condition along the interface of the solid and fluid phases. The kinetic boundary conditions are subsequently derived by employing these kinematic conditions and the balance laws of momentum and energy. A necessary condition for the validity of these interface relations is that they reduce to all known limiting conditions when either of the contacting biphasic materials are taken to the limit of an incompressible viscous or inviscid fluid, a deformable impermeable solid, or a rigid porous solid.

2. Formulation

Continuum mixture models represent idealizations of real porous deformable materials; they assume that fluid and solid particles coexist at every point. The solid and fluid volume fractions are expressed as φ^{s} and φ^{f}, respectively, and for saturated mixtures it follows that $\varphi^{\mathrm{s}} + \varphi^{\mathrm{f}} = 1$. At each point in the mixture, it is possible to define the velocities of the solid and fluid phases as $\mathbf{v}^{\mathrm{s}}$ and $\mathbf{v}^{\mathrm{f}}$, and the stress tensors as $\boldsymbol{\sigma}^{\mathrm{s}}$ and $\boldsymbol{\sigma}^{\mathrm{f}}$, respectively. These stress tensors can also be expressed in terms of the interstitial fluid pressure p and extra stresses as $\boldsymbol{\sigma}^{\mathrm{s}} = -\varphi^{\mathrm{s}} p\mathbf{I} + \boldsymbol{\sigma}^{\mathrm{e}}$, $\boldsymbol{\sigma}^{\mathrm{f}} = -\varphi^{\mathrm{f}} p\mathbf{I} + \boldsymbol{\sigma}^{\mathrm{v}}$, where $\boldsymbol{\sigma}^{\mathrm{e}}$ is usually known as the effective stress in the solid, and $\boldsymbol{\sigma}^{\mathrm{v}}$ is the viscous stress in the fluid. In many applications of porous media theory $\boldsymbol{\sigma}^{\mathrm{v}}$ is often neglected ("inviscid" fluid phase), though in the present analysis it plays an essential role in the viscous friction between sliding biphasic materials.

Let Γ represent the contact interface between two sliding biphasic media and let the regions on both sides of the interface be denoted by 0 and 1, respectively. Clearly, by its definition, Γ follows the solid phase of the biphasic mixture, and hence only fluid may cross this interface. A curvilinear coordinate system is adopted, whereby $\mathbf{n}$ denotes the normal vector to Γ at any point of the contact interface, and $\boldsymbol{\tau}$ is the unit tangent vector in the direction of sliding motion.

For mixture models, the volume fraction of each phase is also equal to the area fraction (Delesses's law [8]). Hence, at the surface Γ, the following relations hold,

$$\varphi^{\mathrm{s}0} + \varphi^{\mathrm{f}0} = 1 \tag{2.1}$$

$$\varphi^{\mathrm{s}1} + \varphi^{\mathrm{f}1} = 1. \tag{2.2}$$

When two solid-fluid mixtures come into contact over Γ, the fraction of the contact area over which solid-to-solid contact occurs is given statistically by the product of the solid area fractions across the interface. Similarly for other interactions, as indicated below:

Solid-to-Solid	$\varphi^{\mathrm{s}0}\varphi^{\mathrm{s}1}$	(2.3)
Solid-to-Fluid	$\varphi^{\mathrm{s}0}\varphi^{\mathrm{f}1}$	(2.4)
Fluid-to-Solid	$\varphi^{\mathrm{f}0}\varphi^{\mathrm{s}1}$	(2.5)
Fluid-to-Fluid	$\varphi^{\mathrm{f}0}\varphi^{\mathrm{f}1}$.	(2.6)

It can easily be shown from Eqs. 2.1–2.2 that these probabilities of interactions add up to unity. These expressions, in particular that of 2.3, can serve to clarify the significance of the interface conditions presented below.

2.1 Kinematic conditions

There are two kinematic conditions which govern the kinematic jump normal to the interface Γ. First, because Γ follows the interface between the opposing immiscible solid phases, the component of solid velocity along the unit normal $\mathbf{n}$ must be continuous,

$$[[v^{\mathrm{s}}_{\mathrm{n}}]] = 0, \tag{2.7}$$

where $[[f]] = f^1 - f^0$ represents the jump condition across Γ for the variable f. Furthermore, from the requirements of the continuity of mass, Hou et al. [15] have derived the general jump requirement for mixture models as

$$[[\varphi^s v_n^s + \varphi^f v_n^f]] = 0. \tag{2.8}$$

Equations 2.8 and 2.7 can be combined to produce

$$[[\varphi^f (v_n^f - v_n^s)]] = 0 \tag{2.9}$$

which represents the continuity of the relative fluid flux normal to the interface.

Along the tangential direction to Γ, because of the viscosity of the interstitial fluid phase, a pseudo-noslip condition needs to be formulated, which generalizes the previously proposed conditions for solid-fluid or biphasic-fluid interfaces. In this study, we propose that this kinematic jump condition tangential to Γ be given by

$$[[(1 - \varphi^{s0}\varphi^{s1})v_\tau^s + \varphi^f (v_\tau^f - v_\tau^s)]] = 0. \tag{2.10}$$

In this pseudo-noslip condition, the expression $1 - \varphi^{s0}\varphi^{s1}$ represents the fraction of the contact area exclusive of the region of solid-to-solid interactions at the contact interface of two sliding biphasic media. Heuristically, it can be explained as the expression nearly equivalent to the jump condition normal to Γ, namely $[[v_n^s + \varphi^f (v_n^f - v_n^s)]] = 0$ derived by summing Eq. 2.7 and Eq. 2.9, except that the region of solid-to-solid contact is non-adhesive. Mathematically, this kinematic condition reduces to the appropriate limiting cases as described further below.

2.2 Interface tractions

To satisfy the balance of linear momentum across Γ, it is necessary to require that $[\boldsymbol{\sigma}^s + \boldsymbol{\sigma}^f - \rho^f \mathbf{v}^f(\mathbf{v}^f - \mathbf{v}^s)]\mathbf{n}$ be continuous across the interface as shown by Hou et al. [15]; the last term of this expression is due to the convection of fluid momentum relative to the solid phase boundary, with ρ^f being the apparent density of the fluid phase, which is related to the true density ρ_T^f by $\rho^f = \varphi^f \rho_T^f$. In most porous media flows with low permeability however, the term $\rho^f \mathbf{v}^f(\mathbf{v}^f - \mathbf{v}^s)\mathbf{n}$ is of higher order and thus negligible. Hence, the following simplified continuity requirements may be satisfied:

$$[[\sigma_{nn}^s + \sigma_{nn}^f]] = 0 \tag{2.11}$$

$$[[\sigma_{n\tau}^s + \sigma_{n\tau}^f]] = 0. \tag{2.12}$$

The balance of energy across Γ requires that

$$\left[\boldsymbol{\sigma}^s \mathbf{v}^s + \boldsymbol{\sigma}^f \mathbf{v}^f - \mathbf{h}^s - \mathbf{h}^f - \rho^f \left(e^f + \frac{1}{2}\mathbf{v}^f \cdot \mathbf{v}^f\right)(\mathbf{v}^f - \mathbf{v}^s)\right] \cdot \mathbf{n} \tag{2.13}$$

be continuous [15]. Here, $\mathbf{h}^{\mathrm{s}}$ and $\mathbf{h}^{\mathrm{f}}$ are the heat fluxes for the solid and fluid, respectively, and e^{f} is the fluid's internal energy density. Neglecting higher order terms and heat fluxes, the energy balance requirement reduces to:

$$[[\sigma^{\mathrm{s}}_{\mathrm{nn}} v^{\mathrm{s}}_{\mathrm{n}} + \sigma^{\mathrm{s}}_{\mathrm{n}\tau} v^{\mathrm{s}}_{\tau} + \sigma^{\mathrm{f}}_{\mathrm{nn}} v^{\mathrm{f}}_{\mathrm{n}} + \sigma^{\mathrm{f}}_{\mathrm{n}\tau} v^{\mathrm{f}}_{\tau}]] = 0 \tag{2.14}$$

This equation can be re-arranged in the form

$$\left[\left[(\sigma^{\mathrm{s}}_{\mathrm{nn}} + \sigma^{\mathrm{f}}_{\mathrm{nn}}) v^{\mathrm{s}}_{\mathrm{n}} + \varphi^{\mathrm{f}}(v^{\mathrm{f}}_{\mathrm{n}} - v^{\mathrm{s}}_{\mathrm{n}}) \frac{\sigma^{\mathrm{f}}_{\mathrm{nn}}}{\varphi^{\mathrm{f}}} + (\sigma^{\mathrm{s}}_{\mathrm{n}\tau} + \sigma^{\mathrm{f}}_{\mathrm{n}\tau}) v^{\mathrm{s}}_{\tau} + \varphi^{\mathrm{f}}(v^{\mathrm{f}}_{\tau} - v^{\mathrm{s}}_{\tau}) \frac{\sigma^{\mathrm{f}}_{\mathrm{n}\tau}}{\varphi^{\mathrm{f}}}\right]\right] = 0. \tag{2.15}$$

The first term on the left of Eq. 2.15 cancels out by virtue of Eqs. 2.7 and 2.11. The second term on the left of Eq. 2.15 is the only remaining one containing velocity components normal to Γ. For the special case of one-dimensional flow normal to Γ, the jump condition on this term must be zero, and by virtue of Eq. 2.9 the continuity condition which must be satisfied under these conditions is

$$[[\frac{\sigma^{\mathrm{f}}_{\mathrm{nn}}}{\varphi^{\mathrm{f}}}]] = 0 \tag{2.16}$$

Though this relation was derived for a special case of one-dimensional flow normal to Γ, it is assumed to hold for all other conditions [15]. The remaining terms in Eq. 2.15 are

$$[[(\sigma^{\mathrm{s}}_{\mathrm{n}\tau} + \sigma^{\mathrm{f}}_{\mathrm{n}\tau}) v^{\mathrm{s}}_{\tau} + \varphi^{\mathrm{f}}(v^{\mathrm{f}}_{\tau} - v^{\mathrm{s}}_{\tau}) \frac{\sigma^{\mathrm{f}}_{\mathrm{n}\tau}}{\varphi^{\mathrm{f}}}]] = 0 \tag{2.17}$$

Multiplying Eq. 2.10 with some constant σ and subtracting it from Eq. 2.17 produces

$$\left[\left[\left(\sigma^{\mathrm{s}}_{\mathrm{n}\tau} + \sigma^{\mathrm{f}}_{\mathrm{n}\tau} - (1 - \varphi^{\mathrm{s}0}\varphi^{\mathrm{s}1})\sigma\right) v^{\mathrm{s}}_{\tau} + \varphi^{\mathrm{f}}(v^{\mathrm{f}}_{\tau} - v^{\mathrm{s}}_{\tau}) \left(\frac{\sigma^{\mathrm{f}}_{\mathrm{n}\tau}}{\varphi^{\mathrm{f}}} - \sigma\right)\right]\right] = 0 \tag{2.18}$$

Now it is argued that since the solid sliding velocities are not continuous across Γ, $[[v^{\mathrm{s}}_{\tau}]] \neq 0$, and since $[[\varphi^{\mathrm{f}}(v^{\mathrm{f}}_{\tau} - v^{\mathrm{s}}_{\tau})]] \neq 0$ in general (see Eq. 2.10), the necessary and sufficient conditions to satisfy Eq. 2.18 are

$$\sigma = \frac{\sigma^{\mathrm{f}0}_{\mathrm{n}\tau}}{\varphi^{\mathrm{f}0}} = \frac{\sigma^{\mathrm{f}1}_{\mathrm{n}\tau}}{\varphi^{\mathrm{f}1}} \tag{2.19}$$

and

$$\sigma^{\mathrm{s}0}_{\mathrm{n}\tau} + \sigma^{\mathrm{f}0}_{\mathrm{n}\tau} - (1 - \varphi^{\mathrm{s}0}\varphi^{\mathrm{s}1}) \frac{\sigma^{\mathrm{f}0}_{\mathrm{n}\tau}}{\varphi^{\mathrm{f}0}} = \sigma^{\mathrm{s}1}_{\mathrm{n}\tau} + \sigma^{\mathrm{f}1}_{\mathrm{n}\tau} - (1 - \varphi^{\mathrm{s}0}\varphi^{\mathrm{s}1}) \frac{\sigma^{\mathrm{f}1}_{\mathrm{n}\tau}}{\varphi^{\mathrm{f}1}} = 0 \tag{2.20}$$

In effect, Eq. 2.20 imposes that the solid-to-solid tangential traction on either layer should reduce to zero (i.e., "frictionless" solid-to-solid sliding) in order for Γ to be a non-dissipative interface. This requirement is entirely compatible with Eqs. 2.12 and 2.19.

In summary, the boundary conditions at the contacting interface of sliding biphasic media with viscous fluid phase are given by:

$$\varphi^{s0} v_n^{s0} + \varphi^{f0} v_n^{f0} = \varphi^{s1} v_n^{s1} + \varphi^{f1} v_n^{f1} \tag{2.21}$$

$$v_n^{s0} = v_n^{s1} \tag{2.22}$$

$$(1 - \varphi^{s0}\varphi^{s1}) v_\tau^{s0} + \varphi^{f0}(v_\tau^{f0} - v_\tau^{s0}) = (1 - \varphi^{s0}\varphi^{s1}) v_\tau^{s1} + \varphi^{f1}(v_\tau^{f1} - v_\tau^{s1}) \tag{2.23}$$

$$\sigma_{nn}^{s0} + \sigma_{nn}^{f0} = \sigma_{nn}^{s1} + \sigma_{nn}^{f1} \tag{2.24}$$

$$\sigma_{n\tau}^{s0} + \sigma_{n\tau}^{f0} = \sigma_{n\tau}^{s1} + \sigma_{n\tau}^{f1} \tag{2.25}$$

$$\frac{\sigma_{nn}^{f0}}{\varphi^{f0}} = \frac{\sigma_{nn}^{f1}}{\varphi^{f1}} \tag{2.26}$$

$$\frac{\sigma_{n\tau}^{f0}}{\varphi^{f0}} = \frac{\sigma_{n\tau}^{f1}}{\varphi^{f1}} \tag{2.27}$$

$$\sigma_{n\tau}^{s0} + \sigma_{n\tau}^{f0} - (1 - \varphi^{s0}\varphi^{s1}) \frac{\sigma_{n\tau}^{f0}}{\varphi^{f0}} = \sigma_{n\tau}^{s1} + \sigma_{n\tau}^{f1} - (1 - \varphi^{s0}\varphi^{s1}) \frac{\sigma_{n\tau}^{f1}}{\varphi^{f1}} = 0 \tag{2.28}$$

3. Limiting cases

The boundary conditions developed for the general case of sliding contact between biphasic media with viscous interstitial fluid must reduce to all known limiting cases in order to be valid. For example, in the simplest cases, these equations should reduce to the proper boundary conditions for the sliding contact of impermeable (single-phase) solids, the flow of a viscous fluid against an impermeable solid, or the interface between two viscous fluids. Furthermore, these boundary conditions should subsume the conditions for viscous flow against a rigid porous solid as derived by Beavers and Joseph [6], viscous flow against a porous deformable biphasic material, as provided by Hou and co-workers [15], and frictionless sliding contact of biphasic media with inviscid interstitial fluid as employed by Ateshian and Wang [1]. The way to achieve these limiting conditions is to repeat the above analysis by removing irrelevant terms from the outset (e.g., the fluid velocity and fluid traction components in the case of a single-phase impermeable solid) and letting the solid and fluid fractions on either side of the interface Γ reduce to zero or one, depending on the case being analyzed. These special cases are described below; note that Hou et al. [15] have provided a detailed analysis demonstrating that their interface boundary conditions reduce to the results of Beavers and Joseph [6] and Taylor [18], whose special cases are therefore omitted.

3.1 Solid-solid interface

The interface conditions for the contact of solids are well established. In this case, $\varphi^{s0} = \varphi^{s1} = 1$ and $\varphi^{f0} = \varphi^{f1} = 0$, and the fluid velocity and fluid traction components no longer appear in the interface conditions which, for a non-dissipative (frictionless) interface, reduce to

$$v_{\mathrm{n}}^{s0} = v_{\mathrm{n}}^{s1}, \quad \sigma_{\mathrm{nn}}^{s0} = \sigma_{\mathrm{nn}}^{s1}, \quad \sigma_{\mathrm{n}\tau}^{s0} = \sigma_{\mathrm{n}\tau}^{s1} = 0. \tag{3.1}$$

3.2 Fluid-fluid interface

This next well-known case represents the interface between two viscous fluids. Here, $\varphi^{s0} = \varphi^{s1} = 0$ and $\varphi^{f0} = \varphi^{f1} = 1$, and the solid velocity and solid traction components no longer appear in the interface conditions which reduce to

$$v_{\mathrm{n}}^{f0} = v_{\mathrm{n}}^{f1}, \quad v_{\tau}^{f0} = v_{\tau}^{f1}, \quad \sigma_{\mathrm{nn}}^{f0} = \sigma_{\mathrm{nn}}^{f1}, \quad \sigma_{\mathrm{n}\tau}^{f0} = \sigma_{\mathrm{n}\tau}^{f1}. \tag{3.2}$$

3.3 Solid-fluid interface

This is the classical case of the noslip condition between a fluid and a solid. For this case, let $\varphi^{s0} = \varphi^{f1} = 1$ and $\varphi^{s1} = \varphi^{f0} = 0$, then the interface conditions become

$$v_{\mathrm{n}}^{s0} = v_{\mathrm{n}}^{f1}, \quad v_{\tau}^{s0} = v_{\tau}^{f1}, \quad \sigma_{\mathrm{nn}}^{s0} = \sigma_{\mathrm{nn}}^{f1}, \quad \sigma_{\mathrm{n}\tau}^{s0} = \sigma_{\mathrm{n}\tau}^{f1}. \tag{3.3}$$

Because of the absence of a solid phase on one side of the interface, the non-dissipative condition of Eq. 2.20 does not apply to this case.

3.4 Biphasic-fluid interface

This is the case previously presented by Hou et al. [15]. Here, let $\varphi^{s0} = 0$ and $\varphi^{f0} = 1$, then the interface conditions reduce to

$$\varphi^{s0} v_{\mathrm{n}}^{s0} + \varphi^{f0} v_{\mathrm{n}}^{f0} = v_{\mathrm{n}}^{f1}, \quad \varphi^{s0} v_{\tau}^{s0} + \varphi^{f0} v_{\tau}^{f0} = v_{\tau}^{f1}$$

$$\sigma_{\mathrm{nn}}^{s0} + \sigma_{\mathrm{nn}}^{f0} = \sigma_{\mathrm{nn}}^{f1}, \quad \sigma_{\mathrm{n}\tau}^{s0} + \sigma_{\mathrm{n}\tau}^{f0} = \sigma_{\mathrm{n}\tau}^{f1}, \quad \frac{\sigma_{\mathrm{nn}}^{f0}}{\varphi^{f0}} = \sigma_{\mathrm{nn}}^{f1}, \quad \frac{\sigma_{\mathrm{n}\tau}^{f0}}{\varphi^{f0}} = \sigma_{\mathrm{n}\tau}^{f1}. \tag{3.4}$$

3.5 Biphasic-solid interface

This is the case of an impermeable solid sliding against a biphasic layer, under non-dissipative conditions. Let $\varphi^{s0} = 0$ and $\varphi^{f0} = 1$, then the interface conditions become

$$\varphi^{s0} v_{\mathrm{n}}^{s0} + \varphi^{f0} v_{\mathrm{n}}^{f0} = v_{\mathrm{n}}^{s1}, \quad v_{\mathrm{n}}^{s0} = v_{\mathrm{n}}^{s1}, \quad v_{\tau}^{f0} = v_{\tau}^{s1}$$

$$\sigma_{\mathrm{nn}}^{s0} + \sigma_{\mathrm{nn}}^{f0} = \sigma_{\mathrm{nn}}^{s1}, \quad \sigma_{\mathrm{n}\tau}^{f0} = \sigma_{\mathrm{n}\tau}^{s1}, \quad \sigma_{\mathrm{n}\tau}^{s0} = 0. \tag{3.5}$$

It appears that this particular case has not been reported previously.

3.6 Inviscid fluid phase

The final limiting case to be presented is that of the interface conditions between two biphasic layers where the fluid phase is assumed to be "inviscid" ($\boldsymbol{\sigma}^{\mathrm{v}} = \mathbf{0}$). These boundary conditions have been reported by Biot [7] and Deresiewicz and Skalak [12], and presented in the context of mixture theory by Ateshian and Wang [1]. For this case, there is no pseudo-noslip boundary condition on the tangential components of velocity, and the general interface conditions reduce to

$$\begin{aligned} \varphi^{s0} v_{n}^{s0} + \varphi^{f0} v_{n}^{f0} = \varphi^{s1} v_{n}^{s1} + \varphi^{f1} v_{n}^{f1}, \quad v_{n}^{s0} = v_{n}^{s1} \\ \sigma_{nn}^{s0} + \sigma_{nn}^{f0} = \sigma_{nn}^{s1} + \sigma_{nn}^{f1}, \quad \frac{\sigma_{nn}^{f0}}{\varphi^{f0}} = \frac{\sigma_{nn}^{f1}}{\varphi^{f1}} \\ \sigma_{n\tau}^{s0} = \sigma_{n\tau}^{f0} = \sigma_{n\tau}^{s1} = \sigma_{n\tau}^{f1} = 0. \end{aligned} \tag{3.6}$$

4. Governing equations

The governing equations for biphasic solid-fluid mixtures under quasi-static conditions are

$$\nabla\cdot\left(\varphi^{s}\mathbf{v}^{s} + \varphi^{f}\mathbf{v}^{f}\right) = 0 \tag{4.1}$$
$$\nabla\cdot\boldsymbol{\sigma}^{s} + \boldsymbol{\pi} = \mathbf{0} \tag{4.2}$$
$$\nabla\cdot\boldsymbol{\sigma}^{f} - \boldsymbol{\pi} = \mathbf{0} \tag{4.3}$$

which represent the continuity equation and the momentum equations for the solid and fluid phases, respectively. In these expressions, $\boldsymbol{\pi}$ is the momentum exchange between the solid and fluid phases. Constitutive equations for the linear biphasic theory are

$$\boldsymbol{\sigma}^{s} = -\varphi^{s} p\mathbf{I} + \lambda_{s}(\mathbf{tr}\boldsymbol{\epsilon}) + 2\mu_{s}\boldsymbol{\epsilon} \tag{4.4}$$
$$\boldsymbol{\sigma}^{f} = -\varphi^{f} p\mathbf{I} + \lambda_{f}(\mathbf{tr}\mathbf{D}) + 2\mu_{f}\mathbf{D} \tag{4.5}$$
$$\boldsymbol{\pi} = \frac{\varphi^{f^{2}}}{k}(\mathbf{v}^{f} - \mathbf{v}^{s}). \tag{4.6}$$

In these expressions, λ_s, μ_s are the Lamé constants for the solid phase, λ_f, μ_f are the apparent viscosity coefficients of the fluid phase, k is the hydraulic permeability of the mixture, $\boldsymbol{\epsilon}$ is the strain tensor in the solid phase, $\boldsymbol{\epsilon} = [(\nabla\mathbf{u}) + (\nabla\mathbf{u})^{T}]/2$, and $\mathbf{D}$ is the rate of deformation tensor in the fluid phase, $\mathbf{D} = [(\nabla\mathbf{v}^{f}) + (\nabla\mathbf{v}^{f})^{T}]/2$. These equations, along with the boundary conditions presented above, can be solved for any given problem involving sliding contact at an interface.

5. Discussion

A set of boundary conditions have been derived for the interface of sliding biphasic media with viscous interstitial fluid, which further generalize the conditions proposed by Hou et al. [15] for the interface between a viscous fluid and a biphasic medium. In the present approach, a generalized pseudo-noslip kinematic condition is introduced for the contact interface, using a heuristic concept of statistical interactions [3]. Hou et al. [15] also derived a set of boundary conditions at the *adhesive* interface of two biphasic media, i.e, one where $v_\tau^{s0} = v_\tau^{s1}$, in contrast to the present model where, in general, $v_\tau^{s0} \neq v_\tau^{s1}$; however, the adhesive case is also subsumed by the present analysis.

Substituting the pseudo-noslip condition into the equations for momentum and energy jump across the interface yielded a set of jump conditions on the interface traction components for both solid and fluid phases; interestingly, this process also yielded a constraint given in Eq. 2.28, which enforces the non-dissipative nature of the interface. This term represents the total tangential traction, $\sigma_{n\tau}^{s} + \sigma_{n\tau}^{f}$, minus the viscous interaction term, $\sigma_{n\tau}^{f}/\varphi^{f}$, acting over the fluid-fluid, fluid-solid and solid-fluid interfaces ($\varphi^{f0}\varphi^{f1} + \varphi^{s0}\varphi^{f1} + \varphi^{f0}\varphi^{s1} = 1 - \varphi^{s0}\varphi^{s1}$). Therefore, this term can be interpreted as the solid-to-solid frictional traction [3]. In many problems, the condition that this traction be equal to zero can be relaxed, with the understanding that a non-zero friction would cause heat dissipation at the interface. For example, Coulomb's law can be used to make the solid-to-solid tangential traction proportional to the corresponding normal traction component [3].

With the availability of these interface boundary conditions, it is now possible to solve problems of sliding contact of biphasic mixtures which can be used to better understand the frictional response of articular cartilage in relation to the solid and fluid interactions occurring at the contact interface.

References

1. Ateshian, G.A., Wang, H. (1995): A theoretical solution for the frictionless rolling contact of cylindrical biphasic articular cartilage layers. J. Biomech. **28**, 1341–1355
2. Ateshian, G.A. (1997): A theoretical formulation for boundary friction in articular cartilage. J. Biomech. Eng. **119**, 81-86
3. Ateshian, G.A., Wang, H., Lai, W.M. (1998): The role of interstitial fluid pressurization and surface porosities on the boundary friction of articular cartilage. J. Tribology **120**, 241–251
4. Ateshian, G.A., Wang, X. (1998): Sliding tractions on a porous deformabel layer. J. Tribology **120**, 89–96
5. Barry, S.I., Parker, K.H., Aldis, G.K. (1991): Fluid-flow over a thin deformable porous layer. Zeitschrift fur angewandte Mathematik und Physik **42**, 633–648
6. Beavers, G.S., Joseph, D.D. (1967): Boundary conditions at a naturally permeable wall. J. Fluid Mech. **30**, 197–207

7. Biot, M.A. (1962): Mechanics of deformation and acoustic propagation in porous media. J. Appl. Phys. **33**, 1482–1498
8. de Boer, R. (1996): Highlights in the historical development of the porous media theory: Toward a consistent macroscopic theory. Appl. Mech. Rev. **49**(4), 201–262
9. Bowen R (1976) Theory of Mixtures. In: Continuum Physics, A.C. Eringen (ed.), Academic Press, New York, **3**, 1–127
10. Brinkman, H.C. (1947): A calculation of the viscous force exerted by a flowing fluid on a dense swarm of particles. Appl. Scien. Res **A1**, 27–34
11. Charnley, J. (1959): The lubrication of animal joints. In: Symposium on Biomechanics, Inst Mech. Eng., London, 12–22
12. Deresiewicz, H., Skalak, R. (1963): On uniqueness in dynamic poroelasticity. Bull. Seismol. Soc. Am. **53**, 783–788
13. Hlavacek, M. (1993): The role of synovial fluid filtration by cartilage in lubrication of synovial joints–I. Mixture model of synovial fluid. J. Biomech. **26**, 1145–1150
14. Hlavacek, M. (1993): The role of synovial fluid filtration by cartilage in lubrication of synovial joints– II. Squeeze-film lubrication: homogeneous filtration. J. Biomech. **26**, 1151–1160
15. Hou, J.S., Holmes, M.H., Lai, W.M., Mow, V.C. (1989): Boundary conditions at the cartilage-synovial fluid interface for joint lubrication and theoretical verification. J. Biomech. Eng. **111**, 78-87
16. Hou, J.S., Mow, V.C., Lai, W.M., Holmes, M.H. (1992): An analysis of the squeeze film lubrication mechanism for articular cartilage. J. Biomech. **25**, 247–259
17. Mow, V.C., Kuei, S., Lai, W.M., Armstrong, C.G. (1980): The biphasic theory for articular cartilage. J. Biomech. Eng. **28**, 123–134
18. Taylor, G.I. (1971): A model for the boundary condition of a porous material, Part I. J. Fluid Mech. **49**, 319–326
19. Wang, W., Parker, K.H. (1995): The effect of deformable porous surface layers on the motion of a sphere in a narrow cylindrical tube. J. Fluid Mech. **283**, 287–305

Computational Experience from the Solution of Coupled Problems in Ship Dynamics

Frank Ihlenburg

Germanischer Lloyd, Hamburg, Germany; e–mail: ihl@germanlloyd.org

Summary. Ideas of modern Finite-Element analysis are employed to increase the reliability in computational predictions of the vibrational response of ships. The fluid-structure problem is solved by coupled FE-BEM. Techniques of homogenization, model adaptation and mesh refinement are used to investigate the numerical stability. The computational results are compared to data from large-scale measurements.

1. Introduction

Coupled problems of fluid-structure interaction arise naturally in the computational simulation of ship dynamics. Obviously, when embarking on the solution of fluid-structure problems, the primal interest of the structural engineer lies not in the fluid per se, but in its influence on the structural behavior. What is needed is therefore an efficient but also sufficiently accurate way of including the fluid response into the structural computations. It is a common practice to use simplified methods towards this goal. Part of this investigation deals with the comparison of a simplified method to a more realistic (but also more expensive) numerical solution with coupled finite and boundary elements. Beyond that, different variants of structural modeling are also considered. The overall goal consists in a quantitative assessment of the effects of both model adaptation and numerical adaptation for large-scale Finite-Element ship vibration analysis in the frequency range 5 ... 25Hz. The model variants include the comparison of membrane to bending shell theory in modeling the ship structure (model adaptation) and a check of numerical convergence by mesh refinement. The corresponding computational results are compared to the data from measurements on a container vessel.

2. Physical assumptions and governing equations

We consider a ship partially embedded in water as sketched in Fig. 2.1. The solid-fluid interaction effects include spring-like effects, damping effects, and mass-like effects which are, respectively, proportional to displacement, velocity, or acceleration. In the present paper, only the inertia forces of the water are considered. The structural displacements are small, allowing for the application of a linear theory in the solid. The fluid is assumed ideal

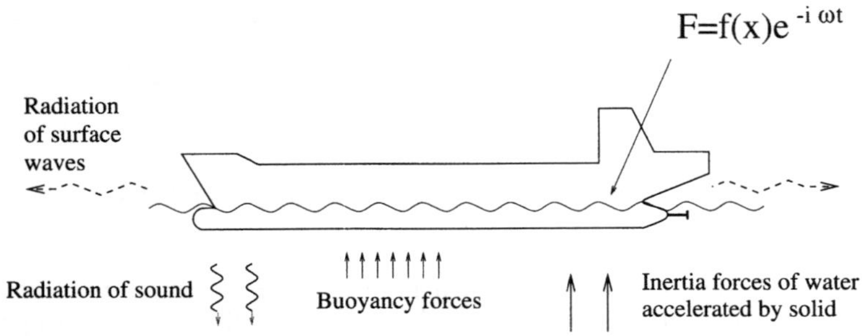

Fig. 2.1. Physical effects of fluid-structure interaction at ships

and incompressible, thus neglecting the acoustic radiation effects. Surface radiation is neglected as well, and hence the fluid equations are also linear.

With these assumptions, the governing equations of fluid-solid interaction can be written in the general form

$$\Delta^* \mathbf{u} - ic\,\Omega \mathbf{u} + \rho_s \Omega^2 \mathbf{u} = -\mathbf{f} \quad \text{in } D_s\,, \tag{2.1}$$

$$\mathbf{u} \cdot \mathbf{n} - \frac{\partial \phi}{\partial n} = 0 \quad \text{on } S_w\,, \tag{2.2}$$

$$\sigma_{ij}(\mathbf{u}) n_i n_j + \rho_f \Omega^2 \phi = 0 \quad \text{on } S_w\,, \tag{2.3}$$

$$\Delta \phi = 0 \quad \text{in } D_f\,, \tag{2.4}$$

where $\mathbf{u}$ is the solid displacement vector, ϕ the fluid displacement potential, σ_{ij} denotes the components of the solid stress tensor, and $\mathbf{n}$ is the normal vector on the wet surface S_w. In particular, (2.1) is the equation of elastodynamic equilibrium, (2.2, 2.3) are coupling conditions, and ϕ satisfies the potential equation in the semi-unbounded fluid domain D_f. The second term in (2.1) accounts for viscous damping of the structure, while the effect of material or structural damping is accounted for in the form of a complexified material law in the elasticity operator Δ^*. In addition to the conditions above, the potential function satisfies a Dirichlet condition on the free water surface S_f and decay conditions in the far field. A more detailed outline of these equations can be found in [1]. The weak form of the coupled system can be written as

$$\int_{D_s} \left[\sigma_i \cdot \nabla v_i - \rho_s \Omega^2 (1 + i\eta_E) \mathbf{u} \cdot \mathbf{v}\right] - \rho_f \Omega^2 \int_{S_w} \phi(\mathbf{v} \cdot \mathbf{n}) = \int_{D_s} \mathbf{f} \cdot \mathbf{v} \tag{2.5}$$

$$\rho_f \int_{D_f} \nabla \phi \cdot \nabla \psi - \rho_f \int_{S_w} (\mathbf{u} \cdot \mathbf{n}) \psi = 0\,, \tag{2.6}$$

with test functions $\mathbf{v}$, ψ. The first equation describes the equilibrium in the solid in terms of continuum mechanics. Different structural equations can be derived from here by inserting the appropriate relations of structural mechanics. Among these, membrane and bending shell models will be considered in this paper. The term $\rho_f \int_{S_w} \phi(\mathbf{v} \cdot \mathbf{n})\, dS$ in (2.5), resulting from the insertion of the coupling condition (2.3) as a traction boundary condition, is called the added hydrodynamic mass.

3. Numerical method and computational implementation

The coupled system (2.5, 2.6) shall be solved by coupling the Finite Element and Boundary Element Methods. Towards this goal, the exterior potential problem is represented by the well-known integral equation

$$\alpha(\mathbf{x})\phi(\mathbf{x}) - V^0(\mathbf{x}, \partial_n\phi, S_\mathrm{f}) + V^1(\mathbf{x}, \phi, S_\mathrm{w}) = V^0(\mathbf{x}, u_n, S_\mathrm{w}), \tag{3.1}$$

where V^0 or V^1 denote, respectively, the single or double layer potential, and $4\pi\alpha$ is the solid angle of the domain in point $\mathbf{x}$. A boundary element method for the solution of (3.1) has been implemented at Rostock University [3] based on the traditional collocation approach [4], but the order of polynomial approximation within the elements may vary from $0 \ldots p$. The finite element method is used for the discretization of the solid equation(2.5). The commercial package ANSYS [5] has been employed within the present project. After discretization, the coupled equations can be written as a system of algebraic equations

$$\begin{aligned} [K - i\Omega C - \Omega^2 M]\{u\} - \Omega^2 A^T\{\phi\} &= \{f\} \\ B\{\phi\} &= A\{u\}, \end{aligned}$$

in which K, C or M are, respectively, the solid stiffness, damping, and mass matrices, B is the fluid matrix, and A is a coupling matrix. The solution of the second system by the BEM formally amounts to the inversion of the matrix B, and hence we can rewrite the first equation as

$$[K - i\Omega C - \Omega^2(M + M_H)]\{u\} = \{f\},$$

where $M_H = A^T B^{-1} A$ is the hydrodynamic mass matrix. The computation of the matrix $M_H = A^T B^{-1} A$ can be interpreted as forming the Schur complement of the system matrix. This evident similarity to the static condensation of FE equations (substructuring) allows to include the inertia response of the fluid as a superelement into the solid finite element model. For technical details, we refer to [1, 2]

4. Computational results

The numerical experiments shall show the quantitative influence of various model adaptations that are aimed at a better mathematical or numerical representation of the physical process considered. Three container ships have been investigated. For each vessel, the base model consists in a relatively coarse finite-element discretization of a geometrical model created from the blueprints. Membrane shell elements are used in the structure, and the added hydrodynamic masses are computed by the traditional method of Lewis [6]. Various upgrades to the base model are considered in the following.

4.1 Fluid-structure interaction

Lewis proposed a simplified computation of the hydrodynamic masses from the analytical solution of 2-D potential problems defined in the exterior of the ship's cross-sections. The cross-sections are assumed rigid. Consequently, there is no coupling between the intertia response in different collocation points neither around nor across the ship hull, and hence the resulting mass matrix is diagonal. As opposed to this simple structure, all collocation points used in the BEM are coupled via the fundamental solution, leading to a fully populated added mass matrix.

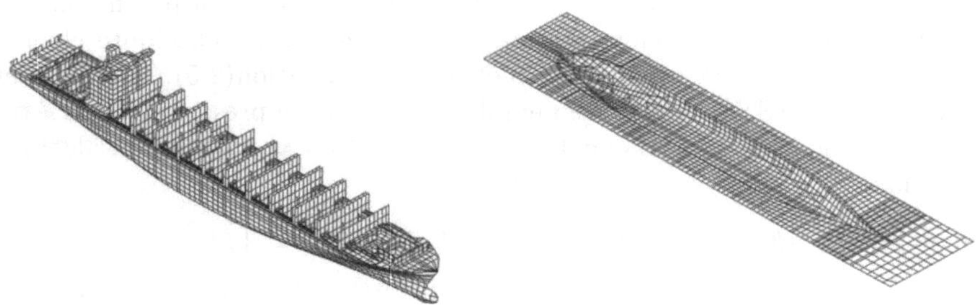

Fig. 4.1. FE model of ship and BE-panelling of water surface

In Fig. 4.1 we show the FE-model of the container vessel together with the BE model of the fluid. Infinite boundary elements (not shown) are attached to the finite boundary elements on the free surface to approximate the far-field behavior of the fluid potential.

Selected computational results together with measured data are shown in Figs. 4.2 and 4.3, where the forced vibrational response (magnitudes of point velocities over a frequency range f=5... 25Hz) is displayed. The measurements were carried out on a normal ocean voyage. The main engine was shut of at the time of the measurements, and the ship structure was excited by a controlled imbalance using a shaker. Two series of measurements were taken for imbalance forces acting, respectively, in vertical or horizontal (w.r. to the deck surfaces) direction.

The velocities were measured using transducers. The plots show the velocities at two measuring points on the top of the deckshouse. The transducers measure, respectively, the velocities in longitudinal (MP11) or horizontal (MP13) direction.

Consider first the results for vertical excitation. In MP11, the results computed with FE-BEM clearly reflect the measured response much better than those obtained with FE-Lewis. This concerns in particular the position of peaks and lows in the whole frequency band. Also the agreement of amplitudes is quite reasonable. The comparison of results for MP13 is less conclu-

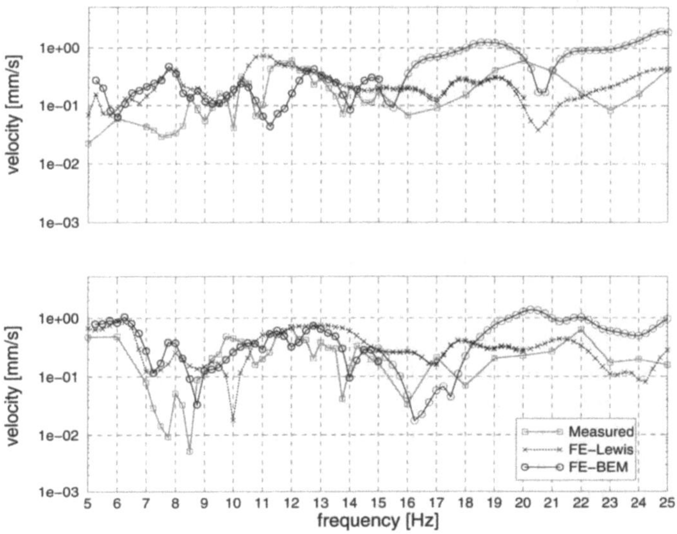

Fig. 4.2. Measured and computed velocities at top of deckshouse for vertical excitation. Comparison of methods for computation of added hydrodynamic mass in measuring points MP11 (below) and MP13 (above).

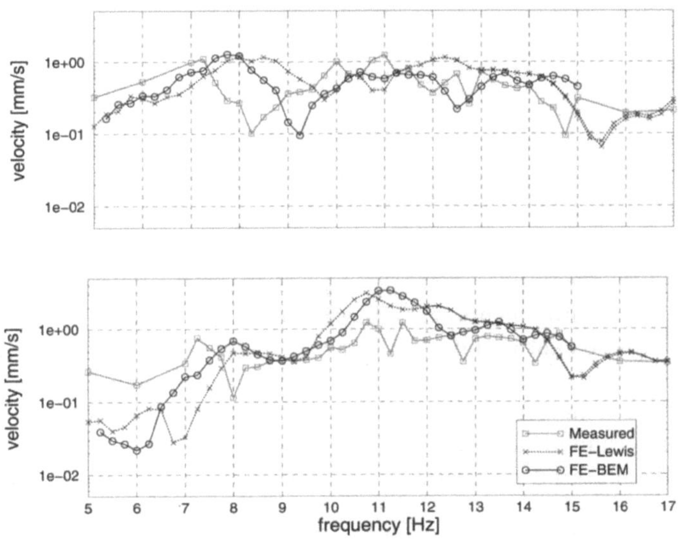

Fig. 4.3. Measured and computed velocities at top of deckshouse, horizontal excitation. Comparison of methods for computation of added hydrodynamic mass in measuring points MP11 (below) and MP13 (above).

sive. Especially in the lower frequency interval 7...9Hz there is a mismatch between the computed results and the measurements. However, in the mid-frequency range from 9...16Hz, the FE-BEM is again in better agreement with the measured data than FE-Lewis.

Regarding the results for horizontal excitation, the influence of the hydrodynamic masses is relatively less significant. Still the agreement with the measured data is improved using FE-BEM.

4.2 Structural model adaptation

In the lower frequency range, the structural response is dominated by the so-called global modes where the ship structure is deformed as a whole. It is safe to assume that the local deformations corresponding to these mode shapes can be well approximated using membrane shell elements. However, the number of eigenmodes per frequency interval rises quickly with the frequency. In Fig. 4.4, the density of the eigenfrequency spectrum for the present vessel is depicted.

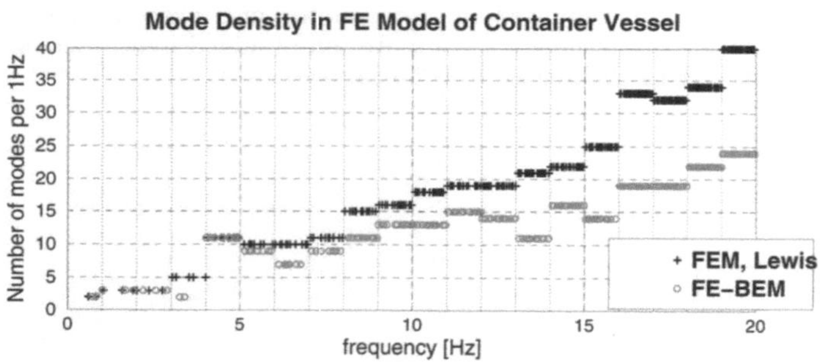

Fig. 4.4. Density of eigenspectrum for container vessel. The magnitudes on the y-axis indicate the number of eigenfrequency per interval of 1Hz.

The higher eigenmodes generally represent local deformations where the structure may exhibit bending stiffness within the scale of single elements. This effect can be captured if membrane shell elements are replaced by bending shell elements. As a further step towards a subsequent mesh refinement, the model was homogenized by "smearing" the stiffnesses and masses of supporting beams and girders onto the shell elements via adequate modifications of the shell's material properties. The cost comparison between the models is shown in Table 4.1. Interestingly, the number of eigenmodes up to 20Hz decreased despite the increase in the number of DOF.

A comparison of results is shown in Figs. 4.5, 4.6. As expected, the computed results from both shell models do not differ much in the lower frequency range. While the difference in the higher frequency range is larger, the bending shell model does not lead to a significantly better agreement between

Table 4.1. Comparison of shell models. Number of DOF and eigenmodes.

Theory	DOF	Modes	
		Lewis	BEM
Membrane	35000	355	243
Bending	66000	308	257

measured and computed data. Especially on the background of the cost comparison displayed in Table4.1, the membrane shell theory can be considered sufficient for the frequency range under investigation.

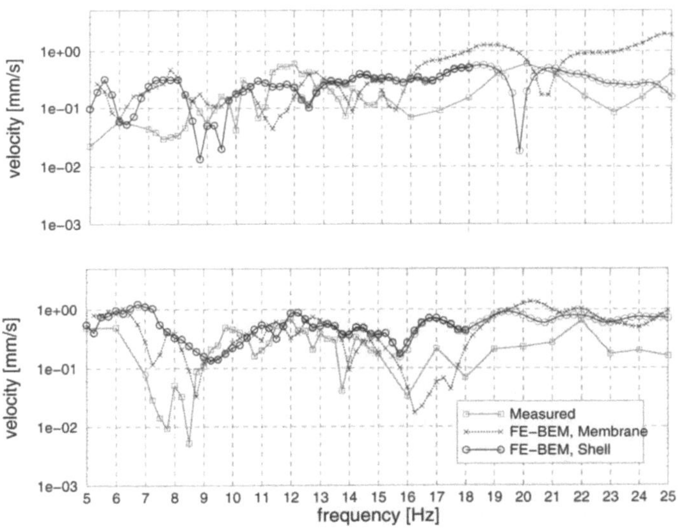

Fig. 4.5. Measured and computed velocities at top of deckshouse, vertical excitation. Comparison of membrane and bending shell models in measuring points MP11 (below) and MP13 (above).

Table 4.2. Mesh parameters of coarse model. L_x - overall length of ship, l_x - typical mesh size in longitudinal direction, $h_x = L_x/l_x$ - mesh size (nondimensional).

l_x	l_y	L_x	L_y	h_x	h_y	DOF
4m	3.8m	254m	53m	0.016	0.08	66000
2m	1.9m	·	·	0.08	0.04	321000

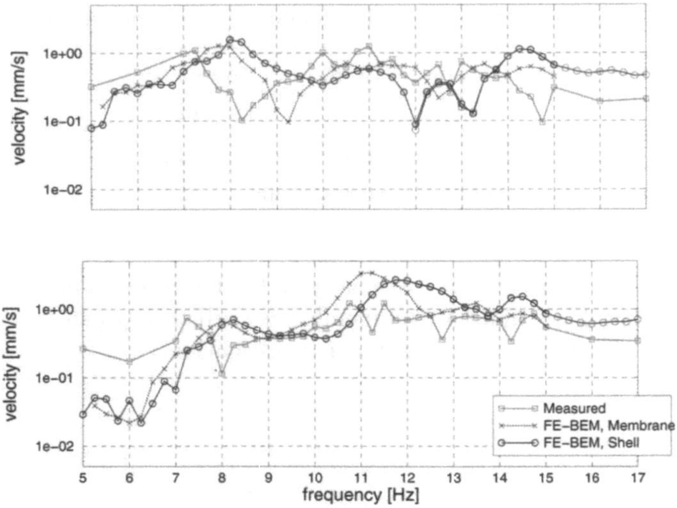

Fig. 4.6. Measured and computed velocities at top of deckshouse, horizontal excitation. Comparison of membrane and bending shell models in measuring points MP11 (below) and MP13 (above).

4.3 Structural mesh refinement

To check the influence of the discretization error, a global mesh refinement was carried out in the bending shell model. Table 4.2 shows typical physical and nondimensional mesh sizes and a cost comparison of the models. The method of Lewis was used for the hydrodynamic masses, since FE-BEM was prohibitively expensive on the fine mesh. Results are displayed in Figs. 4.7, 4.8.

Comparing the coarse and the fine shell models, we observe that the results are almost on top of each other up to a frequency of about 14Hz. This indicates the frequency range where the model is stable w.r. to the numerical parameters. In the higher frequency range, the results differ considerably for both kinds of excitation. A conclusive statement about which result is closer to the measured data is not possible. Thus the numerical data further confirm that for the present applicaton the bending shell model is not superior (regarding the agreement with the measured data) to the membrane shell model.

5. Conclusions

Several ideas from modern FE analysis were applied to the practice of dynamic ship computation in order to increase the reliability in the prediction of the vibrational response. Based on a comparison to measured data,

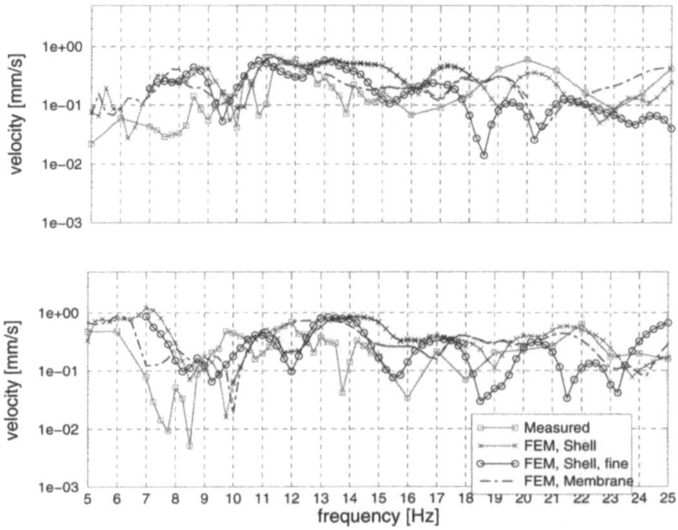

Fig. 4.7. Measured and computed velocities at top of deckshouse, vertical excitation. Comparison of coarse and fine model in measuring points MP11 (below) and MP13 (above).

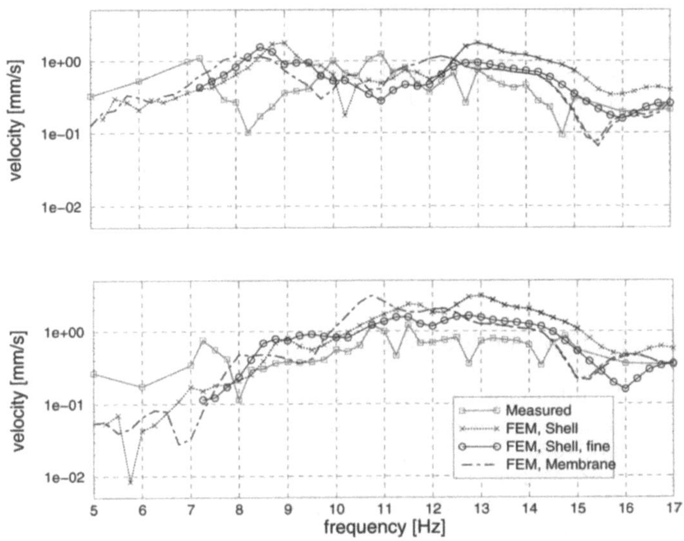

Fig. 4.8. Measured and computed velocities at top of deckshouse, horizontal excitation. Comparison of coarse and fine model in measuring points MP11 (below) and MP13 (above).

the major improvement has been achieved by solving the three-dimensional fluid-structure interaction problem with coupled FE-BEM. A check of the numerical error by means of uniform mesh refinement showed that the meshsize of the original structural model can be considered adequate for the frequency range up to 20Hz. Structural homogenization and subsequent model adaptation was used to facilitate the mesh refinement (which would not have been possible for structural details of the coarse model). These intermediate steps, while also proving the numerical stability of the model in the lower frequency range, did not lead to improved matching with the measurements for higher frequencies. This should not be surprising, since homogenization is in essence contraproductive to mesh refinement. In the future, more powerful computational facilities and improved tools for geometrical modeling will lead to more detailed structural models, the creation of which will be further supported by efficient tools of geometrical modeling. Several other topics of further research suggest themselves. At present, the dynamic forces acting on a vessel in operation are generally computed from empirical models. A more realistic modeling only of the propeller-induced forces will include the solution of challenging fluid-structure interaction problems. Furthermore, the ship is moving when in operation, so the influence of forward velocity must be taken into account in a more general formulation of the fluid-structure problem. Regarding the application of coupled FE-BEM, efficient solution methods for fully populated submatrices will be applied. Finally, the reliability of the measured data should be improved when repeated measurements under relatively constant environmental conditions will be possible.

Acknowledgements The measurements were carried out at Germanischer Lloyd by U. Behrens, H. Kreinath, and H. Mumm. The cooperation with Ch. Cabos, Ch. Beiersdorf, and H. Eisen, Hamburg, as well as I. Runkel, London, is gratefully acknowledged. The research was in part supported by the German Ministry of Research (BMBF) within the project MTK 586.

References

1. Ch. Cabos, F. Ihlenburg, Vibrational Analysis of Ships with Coupled Finite and Boundary Elements, J. Comp. Acoustics, submitted
2. Ch. Cabos, F. Ihlenburg, Verbesserung der Dämpfungsansätze für die Berechnung von Schiffsschwingungen, Technical Report FI 99.139 (178p.), Germanischer Lloyd, Hamburg 1999
3. U. Röhr and P. Möller, Elastische Schiffskörperschwingungen im begrenzten Fahrwasser, *Hauptversammlung der Schiffbautechnischen Gesellschaft*, Proceedings, Hamburg, 1997
4. W.L. Wendland: Die Behandlung von Randwertaufgaben im R^3 mit Hilfe von Einfach- und Doppelschichtpotentialen. Numer. Math. 11, 380-404, 1968
5. ANSYS Theory Manual, 8th ed., SAS IP Inc., Houston PA, 1998
6. F.M. Lewis, The inertia of the water surrounding a vibrating ship, Trans. SNAME, Vol. 37, 1929

On the Adaptive Computation of Shear Bands in Frictional Geomaterials

Wolfgang Ehlers and Peter Ellsiepen

Insitute of Applied Mechanics, University of Stuttgart, 70 550 Stuttgart, Germany

Summary. In frictional geomaterials, shear bands occur as a result of local concentrations of plastic strains in small bands of finite width. Since in practice as well as in numerical simulations both the location of the onset and the direction of shear bands are generally unknown, time- and space-adaptive methods are an excellent tool to detect and to solve shear band problems. In the present contribution, the ill-posedness of the numerical computation of shear band phenomena is overcome by extending the standard continuum mechanical approach by the inclusion of micropolar degrees of freedom in the sense of the *Cosserat* brothers.

1. Introduction

In the present contribution, granular elasto-plastic frictional geomaterials are considered as empty porous solid skeletons within the well-founded framework of the Theory of Porous Media (TPM). Concerning a broad review of the TPM approach and the application of saturated as well as non-saturated porous media to various situations, the reader is referred to the work by *Bowen* [1], *de Boer* and *Ehlers* [2] and *Ehlers* and coworkers [3, 4, 5, 6, 7]. Elasto-plastic frictional geomaterials are characterized by the fact that standard loading situations generally lead to plastic dilatation together with variations of the *Lode* angle. As a result, shear bands, i. e. local concentrations of plastic strains in small bands of finite width, occur, which have to be found and computed by numerical tools as, e. g., the finite element method (FEM). However, it is well known that the application of numerical methods to shear band phenomena reveals two severe problems, namely (1) the existence of a mathematically ill-posed problem and (2) the necessity to apply adaptive methods within the numerical tools. To overcome the ill-posedness of the problem, the standard continuum formulation is extended by including additional and independent micropolar degrees of freedom in the sense of the *Cosserat* brothers. Apart from the regularizing effect, micropolar continua represent the most natural approach to granular frictional materials, since the micropolar degrees of freedom can be interpreted as the local average grain rotations, which cannot be computed within the standard continuum mechanical formulation. Furthermore, since both the location of the onset and the development of shear bands cannot be foreseen in general, it is very convenient to include time- and space-adaptive methods into the numerical procedure. Using adaptive methods leads to a homogeneous distribution of the error both in the time and in the space domain. Following this, one succeeds in finding the correct onset of the shear band as well in refining the

FE mesh in the shear band domain [8], thus additionally proving the regularization of the micropolar formulation by the fact that the shear band width clearly exceeds the element size.

2. Governing equations

By use of an extended micropolar continuum mechanical formulation within the small strain approach, the governing basic kinematical variables of the granular material under study are the displacement vector $\mathbf{u} = \mathbf{x} - \mathbf{X}$, where $\mathbf{x}$ and $\mathbf{X}$ are the position vectors of the material particles at times t and t_0, and the total rotation vector $\bar{\boldsymbol{\varphi}} = \boldsymbol{\varphi} + \overset{*}{\boldsymbol{\varphi}}$, where $\boldsymbol{\varphi}$ and $\overset{*}{\boldsymbol{\varphi}}$ are the continuum rotation and the additional micropolar rotation. In particular,

$$\boldsymbol{\varphi} := -\tfrac{1}{2}\overset{3}{\mathbf{E}}\,(\mathbf{H}_{\text{skw}}) \tag{2.1}$$

is obtained with the aid of the *Ricci* permutation tensor $\overset{3}{\mathbf{E}}$, thus defining the continuum rotation as the axial vector of the skew-symmetric part of the displacement gradient $\mathbf{H} = \operatorname{Grad}\mathbf{u}$, whereas $\overset{*}{\boldsymbol{\varphi}}$ is independent from $\boldsymbol{\varphi}$. Corresponding to $\mathbf{u}$ and $\bar{\boldsymbol{\varphi}}$,

$$\bar{\boldsymbol{\varepsilon}} = \mathbf{H} + \overset{3}{\mathbf{E}}\,\bar{\boldsymbol{\varphi}}\,, \qquad \bar{\boldsymbol{\kappa}} = \operatorname{Grad}\bar{\boldsymbol{\varphi}} \tag{2.2}$$

are defined as the *Cosserat* strain and curvature tensors, which are connected by the micropolar compatibility condition [7, 11]

$$\bar{\boldsymbol{\kappa}} = \tfrac{1}{2}\overset{3}{\mathbf{E}}\left(\operatorname{Grad}\bar{\boldsymbol{\varepsilon}} + \operatorname{Grad}^{\overset{13}{T}}\bar{\boldsymbol{\varepsilon}} - \operatorname{Grad}^{\overset{23}{T}}\bar{\boldsymbol{\varepsilon}}\right)^{\underline{2}}. \tag{2.3}$$

Therein, the transpositions $(\cdot)^{\overset{ik}{T}}$ indicate an exchange of the i-th and k-th basis systems included into the tensor basis of higher order tensors. The additional superscript $(\cdot)^{\underline{n}}$ defines the included contraction $(\cdot)$ to yield a tensor of n-th order.

Apart from the kinematical relations, the material under study is described by the balance relations of mass, momentum and moment of momentum. Assuming materially incompressible solids, the effective (material) density ρ^R is constant during deformation, whereas the partial (bulk) density $\rho = n^S \rho^R$ can still vary through variations of the solid volume fraction n^S. Following this, n^S is governed by the solid deformation through

$$n^S = n_0^S\,(\det\mathbf{F})^{-1}\,, \tag{2.4}$$

where $\mathbf{F} = \operatorname{Grad}\mathbf{x}$ is the deformation gradient, and n_0^S is the volume fraction at time $t = t_0$. Excluding inertia terms, the momentum and moment of momentum balance equations yield the weak forms

$$\int_\Omega \mathbf{T} \cdot \operatorname{Grad} \delta\mathbf{u} \, dv - \int_\Omega (n^S \rho^R) \, \mathbf{b} \cdot \delta\mathbf{u} \, dv = \int_{\Gamma_t} \bar{\mathbf{t}} \cdot \delta\mathbf{u} \, da \,,$$
$$\int_\Omega (\overset{3}{\mathbf{E}} \mathbf{T}) \cdot \delta\bar{\boldsymbol{\varphi}} \, dv + \int_\Omega \mathbf{M} \cdot \operatorname{Grad} \delta\bar{\boldsymbol{\varphi}} \, dv = 0 \,. \tag{2.5}$$

Therein, $\mathbf{T}$ and $\mathbf{M}$ are the non-symmetric stress and couple stress tensors of micropolar materials, $\mathbf{b}$ is the external volume force per unit mass, and $\delta\mathbf{u}$ and $\delta\bar{\boldsymbol{\varphi}}$ are test functions corresponding to the respective field quantities. Furthermore, $\bar{\mathbf{t}}$ is the external load vector acting on the *Neumann* boundary Γ_t. To obtain $(2.5)_2$, it has been assumed that there is no external loading by surface and volume couples.

In the constitutive range of micropolar elasto-plastic materials, one proceeds from an additive decomposition of the geometrically linear *Cosserat* strain $\bar{\boldsymbol{\varepsilon}}$ into elastic and plastic parts. As has been pointed out by *Ehlers & Volk* [7], this assumption directly includes, by use of the micropolar compatibility condtion (2.3), an additive split of the curcvature tensor $\bar{\boldsymbol{\kappa}}$. Thus,

$$\bar{\boldsymbol{\varepsilon}} = \bar{\boldsymbol{\varepsilon}}_e + \bar{\boldsymbol{\varepsilon}}_p \,, \qquad \bar{\boldsymbol{\kappa}} = \bar{\boldsymbol{\kappa}}_e + \bar{\boldsymbol{\kappa}}_p \,. \tag{2.6}$$

As was discussed in [7], the constitutive equations for $\mathbf{T}$ and $\mathbf{M}$ read

$$\mathbf{T} = 2\,\mu\,\bar{\boldsymbol{\varepsilon}}_{e\,\mathrm{sym}} + 2\,\mu_c\,\bar{\boldsymbol{\varepsilon}}_{e\,\mathrm{skw}} + \lambda\,(\bar{\boldsymbol{\varepsilon}}_e \cdot \mathbf{I})\,\mathbf{I} \,,$$
$$\mathbf{M} = 2\,\mu_c\,(l_c)^2\,\bar{\boldsymbol{\kappa}}_e \,. \tag{2.7}$$

Therein, μ and λ are the *Lamé* constants of the porous skeleton material, whereas μ_c is an additional parameter governing the influence of the skew-symmetric part of the elastic *Cosserat* strain on the stress tensor. Finally, l_c represents the intrinsic length scale parameter relating the couple stress to the elastic curvature tensor. The plastic material behaviour of micropolar cohesive-frictional materials [7] is governed by the yield function

$$\bar{F} = \bar{\Phi}^{1/2} + \beta\,\mathrm{I} + \epsilon\,\mathrm{I}^2 + \tfrac{1}{2}\,k_M\,(\mathbf{M} \cdot \mathbf{M})^{1/2} - \kappa = 0 \,,$$
$$\bar{\Phi} = \mathrm{II}^D_{\mathrm{sym}}\,(1 + \gamma\,\vartheta)^m + k_T\,\mathrm{II}_{\mathrm{skw}} + \tfrac{1}{2}\alpha\,\mathrm{I}^2 + \delta^2\,\mathrm{I}^4 \,, \tag{2.8}$$
$$\vartheta = \mathrm{III}^D_{\mathrm{sym}} / (\mathrm{II}^D_{\mathrm{sym}})^{3/2}$$

together with the plastic potential

$$\bar{G} = \bar{\Gamma}^{1/2} + \beta\,\mathrm{I} + \epsilon\,\mathrm{I}^2 - g\,(\mathrm{I}) \,,$$
$$\bar{\Gamma} = \mathrm{II}^D_{\mathrm{sym}} + \tfrac{1}{2}\alpha\,\mathrm{I}^2 + \delta^2\,\mathrm{I}^4 + k_T\,\mathrm{II}_{\mathrm{skw}} \tag{2.9}$$

and the flow rule

$$(\bar{\boldsymbol{\varepsilon}}_p)'_S = \Lambda\,\frac{\partial \bar{G}}{\partial \mathbf{T}} \,. \tag{2.10}$$

In (2.8)-(2.10), I, $\mathrm{II}^D_{\mathrm{sym}}$ and $\mathrm{III}^D_{\mathrm{sym}}$ are the principal invariants of $\mathbf{T}$ and of the symmetric part of the stress deviator $\mathbf{T}^D$, whereas $\mathrm{II}_{\mathrm{skw}}$ defines the second principal invariant of the skew-symmetric part of $\mathbf{T}$. Furthermore, the parameters α, β, γ, δ, ϵ, m, κ and k_M, k_T govern the non-polar and the micropolar parts of the yield function, whereas $g(\mathrm{I})$ relates the dilatation angle to experimental data; Λ is the plastic multiplier. Note in passing that, once the micropolar compatibility condition (2.3) is given, it is straight forward to compute $(\bar{\boldsymbol{\kappa}}_p)'_S$ from (2.10). Thus, it is unnecessary that $\bar{G}$ depends on $\mathbf{M}$.

3. Spatial discretization and numerical tools

In the framework of the FEM, the spatial discretization of the field equations (2.5) is based on quadratic shape functions for the solid displacement $\mathbf{u}$ and linear shape functions for the total rotation $\bar{\boldsymbol{\varphi}}$ (primary variables $\boldsymbol{u}$). Furthermore, both the evolution equation for the plastic strain tensor and the plastic multiplier Λ from (2.10) (internal variables $\boldsymbol{q}$) are computed at the integration points of the numerical quadrature. Following this, one obtains a semi-discrete system of first order in time that can be solved by time- and space-adaptive methods [3, 8].

In the time domain, one-step methods with an embedded time step control are applied, where the solution at time t_{n+1} only depends on the solution at time t_n. This choice is of essential importance with respect to space-adaptive methods, since the transfer of the numerical solution thus only includes two meshes. In particular, 2-stage singly diagonally implicit *Runge-Kutta* methods (SDIRK) are used, thus allowing for an efficient embedded time-step control through the error estimation

$$ERR \approx \|\boldsymbol{y}_{n+1} - \hat{\boldsymbol{y}}_{n+1}\|, \tag{3.1}$$

where $\boldsymbol{y}_{n+1}$ and $\hat{\boldsymbol{y}}_{n+1}$ are two solutions of $\boldsymbol{y} = (\boldsymbol{u}, \boldsymbol{q})^T$ at time t_{n+1} with convergence orders $p = 2$ and $\hat{p} = 1$ [8]. Using the relative and absolute tolerances ϵ_r and ϵ_a together with the weighted error measures

$$e_u := \left(\frac{1}{N} \sum_{k=1}^{N} \left[\frac{u_{n+1}^k - \hat{u}_{n+1}^k}{\epsilon_r \cdot |u_n^k| + \epsilon_a} \right]^2 \right)^{1/2}, \qquad e_q := \max_k \left| \frac{q_{n+1}^k - \hat{q}_{n+1}^k}{\epsilon_r \cdot |q_n^k| + \epsilon_a} \right|, \tag{3.2}$$

where u^k and q^k are the components of $\boldsymbol{u}$ and $\boldsymbol{q}$ and $N = \dim \boldsymbol{u}$, the time-step is accepted if $e_y := \max\{e_u, e_q\} \leq 1$ and rejected otherwise. In both cases, a new step size is predicted by

$$\Delta t_{\text{new}} := \Delta t_{\text{old}} \cdot \min \left\{ f_{\max},\, \max \left\{ f_{\min},\, f_{\text{safety}} \cdot e_y^{-1/(\hat{p}+1)} \right\} \right\}. \tag{3.3}$$

Therein, $f_{\text{safety}} < 1$ is a safety factor, which prevents an oscillation of the time-step size, whereas $f_{\max} > 1$ and $f_{\min} < 1$ are used to limit the step size variation.

In the space domain, the adaptive procedure is based on an extension of the gradient-based error indicator by *Zienkiewicz* and *Zhu* [12] through the inclusion of the driving quantities of micropolar elasto-plastic materials [8]. Proceeding from the L_2-norm $\|\cdot\|_2$ and the corresponding element-wise norm $\|\cdot\|_{2,e}$, smoothened values $(\cdot)^*$ are computed on the basis of the FE quantities $(\cdot)^h$. Following this, the error indicators

$$\eta_1^{(e)} := \|\mathbf{T}^* - \mathbf{T}^h\|_{2,e}, \qquad \eta_2^{(e)} := \|\bar{\varepsilon}_p^* - \bar{\varepsilon}_p^h\|_{2,e} \tag{3.4}$$

are applied, where $\eta_1^{(e)}$ considers the solid elasticity through the non-symmetric stress tensor and (indirectly) through the couple stress. In case of inelastic deformations, $\eta_2^{(e)}$ considers the accumulated plasticity and (indirectly) the plastic curvature. The domain integrals

$$W_1 := \|\mathbf{T}^h\|_2, \qquad W_2 := \|\bar{\varepsilon}_p^h\|_2 \tag{3.5}$$

serve as reference quantities of the respective error indicators. Given the relative and absolute tolerances ϵ_r and ϵ_a together with the weighting factors $\alpha_i \geq 0$ to be included in the local error measures ξ_e, weighted global error measures ξ and weighted local error measures ξ_e per element can be computed on the actual mesh with N_e elements and $\sum_{i=1}^{2} \alpha_i = 1$:

$$\xi := \left(\sum_{e=1}^{N_e} \xi_e^2 \right)^{1/2}, \qquad \xi_e := \left(\sum_{i=1}^{2} \alpha_i \left[\frac{\eta_i^{(e)}}{\epsilon_r \cdot W_i + \epsilon_a} \right]^2 \right)^{1/2}. \tag{3.6}$$

The solution on the actual mesh is accepted if $\xi \leq 1$ and not accepted else. In order to refine ($\xi > 1$) or to coarsen ($\xi \leq 1$) the mesh, a new element radius h_{new} is be computed via

$$h_{\text{new}} := h_{\text{old}} \cdot \xi_e^{-1/(r+1)} \cdot \left[\sum_{e=1}^{N_e} \xi_e^{2/(r+1)} \right]^{-1/(2r)}. \tag{3.7}$$

The radii h_{new} of the elements in the new mesh define a piece-wise constant density function which can be used to generate a new mesh using a suitable mesh generator (remeshing technique). Alternatively, a hierarchical procedure can be used which refines and coarsens the elements in the current mesh using a tree data structure starting from an initial mesh. In the present contribution, a hierarchical approach is applied, thus leading to the hierarchically organized triangulation based on the algorithm by *Kossaczký* [9]. This algorithm implements the *Newest Vertex Bisection* technique proposed by *Mitchell* [10], where, starting from an initial mesh, mesh refinements as well as coarsenings are computed by the method of recursive bisection. This refinement algorithm is stable in the sense that the minimal angle in any triangle is bounded from below by half of the minimal angle of the initial mesh, independent of the number of refinement steps.

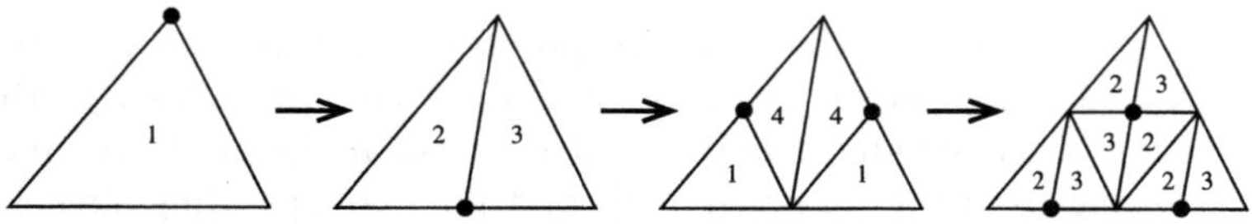

Figure 1. Similarity classes in the *Newest Vertex Bisection* strategy.

In the *Newest Vertex Bisection* strategy, only those edges are allowed for a refinement which are opposite of marked vertices. Furthermore, once a new edge is inserted, the mark is transferred to the newly created vertex. Figure 1 shows the four similarity classes emerging from the refinement process of a triangle.

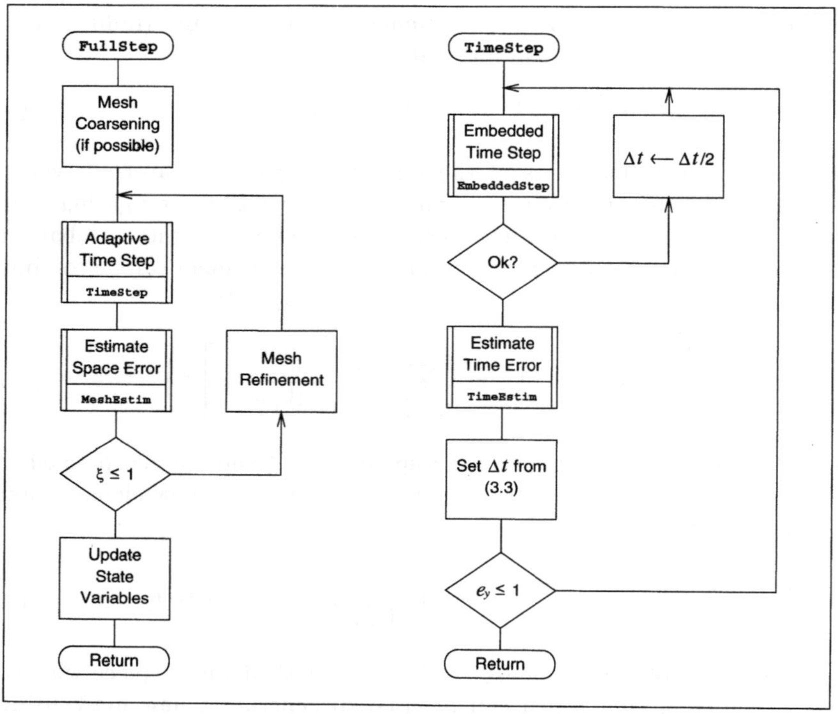

Table 1. Algorithm of a time- and space-adaptive procedure.

In Table 1, the algorithm of a time- and space-adaptive step is shown. The left procedure (**FullStep**) shows the computation of a coupled time- and space-adaptive step while the right procedure (**TimeStep**) collects the sub-steps of the computation of an adaptive time step. Firstly, in a time- and space-adaptive step (procedure **FullStep**), the mesh is coarsened if possible. Thereafter, the outer loop computes an acceptable space discretization, where, in the inner loop, for each new mesh an adaptive time step is computed (procedure **TimeStep**). Thus, after accepting the time step (successful return from **TimeStep**), the space error is computed based on the error measure ξ in (3.6). When the space error is acceptable ($\xi \leq 1$), the computed time- and space-adaptive step is accepted as a whole, and an update of the discrete state variables as well as an update of the current time ($n \leftarrow n + 1$) is carried out. Otherwise, the current mesh is refined, and the complete time step is repeated based on the data at time t_n.

4. Numerical example

The present example exhibits a simulation of a biaxial experiment, where plane strain conditions are prescribed. In particular, a granular elasto-plastic

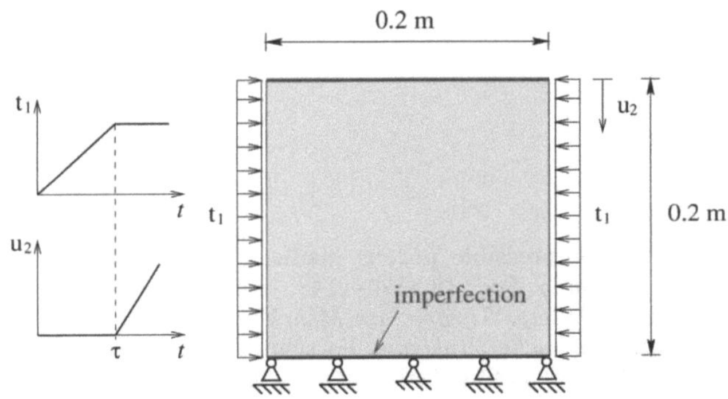

Figure 2. Boundary conditions and loading paths.

micropolar material is considered, where at time $0 \leq t \leq \tau$ a linearly increasing horizontal stress is applied, which is kept constant when $t > \tau$. Simultaneously, the specimen is loaded with a displacement-driven vertical stress. In order to obtain a shear band localization, an imperfection by a local weakening of the elastic constants at the bottom of the material is included, thus initiating the shear band development. It is seen from Figures 3 and 4 that (1) the time- and space-adaptive methods described above are appropriate to correctly locate the shear band and to refine the mesh in the localization zone and that (2) the inclusion of micropolarity leads to a convenient regularization of the shear band problem.

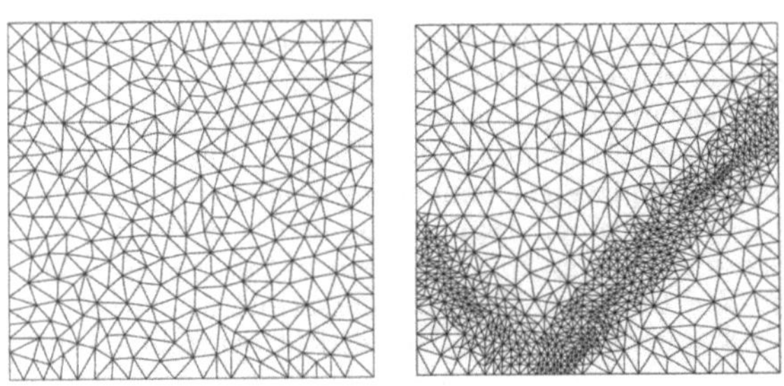

Figure 3. Biaxial experiment: Initial and final mesh.

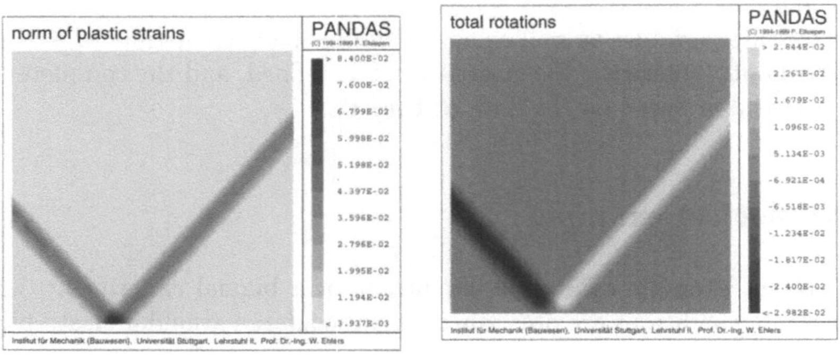

Figure 4. Biaxial experiment: Norm of plastic strains (left) and total rotations (right).

References

1. R. M. Bowen. Incompressible porous media models by use of the theory of mixtures. *Int. J. Engng. Sci.*, 18:1129–1148, 1980.
2. R. de Boer and W. Ehlers. *Theorie der Mehrkomponentenkontinua mit Anwendung auf bodenmechanische Probleme.* Forschungsberichte aus dem Fachbereich Bauwesen, Heft 40, Universität-GH-Essen, 1986.
3. S. Diebels, P. Ellsiepen, and W. Ehlers. Error-controlled *Runge-Kutta* time integration of a viscoplastic hybrid two-phase model. *Technische Mechanik*, 19:19–27, 1999.
4. W. Ehlers. *Poröse Medien – ein kontinuumsmechanisches Modell auf der Basis der Mischungstheorie.* Forschungsberichte aus dem Fachbereich Bauwesen, Heft 47, Universität-GH-Essen, 1989.
5. W. Ehlers. Constitutive equations for granular materials in geomechanical context. In K. Hutter, editor, *Continuum Mechanics in Environmental Sciences and Geophysics*, CISM Courses and Lectures No. 337, pages 313–402. Springer-Verlag, Wien, 1993.
6. W. Ehlers. Grundlegende Konzepte in der Theorie Poröser Medien. *Technische Mechanik*, 16:63–76, 1996.
7. W. Ehlers and W. Volk. On theoretical and numerical methods in the theory of porous media based on polar and non-polar elasto-plastic solid materials. *Int. J. Solids Structures*, 35:4597–4617, 1998.
8. P. Ellsiepen. *Zeit- und ortsadaptive Verfahren angewandt auf Mehrphasenprobleme poröser Medien.* Dissertation, Bericht Nr. II-3 aus dem Institut für Mechanik (Bauwesen), Universität Stuttgart, 1999.
9. I. Kossaczký. A recursive approach to local mesh refinement in two and three dimensions. *J. Comp. Appl. Math.*, 55:275–288, 1994.
10. W. F. Mitchell. Adaptive refinement for arbitrary finite-element spaces with hierarchical bases. *J. Comp. Appl. Math.*, 36:65–78, 1991.
11. W. Volk. *Untersuchung des Lokalisierungsverhaltens mikropolarer poröser Medien mit Hilfe der Cosserat-Theorie.* Dissertation, Bericht Nr. II-2 aus dem Institut für Mechanik (Bauwesen), Universität Stuttgart, 1999.
12. O. C. Zienkiewicz and J. Z. Zhu. A simple error estimator and adaptive procedure for practical engineering analysis. *Int. J. Numer. Methods Eng.*, 24:337–357, 1987.

Intrinsic Viscoelasticity of Porous Materials

Wolfgang Ehlers and Bernd Markert

Institute of Applied Mechanics (Civil Engineering), University of Stuttgart, Pfaffenwaldring 7, D-70569 Stuttgart, Germany

Summary. The classical consolidation problem of a fluid-saturated porous material is governed by the frictional drag force as a result of the pore-fluid flow through the porous solid matrix. This flow-dependent viscoelastic property is overlayed by flow-independent dissipative effects, if the porous solid matrix exhibits viscoelastic properties itself.

The goal of the present paper is the description of this complex material behaviour by a viscoelastic two-phase model accounting for both the flow-dependent fluid-solid interaction and the intrinsic skeleton viscoelasticity. Therefore, the macroscopic Theory of Porous Media (TPM) is applied, where the effective skeleton stress is determined by an appropriate viscoelasticity law. The fundamental approach of the viscoelasticity law is based on the thermodynamics with internal state variables which can be descriptively motivated in the small strain regime by rheological considerations. Thereby, the solid skeleton is modelled materially incompressible and the pore-fluid can either be materially compressible or incompressible.

The viscoelastic two-phase model is implemented into PANDAS[1], an adaptive finite element code for multi-phase problems. To show the capability of this approach, the model is applied to cartilage tissues, since they exhibit significant viscoelastic properties governed by both the flow-dependent and the flow-independent dissipative behaviour. Two representative initial boundary value problems have been computed, where the influence of the intrinsic viscoelasticity of the solid skeleton is studied. In addition, the problem of separating the flow-independent dissipative behaviour from the flow-dependent consolidation process is discussed.

1. Introduction

Fluid-saturated porous solids, e. g. polymer foams, synthetic sponges or biological tissues are subject to large viscoelastic deformations during their daily use. In the framework of macroscopic continuum theories, these types of materials can be described by biphasic viscoelastic models considering both the intrinsic viscoelasticity of the solid skeleton and the dissipative behaviour associated with the viscous pore-fluid flow. Particularly in the field of cartilage modelling, a lot of work has been done concerning the constitutive description of this complex viscoelastic behaviour [8, 12, 13].

In the present contribution, the Theory of Porous Media (TPM) is applied to describe the strongly coupled solid-fluid problem. The Theory of Porous Media is a macroscopic continuum theory based on the theory of mixtures extended by the concept of volume fractions. The reader who is interested in the fundamentals of the TPM approach is referred to Bowen [1, 2], de Boer

[1] **P**orous media **A**daptive **N**onlinear finite element solver based on **D**ifferential **A**lgebraic **S**ystems [11]

and Ehlers [3] and Ehlers [5, 6, 7]. The energy absorbing behaviour of the porous solid matrix is described by a new finite viscoelasticity law based on the thermodynamics with internal state variables [4], where a hyperelastic material formulation is used. In the small strain regime, the respective linear form of the viscoelasticity law can be descriptively motivated by rheological considerations [9]. Under moderate levels of pressure, the porous solid matrix can be assumed to be materially incompressible. Therefore, the so-called point of compaction exists, defined as that deformation state, where all pores are closed and no further volume deformation is possible due to the skeleton incompressibility. Furthermore, the permeability of the solid skeleton depends on the deformation state and is assumed to be a function of the actual porosity, since permeability changes can have an enormous influence on the pore-fluid flow within the large strain regime [10].

2. Governing equations

In the framework of the Theory of Porous Media, a fluid-saturated porous medium can be treated as an immiscible mixture of constituents φ^α ($\alpha = S$: solid skeleton; $\alpha = F$: pore-fluid). Based on the assumption of superimposed continua, the constituents are averaged over a representative elementary volume (REV) occupied by the whole mixture (homogenized model). Starting from different reference positions $\mathbf{X}_\alpha$ at time t_0, each constituent follows its individual motion function and has its own velocity field:

$$\mathbf{x} = \boldsymbol{\chi}_\alpha(\mathbf{X}_\alpha, t) \,, \qquad \overset{\prime}{\mathbf{x}}_\alpha = \partial \boldsymbol{\chi}_\alpha / \partial t \,. \tag{2.1}$$

In the biphasic macro model, the local structure of the mixture is represented by scalar variables, the volume fractions n^α. Thus, to avoid any vacant space, the saturation constraint $n^S + n^F = 1$ is assumed.

The solid motion is described by the solid displacement vector $\mathbf{u}_S$, while the pore-fluid is described relatively to the deforming skeleton by the so-called seepage velocity $\mathbf{w}_F$:

$$\mathbf{u}_S = \mathbf{x} - \mathbf{X}_S \,, \qquad \mathbf{w}_F = \overset{\prime}{\mathbf{x}}_F - \overset{\prime}{\mathbf{x}}_S \,. \tag{2.2}$$

Thus, the solid deformation gradient is defined by

$$\mathbf{F}_S = \frac{\partial \mathbf{x}}{\partial \mathbf{X}_S} =: \mathrm{Grad}_S \, \mathbf{x} = \mathbf{I} + \mathrm{Grad}_S \, \mathbf{u}_S \,. \tag{2.3}$$

For the numerical treatment of the problem within the finite element method (FEM), weak forms of the governing field equations, i. e. the mixture balance of momentum and the fluid balance of mass, are required. Therefore, after eliminating the seepage velocity by use of the Darcy filter law, the balance relations weighted by independent test functions and integrated over the spatial domain Ω with surface $\partial\Omega$ result in the respective weak formulations $\mathcal{G}_{\mathrm{MM}}$ and $\mathcal{G}_{\mathrm{MF}}$:

$$\begin{aligned} \mathcal{G}_{\mathrm{MM}} \equiv \int_{\Omega} \operatorname{grad} \delta\mathbf{u}_S \cdot (\mathbf{T}_E^S - p\,\mathbf{I})\, dv \; - \\ - \int_{\Omega} \delta\mathbf{u}_S \cdot (\rho^S + \rho^F)\, \mathbf{b}\, dv - \int_{\partial\Omega} \delta\mathbf{u}_S \cdot \bar{\mathbf{t}}\, da \;=\; 0\,, \end{aligned} \tag{2.4}$$

$$\begin{aligned} \mathcal{G}_{\mathrm{MF}} \equiv \int_{\Omega} \delta p \left[n^F\,(\rho^{FR})'_S + \rho^{FR} \operatorname{div}\,(\mathbf{u}_S)'_S \right] dv \; + \\ + \int_{\Omega} \operatorname{grad} \delta p \cdot \left[\frac{k^F}{g} \left(\operatorname{grad} p - \rho^{FR}\,\mathbf{b} \right) \right] dv + \int_{\partial\Omega} \delta p\, \bar{q}\, da \;=\; 0\,. \end{aligned} \tag{2.5}$$

Herein, $\delta\mathbf{u}_S$ and δp are the test functions corresponding to the solid displacement $\mathbf{u}_S$ and the pore pressure p, $\bar{\mathbf{t}} = (\mathbf{T}_E^S - p\,\mathbf{I})\,\mathbf{n}$ is the external load vector acting on both constituents, where $\mathbf{T}_E^S$ is the solid Cauchy extra (effective) stress, $\mathbf{I}$ is the identity tensor, and $\mathbf{n}$ is the outward oriented unit surface normal. Furthermore, $\bar{q} = n^F \rho^{FR}\, \mathbf{w}_F \cdot \mathbf{n}$ denotes the filter mass flow of the fluid draining through the surface $\partial\Omega$, $\mathbf{b}$ is the body force density, ρ^S and ρ^F are the partial densities, k^F is the macroscopic Darcy permeability coefficient, and g is the scalar value gravity. In this representation, the pore-fluid can either be compressible, i. e. the effective (material) fluid density is a function of the pore pressure, $\rho^{FR} = \rho^{FR}(p)$, or incompressible, i. e. $\rho^{FR} = \text{const.}$

In general, the permeability parameter k^F is deformation-dependent, i. e. the value of k^F decreases when the pore volume decreases. Thus, it is assumed that the permeability depends on the porosity n^F of the solid skeleton which can be described by a power function

$$k^F(n^F) = k_0^F \left(\frac{n^F}{n_{0S}^F} \right)^{\kappa}, \tag{2.6}$$

where k_0^F is the initial permeability and the exponent κ governs the deformation dependency [10].

3. Finite viscoelasticity law

To describe the intrinsic dissipative phenomenon of the porous solid skeleton structure, an adequate viscoelastic material formulation is required including the property of structural hardening towards the point of compaction. In the framework of a finite deformation theory of materials with elastic and inelastic (here: viscous) behaviour, we proceed from a multiplicative split of the solid deformation gradient (Sidoroff [17])

$$\mathbf{F}_S = (\mathbf{F}_{Se})_n \, (\mathbf{F}_{Si})_n \tag{3.1}$$

into elastic parts $(\mathbf{F}_{Se})_n$ and viscous (inelastic) parts $(\mathbf{F}_{Si})_n$. From elastoplasticity, it is commonly known that the concept of the multiplicative decomposition of deformation gradients is connected with the suggestion of a stress-free, geometrically incompatible intermediate configuration (Fig. 3.1), where the purely inelastic state of deformation is frozen into the memory of the material. Furthermore, from thermodynamics with internal state variables and rheological considerations [16], one obtains the ansatz of a decomposed solid Helmholtz free energy density and a decomposed solid extra stress,

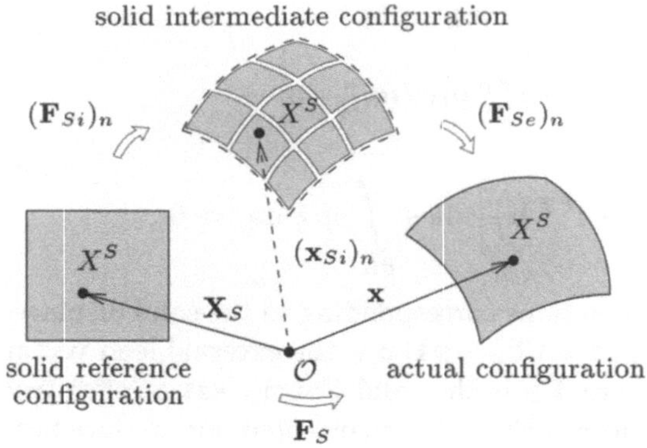

Fig. 3.1. Solid configurations.

$$\psi^S[\mathbf{F}_S, (\mathbf{F}_{Se})_1, ..., (\mathbf{F}_{Se})_n] = \psi^S_{\text{EQ}}[\mathbf{F}_S] + \psi^S_{\text{NEQ}}[(\mathbf{F}_{Se})_1, ..., (\mathbf{F}_{Se})_n] \,, \tag{3.2}$$

$$\boldsymbol{\tau}^S_E[\mathbf{F}_S, (\mathbf{F}_{Se})_1, ..., (\mathbf{F}_{Se})_n] = \boldsymbol{\tau}^S_{\text{EQ}}[\mathbf{F}_S] + \boldsymbol{\tau}^S_{\text{NEQ}}[(\mathbf{F}_{Se})_1, ..., (\mathbf{F}_{Se})_n] \,, \tag{3.3}$$

into equilibrium parts (Index *EQ*) and non-equilibrium parts (Index *NEQ*) which vanish in the thermodynamic equilibrium. Note that the Kirchhoff extra stress is a weighted Cauchy stress $\boldsymbol{\tau}^S_E = J_S \, \mathbf{T}^S_E$, where $J_S = \det \mathbf{F}_S$ is the Jacobian determinant.

Assuming isotropic material behaviour, we proceed from a Neo-Hookean type strain energy function which was formulated by Eipper [10] for purely elastic material behaviour,

$$\psi^S_{\text{EQ}} \;=\; \frac{1}{\rho^S_{0S}} \left[\tfrac{1}{2}\, \mu^S_0 \, (I_S - 3) - \mu^S_0 \, \ln J_S + U^S_0[J_S] \right] , \tag{3.4}$$

and modified Neo-Hookean type strain energy functions describing the viscous part of the skeleton deformation

$$\psi^S_{\text{NEQ}} = \frac{1}{\rho^S_{0S}} \sum_{n=1}^{N} \left[\tfrac{1}{2}\, \mu^S_n \left[(I_{Se})_n - 3 \right] - \mu^S_n \, \ln (J_{Se})_n + U^{Se}_n[(J_{Se})_n] \right] . \tag{3.5}$$

In Eqs.(3.4) and (3.5), I_S and $(I_{Se})_n$ are the first principle invariants of the Cauchy-Green deformation tensors $\mathbf{C}_S = \mathbf{F}^T_S \, \mathbf{F}_S$ or $\mathbf{B}_S = \mathbf{F}_S \, \mathbf{F}^T_S$, respectively, and $(\mathbf{C}_{Se})_n = (\mathbf{F}^T_{Se})_n \, (\mathbf{F}_{Se})_n$ or $(\mathbf{B}_{Se})_n = (\mathbf{F}_{Se})_n \, (\mathbf{F}^T_{Se})_n$, respectively, where μ_1^S and μ_n^S are the first macroscopic Lamé constants. The extension terms $U_0^S[J_S]$ and $U_n^{Se}[(J_{Se})_n]$ describe the finite volume change including the concept of the compaction point:

$$U_0^S[J_S] = \lambda_0^S \left[1 - n_{0S}^S\right]^2 \left(\frac{J_S - 1}{1 - n_{0S}^S} - \ln \frac{J_S - n_{0S}^S}{1 - n_{0S}^S}\right), \tag{3.6}$$

$$U_n^{Se}[(J_{Se})_n] = \lambda_n^S \left[1 - (n_i^S)_n\right]^2 \left(\frac{(J_{Se})_n - 1}{1 - (n_i^S)_n} - \ln \frac{(J_{Se})_n - (n_i^S)_n}{1 - (n_i^S)_n}\right). \tag{3.7}$$

Therein, λ_0^S and λ_n^S are the second Lamé constants, $(J_{Se})_n = \det(\mathbf{F}_{Se})_n$ are the determinants of the elastic parts of the deformation gradient, n_{0S}^S is the initial solidity, and $(n_i^S)_n = n_{0S}^S \det(\mathbf{F}_{Si})_n^{-1}$ are the inelastic solid volume fractions with respect to the intermediate configuration. Thus, for the Neo-Hookean model, the solid extra stresses can be obtained by

$$\boldsymbol{\tau}_{\text{EQ}}^S = \mu_0^S \left[\mathbf{B}_S - \mathbf{I}\right] + \lambda_0^S \left[1 - n_{0S}^S\right]^2 \left(\frac{J_S}{1 - n_{0S}^S} - \frac{J_S}{J_S - n_{0S}^S}\right) \mathbf{I}, \tag{3.8}$$

$$\begin{aligned} \boldsymbol{\tau}_{\text{NEQ}}^S = \sum_{n=1}^{N} \Big[\quad & \mu_n^S \left[(\mathbf{B}_{Se})_n - \mathbf{I}\right] + \\ & + \lambda_n^S \left[1 - (n_i^S)_n\right]^2 \left(\frac{(J_{Se})_n}{1 - (n_i^S)_n} - \frac{(J_{Se})_n}{(J_{Se})_n - (n_i^S)_n}\right) \mathbf{I} \Big]. \end{aligned} \tag{3.9}$$

The inelastic strains, as internal variables, are obtained from evolution equations which have to fulfill the residual dissipation inequality:

$$(\overset{\triangle}{\mathbf{A}}_{Si})_n = \overset{4}{\mathbf{D}}{}_n^{-1} (\boldsymbol{\tau}_{\text{NEQ}}^S)_n, \qquad \overset{4}{\mathbf{D}}{}_n^{-1} = \frac{1}{2\eta_n^S} [\overset{4}{\mathbf{I}} - \tfrac{1}{3} (\mathbf{I} \otimes \mathbf{I})] + \frac{1}{9\zeta_n^S} (\mathbf{I} \otimes \mathbf{I}). \tag{3.10}$$

Therein, $\overset{\triangle}{\mathbf{A}}_{Si}$ is the upper Oldroyd derivative of the Almansian strain [6] and $\overset{4}{\mathbf{D}}{}_n^{-1}$ are positive definite, isotropic fourth order viscosity tensors, where η_n^S and ζ_n^S are macroscopic viscosity parameters.

4. Examples

In order to validate the presented viscoelastic biphasic model, two canonical examples have been computed. On this occasion, the model is applied to articular cartilage as a representative viscoelastic fluid-saturated porous material consisting of two material incompressible phases (extracellular matrix and interstitial fluid) [15]. Note that the numerical treatment of the incompressible model represents the more difficult case compared to the hybrid model with a compressible fluid-phase.

The first example is a torsion test, where a cylindrical sample is twisted by a jump and the torsional moment is recorded in order to determine the intrinsic viscoelastic behaviour of the extracellular matrix under shear [18]. In this test, the volume is preserved. Therefore, no pore pressure is developed and no fluid-flow occurs. Thus, the flow-independent dissipative effect is decoupled. Figure 4.1 shows the tangential displacement after loading and the progression of the torsional moment indicating the characteristic fast stress relaxation of the organic solid matrix of cartilage tissues.

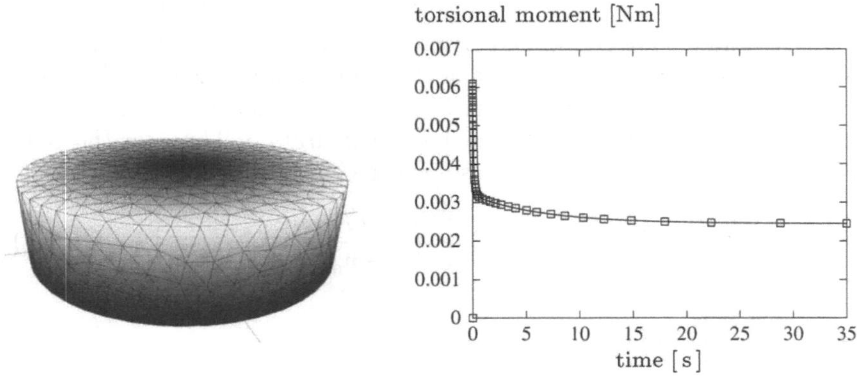

Fig. 4.1. Tangential displacement (left) (white = maximum) and torsional moment of the computed infinitesimal torsion test (right). Squares mark the time steps of the time-adaptive computation done with PANDAS [11].

The second example is a numerical simulation of a hydrostatic compression test, which reveals the problem of isolating the volumetric dissipative response of the solid skeleton. The idea is to consolidate a cubic cartilage specimen due to an external load which is displacement driven increased to its maximum value while recording the hydrostatic pressure. When the maximum load is applied, the test is finished, when, additionally, there is no further fluid exudation and the pore pressure decayed to its initial value. Reaching this equilibrium state, the occurring hydrostatic pressure is only governed by the elastic behaviour of the matrix material. Since no experimental data of such compression tests are available up to now, numerical simulations are necessary. Thus, two different computations are carried out, one with a viscoelastic solid skeleton as the substitute for the missing experimental data and one with an elastic solid skeleton using the elastic parameters governing the equilibrium state of the viscoelastic computation. The difference between both results (Fig. 4.2) is an indicator of the flow-independent volumetric viscoelastic material behaviour, since there are the same fluid properties included in both computations. The major problem is the large consolidation time of the specimen in order to determine the elastic parameters governing the final state of consolidation. Due to the very low permeability of cartilage (here: $k^F = 2.17 \cdot 10^{-11}\,\mathrm{m/s}$), the computed consolidation process has taken about one billion seconds of model time which is equal to 31.7 years. This shows that the experimental determination of the intrinsic volumetric viscoelastic properties by a hydrostatic test is still a problem, since it is not possible to test the empty (without pore-fluid) organic solid matrix alone.

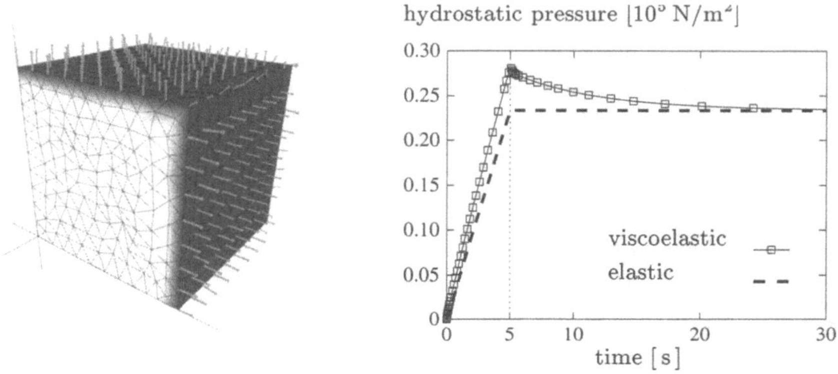

Fig. 4.2. Pore-fluid pressure (white = maximum) and seepage velocity (arrows) (left) and progression of the hydrostatic pressure (right). Squares mark the time steps of the time-adaptive computation done with PANDAS [11].

5. Conclusions

Proceeding from the Theory of Porous Media (TPM), a biphasic viscoelastic model accounting for the flow-dependent frictional drag and the flow-independent (intrinsic) skeleton viscoelasticity has been derived. Under the assumption of a materially incompressible solid skeleton, the structural point of compaction is reached, when all pores are closed and no further volume deformation is possible. To describe this effect, extensions of hyperelastic strain energy functions were introduced, which satisfy all physical requirements for possible finite deformations. Moreover, the pore-fluid can either be materially compressible or incompressible, even an empty viscoelastic solid skeleton can be described. Furthermore, deformation-dependent permeability effects have been considered, since they can have a significant influence on the solution of a boundary value problem within the large strain regime. Therefore, a power function, describing the dependency of the Darcy permeability parameter from the actual porosity of the solid skeleton was used.

The applicability and effectiveness of the presented model was shown in numerical examples, namely the torsion test and the hydrostatic compression test. Moreover, the compression test reveals that further investigations are necessary to measure the volumetric viscoelastic response of a fluid-saturated solid matrix, i. e. to separate the dissipative effects according to the fluid flow at low permeability and according to the properties of the skeleton, respectively.

References

1. Bowen, R. M. (1976): Theory of Mixtures. In: Continuum Physics, Vol. III, Eringen A. C., ed. Academic Press, New York, 1-127
2. Bowen, R. M. (1980): Incompressible porous media models by use of the theory of mixtures. Int. J. Engng. Sci. **18**, 1129-1148
3. de Boer, R., Ehlers, W. (1986): Theorie der Mehrkomponentenkontinua mit Anwendung auf bodenmechanische Probleme. Forschungsberichte aus dem Fachbereich Bauwesen **40**, Universität Essen, Essen
4. Coleman, B. D., Gurtin, M. E. (1967): Thermodynamics with internal state variables. J. Chem. Phys. **47**, 597-613
5. Ehlers, W. (1989): Poröse Medien – ein kontinuumsmechanisches Modell auf der Basis der Mischungstheorie. Forschungsberichte aus dem Fachbereich Bauwesen **47**, Universität Essen, Essen
6. Ehlers, W. (1993): Constitutive equations for granular materials in geomechanical context. In: Continuum Mechanics in Environmental Sciences, CISM Courses and Lectures **337**, Hutter, K., ed. Springer-Verlag, Wien, 313-402
7. Ehlers, W. (1996): Grundlegende Konzepte in der Theorie Poröser Medien. Technische Mechanik **16**, 63-76
8. Ehlers, W., Markert, B. (1999): A linear viscoelastic biphasic model for soft tissues based on the Theory of Porous Media. Berichte aus dem Institut für Mechanik (Bauwesen), Nr. 99-II-3, Universität Stuttgart, Stuttgart
9. Ehlers, W., Markert, B. (1999): On the viscoelastic behaviour of fluid-saturated porous materials. Granular Matter, in press
10. Eipper, G. (1998): Theorie und Numerik finiter elastischer Deformationen in fluidgesättigten porösen Festkörpern. Dissertation, Bericht Nr. II-1 des Instituts für Mechanik (Bauwesen), Universität Stuttgart, Stuttgart
11. Ellsiepen, P. *et al.* (1997): PANDAS - Benutzer- und Referenzhandbuch. Berichte aus dem Institut für Mechanik (Bauwesen), Nr. 97-II-9, Universität Stuttgart, Stuttgart
12. Lai, W. M., Mow, V. C., Zhu, W. (1993): Constitutive modeling of articular cartilage and biomacromolecular solutions. J. biomech. Engng **115**, 474-480
13. Mak, A. F. (1986): The apparent viscoelastic behaviour of articular cartilage - the contributions from the intrinsic matrix viscoelasticity and interstitial fluid flows. J. biomech. Engng **108**, 123-130
14. Mow, V. C., Kuei, S. C., Lai, W. M., Armstrong, C. G. (1980): Biphasic creep and stress relaxation of articular cartilage in compression: theory and experiments J. biomech. Engng **102**, 73-84
15. Mow, V. C., Ratcliff, A. (1997): Structure and Function of Articular Cartilage and Meniscus. In: Basic Orthopaedic Biomechanics, 2nd ed., Mow, V. C., Hayes, W. C., eds. Lipincott-Raven Publishers, Philadelphia, 113-177
16. Reese, S., Govindjee, S. (1998): A theory of finite viscoelasticity and numerical aspects. Int. J. Solids Structures **35**, 3455-3482
17. Sidoroff, F. (1974): Un modèle viscoélastique non linéaire avec configuration intermédiaire. Journal de Mécanique **13**, 679-713
18. Zhu, W. B., Lai, W. M., Mow, V. C. (1986): Intrinsic quasi-linear viscoelastic behavior of the extracellular matrix of cartilage. Trans. Orthop. Res. Soc. **11**, 407

Numerical Simulation of Fluids Interacting with Moving Rigid Bodies

S. Meynen and M. Schäfer

Department of Numerical Methods in Mechanical Engineering,
Darmstadt University of Technology, Germany

Summary. The numerical simulation of fluid-structure interaction has become an important field in engineering science. Since efficient programs exist for the solution of the individual subproblems, coupling schemes have to be developed which allow the use of existing and optimized solution strategies and methods. In this paper two different algorithms are used to couple the computational fluid dynamics (CFD) and the structural dynamics problem: an explicit and an implicit coupling of a predictor-corrector type. In the latter internal iteration are performed to ensure the energy conservation between the structure and the fluid. The grid velocity has to be taken into account within the Navier-Stokes equations applying a space conservation law.

As an example the oscillation of a torsion spring pendulum in a channel flow is presented. Due to the pressure and shear stresses of the fluid the obstacle boundaries are supplied with forces. A damping effect can be observed due to the fluid forces.

1. Introduction

The investigation and optimization of products or processes in engineering practice by numerical simulation is often related to the solution of problems from structural mechanics or fluid mechanics in the framework of continuum mechanics. In many applications one is faced with a coupling of phenomena from both fields. As the state of the art the computational fluid dynamics (CFD) and structural dynamics (CSD) problems can be solved individually with the help of fairly efficient programs, which are optimized for the governing differential equations.

In last years a lot of different approaches were presented for the simulation of fluid-structure interaction. Beside a monolithic treatment, where the solution of the coupled problem is computed in one step, see e. g. [1] or [18], more and more partitioned solution algorithms like staggered schemes are employed which allow the use of existing programs with minor changes inside the CFD part, see e. g. [7], [10], [16]. Here, conservation properties must be guaranteed [9], e.g. through the use of Lagrangian multipliers as proposed by [11] or load projection algorithms [2]. Applications for oscillating rigid bodies can be found in [5], [12] or [14].

In this paper two versions of a coupling algorithm are presented to realize the interaction between CFD and CSD parts: an explicit coupling and an implicit algorithm of predictor-corrector type for which internal iterations are performed. Their behaviour is investigated by means of an exemplary test case concerning an oscillating obstacle in a channel flow.

2. Governing equations and solution methods

First, the governing equations for the fluid and the structural parts as well as the applied solution methods are presented individually. The coupling between the structural part and the fluid part is then described in section 3.

2.1 Navier-Stokes equations for moving grids

The flow of an incompressible Newtonian fluid is described by the Navier-Stokes equations, e. g. [19]. Problems concerning fluid-structure interaction lead to a movement of the boundaries of the fluid domain, so the numerical grid will vary in time. Here, the general transport equation is extended for moving grids [20] and yields the balance of momentum

$$\frac{d}{dt} \int\limits_{V(t)} \rho u_i dV + \int\limits_{S(t)} \left[\rho u_i (u_j - u_j^g) - t_{ij}\right] n_j dS = \int\limits_{V(t)} \rho f_i dV \tag{2.1}$$

and the mass conservation

$$\frac{d}{dt} \int\limits_{V(t)} dV + \int\limits_{S(t)} (u_j - u_j^g) n_j dS = 0 \tag{2.2}$$

in standard notation, see e. g. [17], and the grid velocity u_j^g as an additional variable. To ensure the conservation principle for the movement of the control volumes (CV) the *space conservation law* (SCL) can be derived through the assumption $u_j = 0$:

$$\frac{d}{dt} \int\limits_{V(t)} dV - \int\limits_{S(t)} u_j^g n_j dS = 0 \ . \tag{2.3}$$

The change of the volume of the CV can be computed in different ways as proposed by [3]. The influence of the grid velocity of a CV as described in the above equation is considered by

$$\int\limits_{S(t)} u_j^g n_j dS = \sum_i (u_j^g S_j)_i = \sum_i \frac{\delta V_i}{\Delta t} \ , \qquad i = s, n, e, w \ . \tag{2.4}$$

The grey area in Figure 2.1, for instance, is denoted as swept volume δV_e within the time step Δt for the east side of an arbitrary CV, which can be computed with a small numerical effort.

For the solution of the fluid dynamics problem the finite volume program FASTEST [8] is used. To discretize the Navier-Stokes equations a second-order accurate approximation for the integrals by the midpoint rule is employed, see e. g. [6]. A Crank-Nicolson scheme with a 2nd order of accuracy is used for the time integration of the CFD part [4].

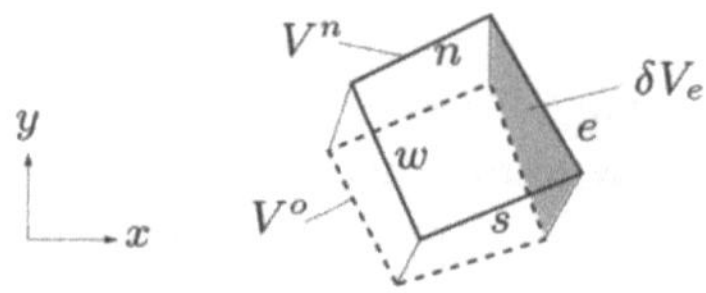

Fig. 2.1. Moving 2D control volume with swept volume δV_e

In the discretized form of the momentum equation (2.1) the correction of the velocity can be considered as a correction of the fluxes through the cell faces of a CV due to the movement of the grid. Using the midpoint rule together with the discrete SCL in equation (2.4) the correction term becomes

$$\int_S \rho u_i u_j^g n_j dS \approx \sum_c \left(\rho u_j^g S_j\right)_c u_i\Big|_c = \sum_c \rho \frac{\delta V_c}{\Delta t} u_i\Big|_c = \sum_c F_c^{\text{corr}} u_i\Big|_c$$

with the fluid velocity $u_i|_c$ in the center of each cell face $c = s, n, e, w$. Considering the SCL within the continuity equation (2.2) yields a divergence free velocity field $u_{i,i} = 0$ since the unsteady term and the integral of u_i^g cancel out each other, so that the mass conservation is ensured.

A SIMPLE-algorithm (Semi-Implicit Method for Pressure Linked Equations) [15] is applied to solve the system of equations arising from the discretization for each time step. Further on, a tremendous reduction of the computation time is achieved by the application of a multigrid technique for the solution of the CFD problem, see e. g. [4].

2.2 Linear rigid body dynamics

For some applications the motion of a rigid body can be sufficiently described with only one degree of freedom, e.g. for the example of a moving obstacle in a channel flow. A sketch of the geometry is shown in Figure 2.2. The movement of the obstacle is given by the solution of the scalar vibration equation

$$\Theta\ddot{\varphi} + \mu\dot{\varphi} + c_T\varphi = M_F(t) \ . \tag{2.5}$$

Here, φ is the angular excursion, Θ the moment of inertia, μ the damping coefficient and c_T the stiffness of the torsion spring. $M_F(t)$ denotes the momentum of the fluid forces due to the pressure and shear stresses.

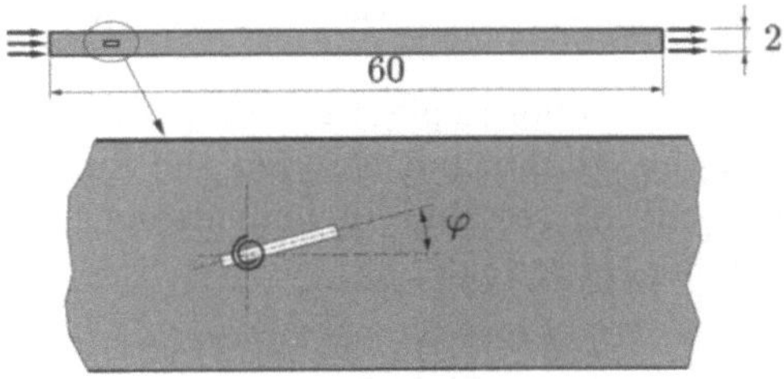

Fig. 2.2. Moving obstacle in a channel, sketch of geometry

In opposite to the fluid domain no spatial discretization is needed for the obstacle since it is considered as a rigid body. The structure dynamics equation is solved by Newmarks method [13] which can be implemented in different forms. In this paper a two-step algorithm [21] for the excursion φ only is used:

$$\begin{aligned}
\left[\Theta + \gamma\Delta t\mu + \beta\Delta t^2 c_T\right]\varphi_{i+1} + \\
\left[-2\Theta + (1-2\gamma)\Delta t\mu + (0.5+\gamma-2\beta)\Delta t^2 c_T\right]\ \varphi_i\ + \\
\left[\Theta + (\gamma-1)\Delta t\mu + (0.5-\gamma+\beta)\Delta t^2 c_T\right]\varphi_{i-1} + \\
= \Delta t^2\left[\beta M_{F,i+1} + (0.5+\gamma-2\beta)M_{F,i} + (0.5-\gamma+\beta)M_{F,i-1}\right]
\end{aligned} \tag{2.6}$$

with the Newmark parameters β and γ. The index i denoting the value at the time level t_i. The Newmark method is unconditionally stable if the Newmark parameters β and γ satisfy $4\beta = (\gamma+0.5)^2 \geq 0.5\gamma$ with $\gamma \geq 0.5$. Here, the parameters are chosen as $\beta = 0.25$ and $\gamma = 0.5$ yielding an asymptotic accuracy of the order $\mathcal{O}(\Delta t^2)$, see [21].

3. Coupling algorithm

The algorithmic approach used in this paper is schematically shown in Figure 3.1. First the CFD part is solved by the finite-volume program FASTEST taking the SCL into account. The shear and pressure forces on the boundary of the obstacle lead to the momentum M_F as the right hand side of equation (2.5). The solution φ of the structural dynamics part is achieved on the basis of the Newmark method as described in (2.6). If the obstacle moves, the fluid mesh is modified by keeping it attached to material points on the wall. Hence, the movement of the obstacle directly determines the position of the fluid nodes on the wall, which is linearly distributed to the internal grid points of the fluid domain to regenerate the fluid mesh.

Here, different implementations of the coupling algorithm are possible: A weak coupling of the structure and the fluid can be easily realized if no internal iterations are performed and the movement of the grid for the fluid domain due to the deflection of the boundaries is considered in the next time step. This explicit treatment of the coupling leads to a limitation of the feasible time step.

In order to achieve a more implicit coupling an iterative process can be carried out to find the equilibrium between the fluid momentum and the structural dynamic effects. No time incrementation is performed until the convergence criteria for a residual

$$r = \frac{\varphi_k(M_F(t_i)) - \varphi_{k-1}(M_F(t_i))}{\varphi_{k-1}(M_F(t_i))} \leq \varepsilon \tag{3.1}$$

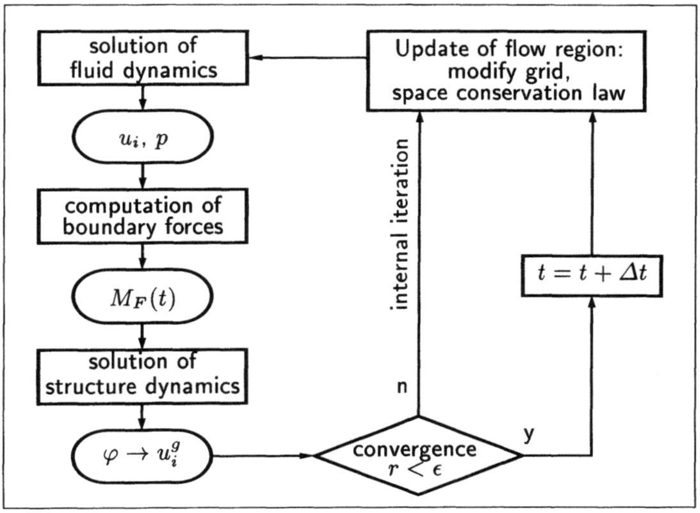

Fig. 3.1. Coupling algorithm for fluid-structure interaction

is reached. k is an internal iteration counter. The iteration stops if the residual is reduced by the factor ε which is typically of the size of 10^{-4}. As usually necessary within CFD simulations an under-relaxation of the excursion $\varphi_k^* = \alpha\ \varphi_k + (1-\alpha)\ \varphi_{k-1}$ with the under-relaxation parameter α can be applied.

4. Numerical example

A laminar channel flow around a horizontal plate is used as a basis for the numerical example. The basic geometry is described in Figure 2.2 (top). After some prescribed sinusoidal oscillations the plate is released at its maximum excursion and oscillates free but supplied with fluid forces. Due to the periodically moved plate an instationary fluid behaviour behind the obstacle is observed.

The channel has a length of 60 m and a width of 2 m, the length l of the obstacle is 1 m and its width d is 0.08 m. The bearing of the plate is located at 1/10 of its length. The coarsest mesh in the initial and deformed configuration at the release point are shown in Figure 4.1. Two refinements of the block-structured grids are used for the multigrid algorithm, which yields approximately 674 400 unknowns on the finest grid.

As fluid parameters the density $\rho_{\text{fluid}} = 1\ \text{kg/m}^3$ and the dynamic viscosity $\eta = 10^{-5}$ Pa s are chosen. At the inflow a Poiseuille flow profile [19] resulting from an undisturbed channel flow with a mean velocity $\bar{u} = 5.0 \cdot 10^{-3}$ m/s is prescribed. A long region behind the plate is used to avoid any feedback from the outflow condition of zero velocity gradients normal to the outlet.

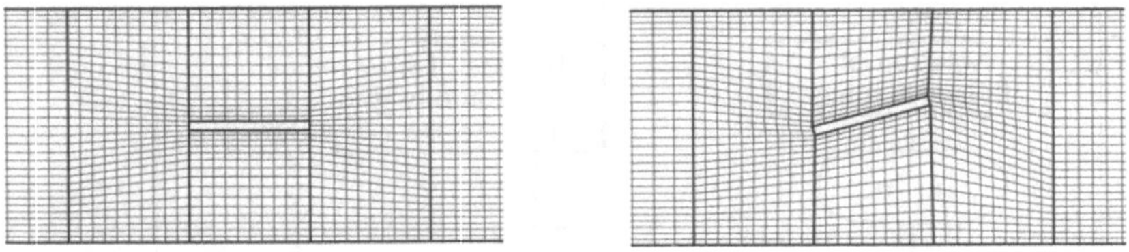

a – initial mesh b – deformed mesh

Fig. 4.1. Meshes for CFD computation

In Table 4.1 four different parameter sets for the structural density, the momentum of inertia and the stiffness of the torsion spring are shown. The Eigen frequency $f = \sqrt{c_T/\Theta}$ of the oscillating obstacle-spring system is the same for all parameter sets. The frequency of the prescribed oscillation of the plate is $f = 1/480\ \mathrm{s}^{-1}$. The influence of the different parameter sets on the oscillation of the obstacle is investigated with the predictor-corrector algorithm using a time step of $\Delta t = 10$s. For set 1 and 2 a nearly undisturbed harmonic oscillation is observed in Figure 4.2 without any noticeable change of amplitude or other values. Due to the high torsion stiffness combined with a high structural density the effect of the fluid momentum is negligible compared to a free vibration of the plate without fluid-structure interaction. A slight reduction of the amplitude and the period of the oscillation is visible for parameter set 3, while a strong damping can be observed when set 4 is applied. Due to the very weak torsion spring and a light material of the plate a reduction of about 80% of the amplitude already occurs within 4 oscillations. Furthermore, a decrease of the period can be recognized. The oscillation of

Table 4.1. Parameter combinations for FSI investigation

		set 1	set 2	set 3	set 4
torsion stiffness c_T	[Nm]	$2.53{\cdot}10^{-2}$	$2.53{\cdot}10^{-3}$	$2.53{\cdot}10^{-4}$	$2.53{\cdot}10^{-5}$
structural density ρ	[kg/m^3]	7800	780	78	7.8
mom. of inertia Θ	[kg m^2]	147.6	14.76	1.476	0.1476

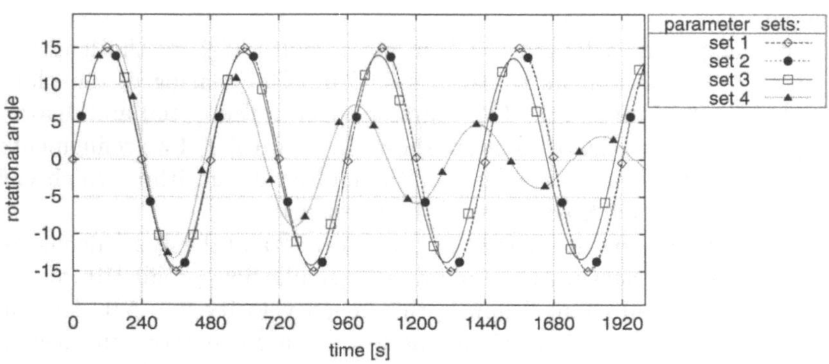

Fig. 4.2. Fluid damping for different parameter sets

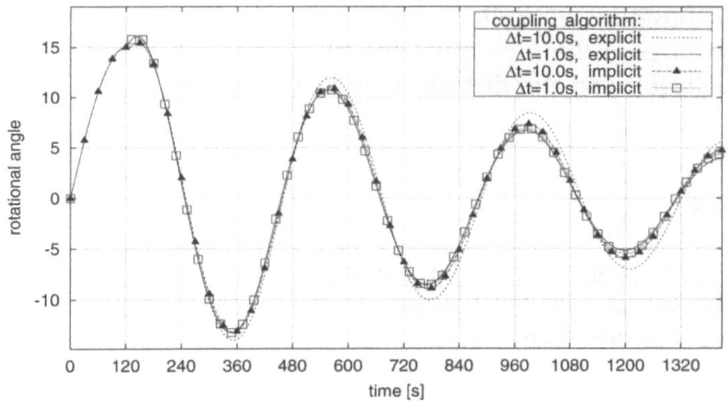

Fig. 4.3. Comparison of the coupling algorithms

the plate is damped out and it gets into a horizontal rest position, while the flow becomes stationary. Due to the low Reynolds number $Re = 40$ no flow induced restart of the oscillation appears.

In Figure 4.3 the influence of the time step is shown for both coupling strategies. For this investigation parameter set 4 and two different time steps $\Delta t = 1$s and 10s are used. The explicit coupling algorithm shows a dependence on Δt, see the lines without markers in Figure 4.3. The angular excursion is too large for time step $\Delta t = 10$ s. This effect becomes even larger if a bigger time step is chosen. Better results are obtained if an implicit coupling is used. Due to the internal iterations the dependence on the used time step is much weaker and the curves for both time steps nearly coincide. Only 2 internal iterations are needed in each time step for the parameter set 1, 2 and 3, for set 4 a maximum of 4 internal iterations is used. A convergence criterion of $\varepsilon = 10^{-4}$ turned out to be sufficient. A relaxation is not necessary, since no convergence problems of the internal iterations have been observed.

5. Conclusions

As an example for fluid-structure interaction the oscillation of a torsion spring pendulum in a channel flow is considered. The dynamics model of a rigid body yields a scalar vibration equation, which is solved by Newmarks method. A finite-volume program is applied for the CFD part. The movement of the obstacle causes a change of the fluid grid, which is taken into account within the Navier-Stokes equations using a *space conservation law*. Additional fluxes are considered in the discrete equations. Although no damping is applied inside the vibration equation the oscillation is damped out by the fluid forces.

The solution of the coupled problem is achieved by an iterative coupling scheme of a predictor-corrector type, which is compared with an explicit coupling approach. Even though second-order accurate time integration schemes are used for the individual subproblems the order of accuracy for the coupled problem is reduced by an explicit coupling. By the use of internal iterations in the predictor-corrector algorithm this effect vanishes. Hence, bigger time steps are applicable and a noticeable reduction of the overall computing time is achieved.

References

1. M. Amabili, F. Pellicano and M.P. Païdoussis. Nonlinear vibrations of simply supported, circular cylindrical shells, coupled to quiescent fluid. *Journal of Fluids and Structures* 12:883–918, 1998.
2. J.R. Cebral and R. Löhner. Conservative load projection and tracking for fluid–structure problems. *AIAA* 35:4, 687–692, 1997.
3. I. Demirdžić and M. Perić. Finite volume method for prediction of fluid flow in arbitrarily shaped domains with moving boundaries. *Int. J. Num. Meth. Eng.*, 10:771–790, 1990.
4. F. Durst and M. Schäfer. A parallel block-structured multigrid method for the prediction of incompressible flows. *Int. J. Num. Meth. Fluids*, 22:549–565, 1996.
5. H. Dütsch, A. Melling, and F. Durst. Coupled numerical solution of the fluid damped oscillations of a lamina. In *FORTWIHR Conference*, Munich, 1998.
6. H. Ferziger and M. Perić. *Comp. Meth. Fluid Dynamics.* Springer, Berlin, 1997.
7. M. Heil. Stokes flow in an elastic tube — a large-displacement fluid-structure interaction problem. *Int. J. Num. Meth. Fluids* 28:243–265, 1998.
8. INVENT Computing GmbH. FASTEST - parallel multigrid solver for flows in complex geometries. 1994.
9. M. Lesoinne and C. Farhat. Geometric conservation laws for flow problems with moving boundaries and deformable meshes, and their impact on aeroelastic computations. *Comp. Meth. Appl. Mech. Eng.* 134:71–90, 1996.
10. M. Lesoinne and C. Farhat. Improved Staggered Algorithms for the Serial and Parallel Solution of Three-Dimensional Nonlinear Transient Aeroelastic Problems. In *Proceedings of WCCM '98*, Buenos Aires, Argentina, 1998.
11. P. Le Tallec. Fluid structure interaction with large structural displacements. *submitted to Elsevier*, 1998.
12. S. Meynen and M. Schäfer. Numerical simulation of fluid-structure interaction for fluid damped oscillations. In *Proceedings of the ECCM '99*, München, 1999.
13. N.M. Newmark. A method of computation for structural dynamics. *Proceedings of ASCE, Journal of Engineering Mechanics (EM3)*, 85:67–94, 1959.
14. T. Nomura and T.J.R. Hughes. An arbitrary lagrangian-eulerian finite element method for interaction of fluid and a rigid body. *Int. J. Num. Meth. Eng.* 95:115–138, 1992.
15. S.V. Patankar and D.B. Spalding. A calculation procedure for heat, mass and momentum transfer in three dimensional parabolic flows. *Int. J. Heat Mass Transfer*, 15:1787–1806, 1972.
16. S. Piperno and C. Farhat. Design and evaluation of staggered partitioned procedures for fluid-structure interaction simulations. INRIA Rep. 3241, 1997.
17. M. Schäfer. *Numerik im Maschinenbau.* Springer-Verlag, 1999.
18. M. Schäfer and I. Teschauer. Numerical simulation of coupled fluid-solid problems. to appear in *Comp. Meth. Appl. Mech. Eng.* 2000.
19. J.H. Spurk. *Fluid mechanics.* Springer-Verlag, 1993.
20. P.D. Thomas and C.K. Lombard. Geometric conservation law and its application to flow computations on moving grids. *AIAA Jour.* 17:1030–1037, 1979.
21. W.L. Wood. *Practical Time-stepping Schemes.* Oxford University Press, 1990.

Partitioned Analysis of Transient Nonlinear Fluid Structure Interaction Problems Including Free Surface Effects

W.A. Wall, D.P. Mok, J. Schmidt and E. Ramm

Institute of Structural Mechanics, University of Stuttgart, Germany

Summary. This paper sketches some of our current efforts towards the development of a general and robust computational procedure for solving time-dependent nonlinear fluid structure interaction problems. In our approach, discretization in space is done solely via the finite element method, and direct integration schemes are employed in time. A stabilized finite element method based upon an arbitrary Lagrangean-Eulerian (ALE) approach has been developed to model the incompressible viscous fluid. Free fluid surfaces are described by a local Lagrange approach. To capture large structural displacements, geometrically nonlinear effects are taken into account. The final three-field coupled problem, with the moving mesh considered as the third field, is solved by partitioned analysis procedures.

1. Introduction

Fluid-structure interaction problems are found in many engineering areas. Some examples are fluid sloshing in a tank, wind-induced vibration of a bridge, vibrating off-shore structures, as well as applications in bioengineering, e.g. cardiovascular blood flow. In such situations, simplified models are often not able to give satisfactory answers. Therefore, the complexity of these coupled multi-physics problems has to be taken into account when modelling the different fields as well as their interaction. In turn, this sophistication of the separate models asks for a computational approach based on partitioned analysis. This allows the independent use of well-established discretization and solution methods within each field, including the application of different time integration schemes and time step sizes. At the same time, software development is simplified and software modularity is preserved.
The next three sections of this paper are devoted to the numerical treatment of the three computational fields, namely fluid, structure, and moving mesh. Our loose and strong coupling algorithms are discussed in Section 5. Numerical examples demonstrate the applicability of our approach.

2. CFD - Computational Fluid Dynamics

The fluid is modelled as an incompressible viscous Newtonian fluid. Its behavior is fully described by the incompressible Navier-Stokes equations, given in terms of velocity $\mathbf{u}$ and kinematic pressure p as

$$\frac{\partial \mathbf{u}}{\partial t} + \mathbf{u} \cdot \nabla \mathbf{u} - 2\nu \nabla \cdot \epsilon(\mathbf{u}) + \nabla p = \mathbf{b} \tag{2.1}$$

$$\nabla \cdot \mathbf{u} = 0 \tag{2.2}$$

with appropriate standard initial and boundary conditions on complementary boundary subsets, namely the Dirichlet boundary Γ_g and the Neumann boundary Γ_h.

2.1 Arbitrary Lagrangean Eulerian (ALE) formulation

Fluid equations are generally given in an Eulerian formulation, as in (2.1), which allows for simple treatment of large deformations. For the structure, however, the proper definition of moving boundaries is essential. Hence, a Lagrangean formulation is used to describe the structural equations, see Section 3. To close the gap between the fluid and structural domains, an Arbitrary Lagrangean Eulerian (ALE) formulation is employed for parts of the fluid domain (Hughes et al. [4]). In this formulation, an additional reference domain Ω_x is introduced with reference coordinates $\mathbf{x}$. Material and spatial domain and coordinates are denoted by (Ω_z, $\mathbf{z}$) and (Ω_y, $\mathbf{y}$), respectively. The reference domain is allowed to move arbitrarily, independent of spatial or material points. The material time derivative of a quantity f, which is needed for the formulation of the conservation laws, also referred to as the 'ALE fundamental equation', is defined as

$$\frac{Df}{Dt} = \frac{\partial f}{\partial t}(\mathbf{z},t)\Big|_z = \frac{\partial f}{\partial t}(\mathbf{x},t)\Big|_x + c_i \frac{\partial f}{\partial y_i}(\mathbf{y},t) \tag{2.3}$$

$\mathbf{c}$ denotes the ALE convective velocity, given as the difference between the particle and the mesh velocities. For the ease of implementation, a formulation written in spatial coordinates $\mathbf{y}$ is used, while keeping the time derivative in reference coordinates $\mathbf{x}$. The conservation of momentum, given in (2.1) for the Eulerian description, can now be written in components as

$$\frac{\partial u_i}{\partial t}\Big|_x + c_j \frac{\partial u_i}{\partial y_j} = \frac{\partial \sigma_{ij}}{\partial y_j} + b_i \quad in\ \Omega_y \quad with\ \ \sigma_{ij} = 2\nu\epsilon_{ij} - p\delta_{ij} \tag{2.4}$$

In the finite element context, the 'arbitrarily' moving finite element mesh represents the reference domain (see Section 4).

2.2 Stabilized finite element method

Solving equations (2.4) and (2.2) by the classical finite element method leads to a number of numerical difficulties. The nonlinearity and presence of the convective term can lead to convergence problems. The hyperbolic character of these equations and the incompressibility constraint can result in spurious oscillations in the finite element solution. Hence, stabilized finite elements are employed in our approach. The variational formulation for our stabilized ALE finite element method can be given as:
Find $\mathbf{u}^h \in \mathcal{V}^h$ and $p^h \in \mathcal{P}^h$ such that $\forall(\mathbf{v}^h, q^h) \in \mathcal{V}^h \times \mathcal{P}^h$

$$\Big(\frac{\partial \mathbf{u}^h}{\partial t}\Big|_x + \mathbf{c}^h \cdot \nabla\mathbf{u}^h, \mathbf{v}^h\Big) + (2\nu\epsilon(\mathbf{u}^h), \epsilon(\mathbf{v}^h)) - (\nabla\cdot\mathbf{v}^h, p^h) - \tag{2.5}$$

$$(\nabla\cdot\mathbf{u}^h, q^h) + ST = (\mathbf{b}, \mathbf{v}^h) + (\mathbf{h}, \mathbf{v}^h)_{\Gamma_h} \tag{2.6}$$

$\mathcal{V}^h$ and $\mathcal{P}^h$ denote appropriate finite-dimensional function spaces, and $\mathbf{v}^h$ and q^h are the test functions for the velocities and the pressure, respectively. The stabilization terms ST are defined as, see Wall [8],

$$ST = \sum_e ((\frac{\partial \mathbf{u}^h}{\partial t}\bigg|_x + \mathbf{c}^h \cdot \nabla \mathbf{u}^h + \nabla p^h - 2\nu\nabla \cdot \epsilon(\mathbf{u}^h) - \mathbf{b}),$$
$$\tau_m(\mathbf{c}^h \cdot \nabla \mathbf{v}^h - \nabla q^h - 2\nu\nabla \cdot \epsilon(\mathbf{v}^h)))_e + (\nabla \cdot \mathbf{u}^h, \tau_c \mathbf{v}^h) \tag{2.7}$$

with appropriately chosen stabilization parameters τ_m for the momentum and τ_c for the continuity equation. The trial and test functions are now expanded in terms of finite element basis functions, yielding the semi-discrete matrix equation

$$\mathbf{M}_f \dot{\mathbf{u}} + \mathbf{N}_f(\mathbf{u})\mathbf{u} + \mathbf{G}_f \mathbf{p} = \mathbf{F}_f \tag{2.8}$$

Different approaches are adopted for time discretization and for treatment of the involved nonlinearities (Wall and Ramm [10]).

2.3 Description of the free surface

Extending our previous research efforts, free surface effects are incorporated into the present formulation. A free surface is generally an interface between a liquid and a gas, e.g. between water and the surrounding air. At this interface, both a kinematic and a dynamic boundary condition have to be satisfied. The kinematic condition enforces mass conservation, i.e. there can be no fluid flow across the free surface. In this paper, a local Lagrange formulation is used for the description of the free surface, for which the kinematic condition simply states that the convective velocity at the free surface Γ_S be equal to zero. Equivalently, the grid velocity $\mathbf{u}^G$ has to be equal to the fluid velocity $\mathbf{u}$ at the free surface:

$$\mathbf{c} = \mathbf{u} - \mathbf{u}^{\mathbf{G}} = \mathbf{0} \quad on \quad \Gamma_S \tag{2.9}$$

The dynamic condition describes conservation of linear momentum, i.e. force equilibrium, at the free surface. For applications considered here, forces at the free surface are due to surface tension only. The dynamic condition then reduces to

$$\sigma \cdot \mathbf{n} = \gamma\kappa\mathbf{n} \quad on \quad \Gamma_S \tag{2.10}$$

Assuming isothermal flow, the surface tension is a function of the constant surface tension coefficient γ and the local radius of curvature κ.

3. CSD - Computational Structural Dynamics

The displacements $\mathbf{d}$ of the geometrically nonlinear structure are governed by the Cauchy equation of motion,

$$\rho\ddot{\mathbf{d}} = \nabla \cdot \mathbf{S} + \rho\mathbf{f} \quad in \ \Omega_s \times (0, T) \tag{3.1}$$

with appropriate boundary and initial conditions. Here, ρ is the structural density and $\mathbf{f}$ is the body force. The second Piola-Kirchhoff stress tensor $\mathbf{S}$

and the Green-Lagrange strain $\mathbf{E}$ are defined as $\mathbf{S} = \mathbf{C} : \mathbf{E}$, $\mathbf{E} = \frac{1}{2}(\mathbf{F}^T \cdot \mathbf{F} - \mathbf{I})$, with $\mathbf{F} = \nabla \mathbf{d}$ and $\mathbf{C}$ denoting the constitutive tensor.
These equations are discretized in space using different types of finite elements. The semidiscrete equation of motion, with an added damping term, can then be written as

$$\mathbf{M}_s \ddot{\mathbf{d}} + \mathbf{D}_s \dot{\mathbf{d}} + \mathbf{N}_s(\mathbf{d}) = \mathbf{F}_s \tag{3.2}$$

where $\mathbf{M}_s$, $\mathbf{D}_s$, $\mathbf{N}_s$ and $\mathbf{F}_s$ are the mass and damping matrices and the internal and external force vectors, respectively. $\mathbf{d}$, $\dot{\mathbf{d}}$ and $\ddot{\mathbf{d}}$ denote the vectors of nodal displacements, velocities and accelerations, respectively. In the present approach, this system ist solved using the 'Generalized-α Method' of Chung and Hulbert [2] along with consistent linearization and a Newton-Raphson iterative scheme. The 'Generalized-α Method' is an implicit, one-step second order accurate time integration scheme based on Newmark approximations in the time domain. An appropriate selection of the involved time integration and Newmark parameters governs its numerical characteristics, i.e. its numerical dissipation of low and high frequency modes and its stability. Thus it is possible to obtain unconditional stability even in the nonlinear regime.

4. CMD - Computational Mesh Dynamics

When solving the three-field problem, a mesh-moving scheme has to be chosen that defines the moving reference domain Ω_x in terms of the grid displacements $\mathbf{r}$. Modelling the moving finite element mesh as a pseudo-structural system with pseudo-stiffness matrix $\mathbf{K}^m$ has proved to be a very powerful approach. For simplicity, nonlinear effects are neglected in this context. To avoid numerical difficulties due to 'artificially oscillating meshes', only static equilibrium is considered. On the fluid-structure interface Γ_I, the mesh is required to follow the structure, i.e. the grid displacements $\mathbf{r}$ are equal to the structural displacements $\mathbf{d}$. On the free surface Γ_S, the grid velocity $\dot{\mathbf{r}} \equiv \mathbf{u}^G$ has to be equal to the fluid velocity $\mathbf{u}$. Hence, in each time step, the grid motion is governed by the elastostatic system of equations

$$\mathbf{K}^m \mathbf{r} = \mathbf{F} \; \textit{with} \; \mathbf{r} = \mathbf{d} \; \textit{on} \; \Gamma_I, \; \dot{\mathbf{r}} = \mathbf{u} \; \textit{on} \; \Gamma_S \; \textit{and} \; \mathbf{r} = \mathbf{0} \; \textit{on} \; \Gamma \setminus (\Gamma_I \cup \Gamma_S) \tag{4.1}$$

where the right hand side vector $\mathbf{F}$ results only from the Dirichlet boundary conditions on $\mathbf{r}$ and $\dot{\mathbf{r}}$.

5. Partitioned staggered analysis approach

Our partitioned analysis approach is based upon a non-overlapping partitioning between the two physical partitions fluid and structure. The three computational fields CFD, CSD and CMD are solved separately within each time step in a sequential manner, and always with the newest available coupling information. The coupling conditions – kinematic and dynamic continuity across the fluid-structure interface – are taken into account by applying the fluid traction as a Neumann boundary condition to the structure, and the movement of the wetted structural surface, as well as the free surface condition described above, as Dirichlet boundary conditions for the deforming fluid mesh.

5.1 Algorithmic setup

The staggered time integration of the coupled system over one time step Δt from time $t^n = n\Delta t$ to t^{n+1} is depicted in Fig. 5.1, with U and D denoting the respective sets of variables. The algorithm follows the following steps,

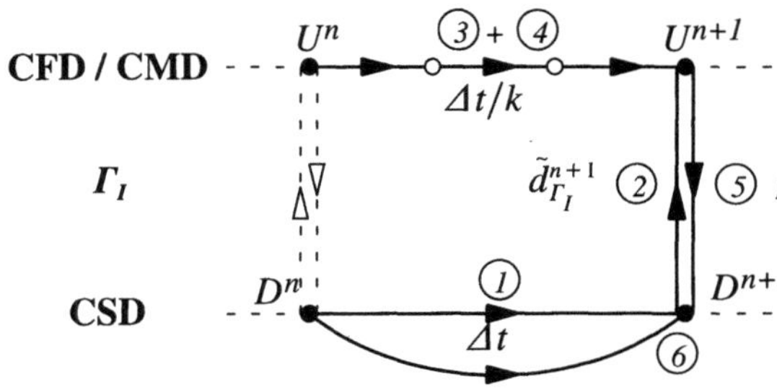

Fig. 5.1. Partitioned time integration algorithm

starting from a known state of motion before and at t^n in all three fields:

1. Predict displacements of wetted structural surface at t^{n+1} with a predictor $\tilde{\mathbf{d}}_{\Gamma_I}^{n+1} = \tilde{\mathbf{d}}_{\Gamma_I}^{n+1}(\mathbf{d}_{\Gamma_I}^n, \dot{\mathbf{d}}_{\Gamma_I}^n, \mathbf{d}_{\Gamma_I}^{n-1}, \dot{\mathbf{d}}_{\Gamma_I}^{n-1})$ further discussed below.
2. Pass over $\tilde{\mathbf{d}}_{\Gamma_I}^{n+1}$ from CSD to CMD as Dirichlet b.c.
3. Solve CMD with Dirichlet b.c. on Γ_I: $\mathbf{r}_{\Gamma_I}^{n+1} = \tilde{\mathbf{d}}_{\Gamma_I}^{n+1}$ and on Γ_S: $\dot{\mathbf{r}}_{\Gamma_S}^{n+1} = \mathbf{u}_{\Gamma_S}^n$ for $\mathbf{r}^{n+1}$ and then compute grid velocity $(\mathbf{u}^G)^{n+1} \equiv \dot{\mathbf{r}}^{n+1}$.
4. Solve CFD (possibly with subcycling: $\Delta t_F = \Delta t/k$, $k \in \mathcal{N}$) on new mesh $(\mathbf{r}^{n+1}, (\mathbf{u}^G)^{n+1})$ for fluid velocity and pressure fields $\mathbf{u}^{n+1}$ and $\mathbf{p}^{n+1}$.
5. Pass over fluid pressure $\mathbf{p}_{\Gamma_I}^{n+1}$ from CFD to CSD as Neumann b.c. and convert it into a consistent nodal load vector $\mathbf{F}_{\Gamma_I}^{n+1}$.
6. Solve CSD with $\mathbf{F}_{\Gamma_I}^{n+1}$ for the new displacements $\mathbf{d}^{n+1}$.
7. Proceed to the next time step by setting $n \to n + 1$.

The quality of the predictor in step 1 has a main impact on the accuracy and stability properties of the staggered algorithm. Choosing the old interface displacements at t^{n+1}, $\tilde{\mathbf{d}}_{\Gamma_I}^{n+1} = \mathbf{d}_{\Gamma_I}^n$ yields the basic sequential staggered scheme dating back to Felippa and Park [3]. In most cases a much better choice is the $O(\Delta t^2)$-accurate predictor suggested by Piperno [7],

$$\tilde{\mathbf{d}}_{\Gamma_I}^{n+1} = \mathbf{d}_{\Gamma_I}^n + \Delta t(\frac{3}{2}\dot{\mathbf{d}}_{\Gamma_I}^n - \frac{1}{2}\dot{\mathbf{d}}_{\Gamma_I}^{n-1}) \tag{5.1}$$

However, due to their explicit character the sequential staggered coupling procedures still cannot exactly fulfill the above stated coupling conditions within every time step. In other words, only a loose coupling between the fields can be achieved and energy is not conserved across the fluid structure interface. Consequently, for such loose coupling schemes classes of examples exist with parameter configurations unable to produce stable and accurate solutions. In fact, such solutions can only be guaranteed a priori by the application of implicit, iterative staggered schemes.
Iterative staggered schemes, also called iterative substructuring schemes, use iterations over the fields within every time step until the coupling conditions are fulfilled, i.e. convergence. Thus a strong algorithmic coupling is obtained. In the algorithmic framework given above, this means that steps 2 to 6 are repeated until convergence, using the most recently computed interface displacements as new predictor $\tilde{\mathbf{d}}^{n+1}_{\Gamma_I,i+1} = \mathbf{d}^{n+1}_{\Gamma_I,i}$. As this fixed-point iteration generally shows poor convergence behavior, a relaxed update of the predictor

$$\tilde{\mathbf{d}}^{n+1}_{\Gamma_I,i+1} = (1-\omega_i)\mathbf{d}^{n+1}_{\Gamma_I,i-1} + \omega_i \mathbf{d}^{n+1}_{\Gamma_I,i} \tag{5.2}$$

should be used to accelerate and ensure convergence (Le Tallec and Mouro [5], Marini and Quarteroni [6]).

5.2 A robust iterative substructuring scheme

Concerning the convergence properties of the iterative scheme described above the key point is the choice of the relaxation parameter ω_i. With constant $\omega > 0$ throughout the whole computation, the scheme can be interpreted as a relaxed Richardson iteration on the interface variables $\tilde{\mathbf{d}}^{n+1}_{\Gamma_I}$. Applied to a linear system, the optimal parameter is determined as $\omega_{opt} = 2/(\lambda_{max} + \lambda_{min})$. However, this requires knowledge of the extreme eigenvalues of the coupled system. In the nonlinear case the spectrum of the linearized operators changes within each iteration and time step, so that the additional cost for repeated eigenvalue analyses becomes prohibitive.
In Wall et al. [9] we proposed the evaluation of optimal relaxation parameters which are iteration dependent and use the method of steepest descent applied to the iterative substructuring scheme. It exhibits the same optimal convergence behavior as the relaxed Richardson iteration with ω_{opt}, even in the nonlinear regime, leading to very small iteration counts. The relaxation parameter is then equivalent to the step length in the method of steepest descent,

$$\omega_i = \frac{\mathbf{g}_i^T\,\mathbf{g}_i}{\mathbf{g}_i^T\,\mathbf{S}_s^{-1}(\mathbf{S}_f + \mathbf{S}_s)\,\mathbf{g}_i} \tag{5.3}$$

where $\mathbf{g}_i = \mathbf{d}^{n+1}_{\Gamma_I,i} - \mathbf{d}^{n+1}_{\Gamma_I,i-1}$ denotes the search direction equal to the actual residual, and $\mathbf{S}_f$ / $\mathbf{S}_s$ are the Schur complement matrices of the fluid / structural field, respectively. In Wall et al. [9] we further describe a procedure to evaluate the relaxation parameters without explicitly computing and storing these Schur complements.

6. Numerical examples

6.1 Floating vertical cylinder in viscous fluid

In this example, a fluid-structure interaction problem incorporating a free surface and gravity effects (g) is discussed. The fluid in a container is excited by a harmonically oscillating surface-piercing rigid body. The ensuing sloshing of the free surface is demonstrated in Fig. 6.1. The results of our approach compare well with detailed studies given in Ananthakrishnan [1].

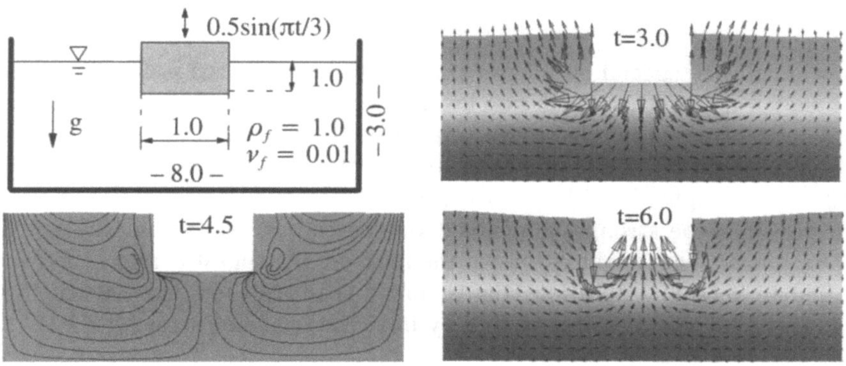

Fig. 6.1. Velocity vectors and streamlines for surface-piercing cylinder

6.2 Cavity with oscillating top plate and flexible bottom

The example shown in Fig. 6.2 is a 2-D simulation of a square liquid-filled tank with a flexible bottom and a harmonically oscillating top plate. The flexible

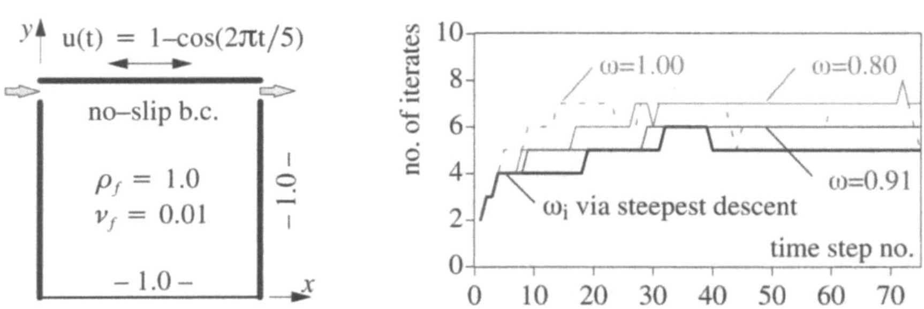

Fig. 6.2. Cavity: problem statement and number of iterates per time step

bottom structure has a thickness of $h = 0.002$, a density of $\rho_s = 500.0$ and a Young's modulus of $E = 250.0$. As in 6.1, Q1Q1 elements are used to model the fluid, and here Q2 elements are used to model the structure. Both fields are implicitly integrated in time with a time step of $\Delta t = 0.1$. Computations for this problem demonstrate the unreliability of both sequential staggered (loose coupling) schemes described in Section 5. Stable solutions could only

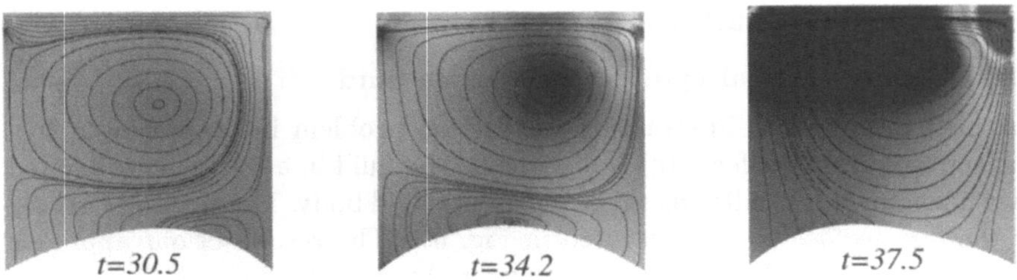

Fig. 6.3. Streamlines on pressure and deformed configurations

be obtained with special parameter configurations (material properties, time step size). On the other hand, employment of the proposed iterative (strong coupling) schemes always led to convergent solutions, even for long-term computations (Fig. 6.3). The number of iterates to reach the required accuracy of 10^{-8} in a weighted residual norm and the respective computational costs were proven to be strongly dependent on the chosen relaxation parameter. Fig. 6.2 shows that ω-evaluation via the method of steepest descent (Section 5.2) led to very low iteration counts, lower than the best fixed relaxation parameter that we were able to find by intricate trial and error ($\omega = 0.91$).

References

1. P. Ananthakrishnan: *Radiation hydrodynamics of floating vertical cylinder in viscous fluid.* J. Eng. Mech. **125**, 836-847 (1999).
2. J. Chung G.M. Hulbert: *A time integration algorithm for structural dynamics with improved numerical dissipation: The Generalized-α method.* J. Appl. Math. **60**, 371-375 (1993).
3. C.A. Felippa, K.C. Park: *Staggered transient analysis procedures for coupled mechanical systems: Formulation.* Comp. Meth. Appl. Mech. Eng. **24** (1980), 61-111.
4. T.J.R. Hughes, W.K. Liu, T.K. Zimmermann: *Lagrangian-Eulerian finite element formulation for viscous flows.* Comp. Meth. Appl. Mech. Eng. **29**, 329-349 (1981).
5. P. Le Tallec, J. Mouro: *Fluid structure interaction with large structural displacements.* To appear in: Comp. Meth. Appl. Mech. Eng.
6. L.D. Marini, A. Quarteroni: *A relaxation procedure for domain decomposition methods using finite elements.* Numerische Mathematik **55** (1989), 575-598.
7. S. Piperno: *Explicit/implicit fluid/structure staggered procedures with a structural predictor and fluid subcycling for 2D inviscid aeroelastic simulations.* Int. J. Num. Meth. Fluids **25** (1997), 1207-1226.
8. W.A. Wall: *Fluid-Struktur-Interaktion mit stabilisierten Finiten Elementen.* Ph.D.-Dissertation, Bericht Nr. 31, Inst. f. Baustatik, Univ. Stuttgart (1999).
9. W.A. Wall, D.P. Mok, E. Ramm: *Partitioned analysis approach for the transient, coupled response of viscous fluids and flexible structures.* Solids, Structures and Coupled Problems in Eng., Proc.ECCM'99, W. Wunderlich (ed.), Munich, 1999.
10. W.A. Wall, E. Ramm: *Fluid-structure interaction based upon a stabilized (ALE) finite element method.* Computational Mechanics - New Trends and Applications, Proc. WCCM IV. S.R. Idelsohn, E. Oñate, E.N. Dvorkin (eds.), CIMNE, Barcelona, 1998.

Section 4

Efficient Solvers and Adaptivity

Edited by O. Steinbach and C. Wieners

Domain Decomposition Methods in the Design of High Power Electronic Devices

R. Hoppe, Y. Iliash, and G. Mazurkevitch

Institute of Mathematics, University of Augsburg, Germany

Summary. The optimal design of high power electronic devices and systems requires a subtle physically consistent modeling of the relevant physical effects as well as the use of state-of-the-art algorithmic methods for an efficient and reliable simulation of the operating behavior. In this contribution, we shall consider two important issues with emphasis on the development, analysis, and implementation of domain decomposition methods as effective simulation tools. The first one addresses electrothermalmechanical coupling phenomena in IHV-modules which involves self-heating due to Joule heat thus causing heat stresses that have an impact on the mechanical behavior. The second one is related to the minimization of parasitic inductivities in converter modules which amounts to the computation of eddy currents as the solution of the quasistationary limit of Maxwell's equations.

1. Introduction

We consider multifield problems both as they arise in the coupling of various physical effects as well as they occur in the numerical simulation of physical processes by the application of domain decomposition methods. There is a wide variety of technological applications where multifield problems play a significant role. Here, we concentrate on one specific area of applications, namely the layout and design of devices and systems in high power electronics. In particular, as two significant applications in this field, in section 2 we present the electrothermomechanical coupling in IHV-modules and the minimization of parasitic inductivities in converter modules, both modules being used as electrical drives for high power electromotors. A common feature of both problems is that the geometrical and physical structure suggests a decomposition of the computational domain into subdomains and the application of corresponding domain decomposition methods. We emphasize on such techniques involving nonmatching grids on the interfaces between adjacent subdomains also known as mortar methods. In particular, we suggest mortar methods based on Lagrangian finite elements for the thermomechanical coupling and mortar methods for edge element discretizations in the computation of eddy currents. The fundamental ideas will be outlined in section 3 including a brief discussion of efficient iterative solvers and a posteriori error estimators as a basic tool for adaptive grid refinement. Finally, section 4 contains some results of numerical simulations.

2. High power electronic devices and systems

In the area of high power electronics significant devices are given by IHV- and Converter-Modules both used as electric drives for high power electromotors. The field of applications ranges from natural power generation by windmills to public transportation systems such as trams and - as the most spectacular example - high speed trains like the German ICE or the French TGV.

2.1 Electrothermomechanical couplings in IHV-Modules

IHV-Modules consist of specific semiconductor devices such as IGBTs (Insulated Gate Bipolar Transistors) and power diodes serving as switches for electric currents ranging roughly between 300 A and 1800 A. The problem is that during operation electric energy is converted into Joule heat thus causing a significant self-heating of the devices. This has an impact on the mechanical behavior due to heat sresses that lead to mechanical deformations and even mechanical damage such as cracks without appropriate cooling mechanisms.

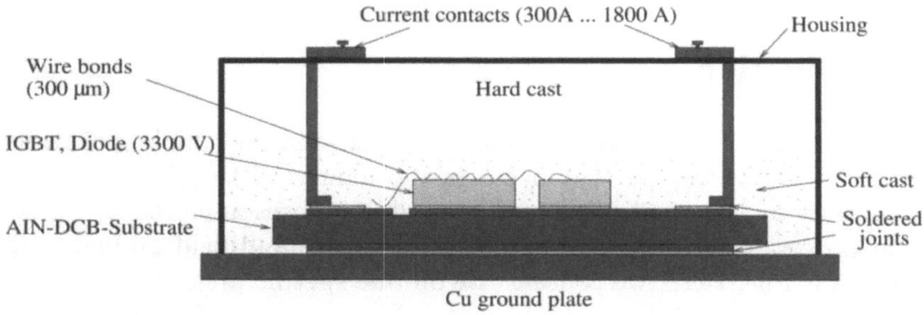

Fig. 1: Schematic representation of an IHV-Module

Figure 1 shows the schematic representation of an IHV-Module as used as an electrical drive in trams. The IGBTs and power diodes are integrated into copper blocks that are mounted on several layers of different materials attached to each other by thin soldered joints. The copper ground plate either features a direct cooling or is fixed on a cooling device.
The transport processes in the semiconductor devices can be modelled by the classical drift-diffusion model consisting of a potential equation for the electric potential ψ and two drift-diffusion equations for the concentrations of the carriers, namely the electrons n and the holes p

$$-\nabla\cdot\varepsilon\nabla\psi + N_{dop}(n,p) + q(n-p) = 0 \qquad (2.1)$$

$$\frac{\partial n}{\partial t} = +q^{-1}\nabla\cdot\mathbf{J_n} + G(\mathbf{J_n},\mathbf{J_p},\mathbf{E}) - R(n,p) \qquad (2.2)$$

$$\frac{\partial p}{\partial t} = -q^{-1}\nabla\cdot\mathbf{J_p} + G(\mathbf{J_n},\mathbf{J_p},\mathbf{E}) - R(n,p) \qquad (2.3)$$

with appropriately chosen initial and boundary conditions. In (2.1), q is the elementary charge and N_{dop} stands for the doping profile whereas in (2.2),(2.3) $\mathbf{E}$ is the electric field $\mathbf{E} = -\nabla\psi$, $\mathbf{J_n}$ and $\mathbf{J_p}$ denote the current densities as given by

$$\mathbf{J_n} = -q\mu_n n \nabla\psi \quad , \quad \mathbf{J_p} = -q\mu_p p \nabla\psi$$

where μ_n and μ_p are the mobilities, and G and R refer to the generation/recombination terms (cf.,e.g.,[19, 21]).
The dominant source of heat generation is the Joule heat

$$H_{\text{Joule}} = \frac{\mathbf{J_n} \cdot \mathbf{J_n}}{q\mu_n n} + \frac{\mathbf{J_p} \cdot \mathbf{J_p}}{q\mu_p p} \quad ,$$

but also other effects can be taken into account (cf.,e.g.,[22, 23]).
For the efficient numerical solution of the drift-diffusion model finite volume techniques like the Scharfetter-Gummel scheme (cf. [19, 21]) or related mixed-hybrid finite element methods (cf. [13, 14]) are available.
In the following we shall concentrate on the thermomechanical coupling process. The temporal and spatial temperature distribution is described by a heat equation in the bounded 3D domain Ω occupied by the module

$$\rho c \frac{\partial T}{\partial t} = \nabla \cdot (\kappa \nabla T) \quad \text{in } Q := \Omega \times (0, t^*) \ , \tag{2.4}$$

$$\mathbf{n} \cdot \kappa \nabla T = J(t) \quad \text{on } \Gamma_0 \times (0, t^*) \ , \tag{2.5}$$

$$\mathbf{n} \cdot \kappa \nabla T = h\,(T^* - T) \quad \text{on } \Gamma_1 \times (0, t^*) \ , \tag{2.6}$$

$$\mathbf{n} \cdot \kappa \nabla T = 0 \quad \text{on } \Gamma_2 \times (0, t^*) \ , \tag{2.7}$$

$$T(x, 0) = T_0(x) \quad \text{in } \Omega \ . \tag{2.8}$$

In particular, the generated Joule heat is modelled by a heat flux $J(t), t \in (0, t^*)$ through the upper boundary $\Gamma_0 \subset \partial\Omega$ of the copper blocks housing the IGBTs and power diodes. At the bottom $\Gamma_1 \subset \partial\Omega$ of the copper ground plate, heat exchange with the attached cooling device is assumed where in (2.6) h stands for the corresponding heat exchange coefficient and T^* denotes the temperature of the ambient medium. Otherwise, i.e., on the rest Γ_2 of the boundary we assume homogeneous Neumann type boundary conditions. Note that in (2.4) the quantities ρ, c, and κ refer to the material dependent densities, heat capacities, and heat conductivities for the various layers of the module.
The change in temperature causes heat stresses leading to mechanical deformations that can be described by the equilibrium equations of linear elasticity

$$\nabla \cdot \sigma(\mathbf{u}) = \frac{\alpha(1-\nu)E}{(1+\nu)(1-2\nu)} \nabla (T - T_0) \quad \text{in } \Omega \ , \tag{2.9}$$

$$\mathbf{u} = \mathbf{0} \quad \text{on } \Gamma_1 \quad , \quad \mathbf{n} \cdot \sigma(\mathbf{u}) = \mathbf{0} \quad \text{elsewhere on } \Gamma \tag{2.10}$$

with Hooke's law as the constitutive equation

$$\sigma(\mathbf{u}) \; = \; \frac{E}{1 \, + \, \nu} \; [\, \mathcal{D}(\mathbf{u}) \; + \; \frac{\nu}{1 \, - \, 2\nu} \; \mathrm{tr}\,\mathcal{D}(\mathbf{u})\,\mathbf{I}\,]$$

where $\mathcal{D}(\mathbf{u})$ is the linearized strain tensor, and E and ν refer to Young's moduli and Poisson's ratios for the different material layers.
Note that the model does not contain damage mechanisms, but the magnitude of the deviatoric stress tensor may serve as an indicator for possible damage (cf. Section 4).

2.2 Minimization of parasitic effects in Converter Modules

Converter Modules are high power electronic devices that are designed to convert AC into DC or vice versa. As shown in Figure 2a, they consist of IGBTs that are connected to each other and with the high voltage source as well as the electromotor to be driven by copper made bus bars. The typical geometry of such a bus bar is displayed in Figure 2b. The bus bars feature several ports where electric currents are either fed to the bar or are supplied to an attached IGBT.

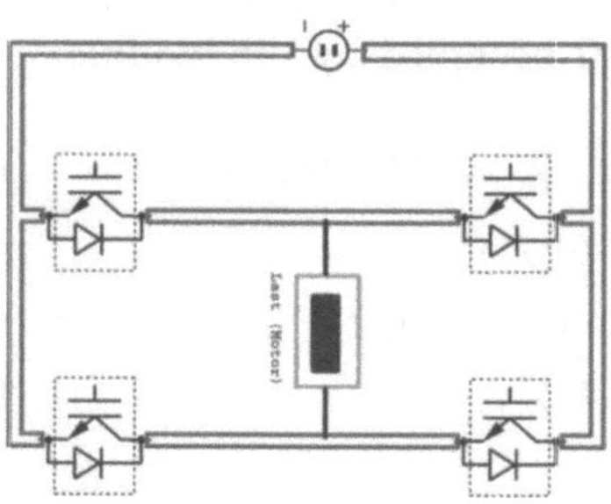

Fig. 2a: Converter Module

Fig. 2b: Geometry of a bus bar

The problem is that due to the high intended voltages in the range of several hundred kilovolts and the fast switching times of the IGBTs of less than 100 nanoseconds, parasitic inductivities caused by eddy currents build up that lead to a significant loss in the power transmission. Assuming a module with N bus bars each bus bar containing N_ν ports $T_{\nu\alpha}$, the ultimate goal is to design both the individual bus bars as well as the entire configuration in such a way that the total inductivity

$$L_{\mathrm{tot}} := \left(\sum_{\nu,\alpha} \sum_{\mu,\beta} \int_0^T |\, L_{\nu\alpha,\mu\beta}(t)\,|^2 \, dt \right)^{1/2} \tag{2.11}$$

is minimized given in terms of the generalized transient inductivity coefficients

$$L_{\nu\alpha,\mu\beta}(t) := \frac{1}{\sigma} \int_{\Omega_\nu} \mathbf{J}_{\nu\alpha}(x) \cdot \mathbf{S}(t) \mathbf{J}_{\mu\beta}(x) \, dx$$

where $J_{\nu\alpha}$ denotes the current associated with the bar Ω_ν at the port $T_{\nu\alpha}$ and $S(t), 0 \le t \le T$, stands for the solution operator of the equation for the associated magnetic vector potential (see (2.13) below).
Since the dielectric time constant is much less than the switching time, the generated electromagnet field is slowly varying in time and can thus be described by the quasistationary limit of Maxwell's equations. For the computation of the eddy currents we resort to a potential formulation and introduce a scalar electric potential φ and a magnetic potential $\mathbf{A}$ according to

$$\mathbf{E} \;=\; -\mathrm{grad}\,\varphi \;-\; \frac{\partial\,\mathbf{A}}{\partial\,t} \quad , \qquad \mathbf{B} \;=\; \mathrm{curl}\,\mathbf{A}$$

where $\mathbf{E}$ and $\mathbf{B}$ denote the electric field and magnetic induction, respectively. We are then lead to the coupled system

$$\mathrm{div}\,(\sigma\;\mathrm{grad}\;\varphi) = 0 \;\;\text{in}\;\Omega, \quad \sigma\mathbf{n}\cdot\mathrm{grad}\,\varphi = \begin{cases} -I_{\nu\alpha}(t) & \text{on}\;T_{\nu\alpha} \\ 0 & \text{else} \end{cases} , \tag{2.12}$$

$$\sigma\,\frac{\partial\,\mathbf{A}}{\partial\,t} \;+\; \frac{1}{\mu}\;\mathrm{curl}\;\mathrm{curl}\;\mathbf{A} = \begin{cases} \mathbf{J}_{\nu\alpha} & \text{in}\;\;\Omega \\ 0 & \text{in}\;\mathbf{R}^3\setminus\Omega \end{cases} \tag{2.13}$$

with σ and μ being the electric conductivity and magnetic permeability, respectively.
Choosing the conductivity σ as a material parameter, the minimization of the total inductivity (2.11) leads to a topology optimization problem. Under some appropriate constraints on σ, a local minimizer is determined by an iterative process featuring the computation of the potentials φ and $\mathbf{A}$ for the actual design followed by the determination of the descent directions by means of a sequential quadratic programming approach.

3. The mortar element methods

Mortar methods have been originally introduced in [7] within the framework of spectral methods. There is a vast amount of literature on such methods for standard finite element discretizations (cf. [4, 8, 9, 18] and the references therein) whereas only a few papers have addressed related techniques for edge elements (cf. [5, 6, 15]). For notational convenience and better readability we shall outline the basic ideas for simply structured model problems.

3.1 Mortar methods for Lagrangian finite elements

We shall outline the basic ideas and concepts for a model problem as given by the following primal variational formulation of a boundary value problem for a linear second order elliptic differential operator

$$\int\limits_{\Omega} (\;a\nabla u\cdot\nabla v \;+\; cuv)\;dx \;=\; \int\limits_{\Omega} fv\;dx \;+\; \int\limits_{\Gamma_N} gv\;d\sigma \quad , \qquad v\in H^1_{0,\Gamma_D}(\Omega) \tag{3.1}$$

under the usual assumptions on the data of the problem where $H^1_{0,\Gamma_D}(\Omega) := \{v \in H^1(\Omega) \mid v\mid_{\Gamma_D} = 0\}$ is the Sobolev space of functions with vanishing trace on the Dirichlet part Γ_D of the boundary. We start from a decomposition

$$\bar{\Omega} = \bigcup_{i=1}^{n} \bar{\Omega}_i \quad , \quad \Omega_i \cap \Omega_j = \emptyset \quad , \quad i \neq j \tag{3.2}$$

of the computational domain into n mutually disjoint subdomains and we assume that this decomposition is geometrically conforming in the sense that the boundary of any subdomain either is part of the boundary of the entire computational domain or corresponds to the boundary of an adjacent subdomain. For later reference, the union

$$S := \bigcup_{i \neq j} \Gamma_{ij} \quad , \quad \Gamma_{ij} = \bar{\Omega}_i \cap \bar{\Omega}_j \tag{3.3}$$

of the interfaces between adjacent subdomains will be denoted as the skeleton of the decomposition. Moreover, according to common notational customs, for a given interface $\Gamma_{ij} \subset S$ we refer to Ω_j as the mortar- and to Ω_i as the nonmortar-side. The idea is to use invidual simplicial triangulations $\mathcal{T}_i$ of the subdomains and corresponding finite element discretizations by means of continuous, piecewise linear finite elements regardless the situation on the interfaces so that typically nonconforming nodal points will arise on the skeleton. Consequently, the resulting discretization is a nonconforming one. In order to guarantee consistency of that nonconforming approach we have to establish coupling constraints on the interfaces making up the skeleton of the decomposition. This can be done by Lagrangian multipliers living in appropriately defined multiplier spaces. For that purpose we consider the product space

$$V_h := \prod_{i=1}^{n} S_{1,\Gamma_D}(\Omega_i; \mathcal{T}_i) \; ,$$

with $S_{1,\Gamma_D}(\Omega_i; \mathcal{T}_i)$ denoting the finite element spaces associated with the P1-approximations of the subdomain problems. Due to the nonconforming situation, for $v_h \in V_h$ the traces on Γ_{ij} in general do not coincide, i.e., v_h experiences a jump across Γ_{ij} given by $[v_h]_J := v_h\mid_{\Gamma_{ij}\cap\Omega_i} - v_h\mid_{\Gamma_{ij}\cap\Omega_j}$. We measure that jump by means of the mesh-dependent weighted L^2-norm

$$\| \, [v_h] \mid_{\Gamma_{ij}} \|_{\frac{1}{2},h,\Gamma_{ij}} := h^{-1/2} \, \| \, [v_h] \mid_{\Gamma_{ij}} \|_{0,\Gamma_{ij}}$$

and equip V_h with the graph norm

$$\|v_h\|_{V_h} := \Big(\sum_{i=1}^{n} \|v_h\|^2_{1,\Omega_i} + \sum_{\Gamma_{ij} \subset S} \| \, [v_h] \mid_{\Gamma_{ij}} \|^2_{\frac{1}{2},h,\Gamma_{ij}} \Big)^{1/2} \; .$$

The multiplier space will be defined with respect to the triangulation $\mathcal{T}_{ij}$ of Γ_{ij} inherited from the triangulation $\mathcal{T}_i$ of the nonmortar-side. Denoting by $S_1(\Gamma_{ij}, \mathcal{T}_{ij})$ the space of continuous, piecewise linear finite elements with respect to $\mathcal{T}_{ij}$ and setting $S_{0,1}(\Gamma_{ij}, \mathcal{T}_{ij}) := \{v_h \in S_1(\Gamma_{ij}, \mathcal{T}_{ij}) \mid v_h |_{\partial\Gamma_{ij}} = 0\}$, we define $M_h(\Gamma_{ij})$ such that

$$S_{1,0}(\Gamma_{ij}; \mathcal{T}_{ij}) \subset M_h(\Gamma_{ij}) \quad , \quad \dim M_h(\Gamma_{ij}) \quad = \quad \dim \mathrm{S}_{1,0}(\Gamma_{ij}; \mathcal{T}_{ij}) \quad .$$

This can be done by a modification of the nodal basis functions associated with those nodal points in the interior of Γ_{ij} having neighboring nodal points on $\partial\Gamma_{ij}$. The modification consists in adding a convex combination of the nodal basis functions that correspond to the neighbors on $\partial\Gamma_{ij}$ (for details we refer to [8]). We define the global multiplier space according to

$$M_h(S) \quad := \quad \prod_{\Gamma_{ij} \subset S} M_h(\Gamma_{ij}) \quad , \quad \|\mu_h\|_{M_h(S)} \; := \; \Big(\sum_{\Gamma_{ij} \subset S} \|\mu_h\|^2_{-\frac{1}{2},h,\Gamma_{ij}} \Big)^{1/2}$$

where $\| \cdot \|_{-\frac{1}{2},h,\Gamma_{ij}}$ stands for the mesh-dependent weighted L^2-norm

$$\|\mu_h\|_{-\frac{1}{2},h,\Gamma_{ij}} \quad := \quad h^{1/2} \; \|\mu_h\|_{0,\Gamma_{ij}} \quad , \quad \mu_h \in M_h(S) \; .$$

A weak continuity across the interfaces $\Gamma_{ij} \subset S$ can be enforced by the bilinear form $b_h : V_h \times M_h(S) \to \mathbf{R}$ given by

$$b_h(v_h, \mu_h) \quad := \quad - \sum_{\Gamma_{ij} \subset S} \int_{\Gamma_{ij}} \mu_h \, v_h \, d\sigma \quad , \quad v_h \in V_h \quad , \quad \mu_h \in M_h(S) \quad .$$

We refer to $B_h : V_H \to (M_h(S))^*$ as the operator associated with b_h and further denote by $a_h : V_h \times V_h \to \mathbf{R}$ the bilinear form

$$a_h(v_h, w_h) \quad := \quad \sum_{i=1}^{n} \int_{\Omega_i} [a \nabla v_h \cdot \nabla w_h \; + \; c v_h w_h] dx \quad , \quad v_h, w_h \in V_h \quad .$$

Then, the mortar finite element approximation of (3.1) amounts to the computation of $(u_h, \lambda_h) \in V_h \times M_h(S)$ as the solution of the discrete saddle point problem

$$a_h(u_h, v_h) \; + \; b_h(v_h, \lambda_h) \quad = \quad \ell_h(v_h) \quad , \quad v_h \in V_h \; , \qquad (3.4)$$

$$b_h(u_h, \mu_h) \quad = \quad 0 \quad , \quad \mu_h \in M_h(S) \qquad (3.5)$$

where $\ell_h(v_h) := \int_\Omega f v_h \, dx + \int_{\Gamma_N} g v_h \, d\sigma$. The saddle point problem (3.4,3.5) admits a unique solution depending continuously on the data. This is a consequence of the following result.

Theorem 3.1. The bilinear form a_h is elliptic on $\ker B_h$, i.e.,

$$a_h(v_h, v_h) \; \geq \; \alpha \|v_h\|^2_{V_h} \quad , \quad v_h \in \ker B_h \; , \; \alpha > 0 \; , \qquad (3.6)$$

and the bilinear form b_h satisfies the LBB-condition

$$\inf_{\mu_h \in M_h(S)} \sup_{v_h \in V_h} \frac{b_h(v_h, \mu_h)}{\|v_h\|_{V_h} \|\mu_h\|_{M_h(S)}} \geq \beta > 0 . \qquad (3.7)$$

A proof of (3.6) and (3.7) as well as a priori estimates for the global discretization error can be found in [8] (cf. also [4]). Note that the stability estimate (3.7) was first proved in [1] under the assumption of quasiuniform meshes within each subdomain.

For the iterative solution of (3.4,3.5) we suggest a preconditioned Lanczos iteration featuring a block diagonal preconditioner $\mathcal{R} := \text{blockdiag}(R_u, R_\lambda)$ The first diagonal block $R_u := \text{diag}(R_1, ..., R_n)$ involves preconditioners R_i, $1 \leq i \leq n$, of the subdomain stiffness matrices whereas the second diagonal block R_λ invokes a preconditioner of the Schur complement resulting from block Gaussian elimination of the unknowns associated with the subdomains. For the development, analysis, and implementation of the preconditioners in a multilevel framework given by an adaptively generated hierarchy of triangulations we refer to [11, 12, 18]. These references also contain the construction and analysis of both residual based and hierarchical type a posteriori error estimators as basic tools for adaptive drid refinement.

3.2 Mortar methods for curl-conforming edge elements

We confine ourselves to the interior domain problem (2.13) for the magnetic vector potential discretized implicitly in time by the backward Euler scheme. For simplicity we suppose vanishing tangential traces on $\Gamma = \partial\Omega$. Then, we are faced with a boundary value problem for the double curl-operator whose weak formulation requires the computation of $\mathbf{j} \in H_0(\text{curl}, \Omega)$ as the solution of

$$\int_\Omega [\text{curl}\mathbf{j} \cdot \text{curl}\mathbf{q} + c\mathbf{j} \cdot \mathbf{q}] dx = \int_\Omega \mathbf{f} \cdot \mathbf{q} dx \quad , \quad \mathbf{q} \in H_0(\text{curl}, \Omega) . \qquad (3.8)$$

Here, $H_0(\text{curl}, \Omega) := \{\mathbf{q} \in H(\text{curl}, \Omega) \mid \mathbf{n} \wedge \mathbf{q} \mid_\Gamma = 0\}$ where $H(\text{curl}, \Omega) := \{\mathbf{q} \in L^2(\Omega)^3 \mid \text{curl}\mathbf{q} \in L^2(\Omega)^3\}$ with norm $\|\mathbf{q}\|_{\text{curl},\Omega} := (\|\mathbf{q}\|_{0,\Omega}^2 + \|\text{curl}\,\mathbf{q}\|_{0,\Omega}^2)^{1/2}$.

Again, we assume a geometrically conforming decomposition (3.2) with skeleton (3.3) and invidual simplicial triangulation $\mathcal{T}_i$ of Ω_i, $1 \leq i \leq n$. For discretization of the subdomain problems we choose Nédélec's lowest order curl-conforming edge elements $Nd_1(T) := \{\mathbf{q} = \mathbf{a} + \mathbf{b} \wedge \mathbf{x} \mid \mathbf{a}, \mathbf{b} \in \mathbf{R}^3\}$, $T \in \mathcal{T}_i$ using the zero order moments $\int_E \mathbf{t}_E \cdot \mathbf{q} ds$ of the tangential component with respect to the edges E of T as the degrees of freedom (cf. [20]). We denote by $Nd_{1,\Gamma}(\Omega_i, \mathcal{T}_i)$, $1 \leq i \leq n$, the corresponding edge element spaces with vanishing tangential traces on $\Gamma \cap \partial\Omega_i$. We introduce the product space

$$V_h := \prod_{i=1}^n Nd_{1,\Gamma}(\Omega_i, \mathcal{T}_i)$$

with norm

$$\|\mathbf{q_h}\|_{V_h} := \Big(\sum_{i=1}^{n} \|\mathbf{q_h}\|^2_{curl;\Omega_i} + \sum_{\Gamma_{ij} \subset S} \|[\mathbf{n} \wedge \mathbf{q_h}] \,|_{\Gamma_{ij}}\, \|^2_{\frac{1}{2},h,\Gamma_{ij}} \Big)^{1/2}$$

which takes into account the jumps $[\mathbf{n} \wedge \mathbf{q_h}] \,|_{\Gamma_{ij}}$ of the tangential traces across $\Gamma_{ij} \subset S$. The multiplier space $M_h(S)$ is specified by means of

$$M_h(S) := \prod_{\Gamma_{ij} \subset S} M_h(\Gamma_{ij}) \quad , \quad \|\boldsymbol{\mu}_h\|_{\mathbf{M_h(S)}} := \Big(\sum_{\Gamma_{ij} \subset S} \|\boldsymbol{\mu}_h\|^2_{-\frac{1}{2},h,\Gamma_{ij}} \Big)^{1/2}$$

with $M_h(\Gamma_{ij})$, $\Gamma_{ij} \subset S$, constructed according to

$$Nd_{1,0}(\Gamma_{ij}; \mathcal{T}_{ij}) \subset M_h(\Gamma_{ij}) \quad , \quad \dim M_h(\Gamma_{ij}) = \dim Nd_{1,0}(\Gamma_{ij}; \mathcal{T}_i)$$

following similar ideas as before (cf. [15]). Defining bilinear forms $a_h : V_h \times V_h \to \mathbf{R}$ and $b_h : V_h \times M_h(S) \to \mathbf{R}$ by

$$a_h(\mathbf{p_h}, \mathbf{q_h}) := \sum_{i=1}^{n} \int_{\Omega_i} [\operatorname{curl} \mathbf{p_h} \cdot \operatorname{curl} \mathbf{q_h} + c\mathbf{p_h} \cdot \mathbf{q_h}]\, dx \quad , \quad \mathbf{p_h}, \mathbf{q_h} \in V_h \; ,$$

$$b_h(\mathbf{q_h}, \boldsymbol{\mu}_h) := \sum_{\Gamma_{ij} \subset S} \int_{\Gamma_{ij}} \boldsymbol{\mu}_h \cdot [\mathbf{n} \wedge \mathbf{q_h}] \,|_{\Gamma_{ij}}\, d\sigma \quad , \quad \mathbf{q_h} \in V_h, \; \boldsymbol{\mu}_h \in M_h(S),$$

the mortar edge element approximation of (3.8) requires the computation of $(\mathbf{j_h}, \boldsymbol{\lambda}_h) \in V_h \times M_h(S)$ as the solution of

$$a_h(\mathbf{j_h}, \mathbf{q_h}) + b_h(\mathbf{q_h}, \boldsymbol{\lambda}_h) = \int_{\Omega} \mathbf{f} \cdot \mathbf{q_h}\, dx \quad , \quad \mathbf{q_h} \in V_h \; , \tag{3.9}$$

$$b_h(\mathbf{j_h}, \boldsymbol{\mu}_h) = 0 \quad , \quad \boldsymbol{\mu}_h \in M_h(S) \; . \tag{3.10}$$

Denoting by B_h the operator associated with b_h the following result can be established.

Theorem 3.2. The bilinear form a_h is elliptic on $\ker B_h$, i.e.,

$$a_h(\mathbf{q_h}, \mathbf{q_h}) \geq \alpha \|\mathbf{q_h}\|^2_{V_h} \quad , \quad \mathbf{q_h} \in \ker B_h \; , \; \alpha > 0$$

and the bilinear form b_h satisfies the LBB-condition

$$\inf_{\boldsymbol{\mu}_h \in \mathbf{M_h(S)}} \sup_{\mathbf{q_h} \in \mathbf{V_h}} \frac{b_h(\mathbf{q_h}, \boldsymbol{\mu}_h)}{\|\mathbf{q_h}\|_{\mathbf{V_h}} \, \|\boldsymbol{\mu}_h\|_{\mathbf{M_h(S)}}} \geq \beta > 0 \; .$$

For the proof of Theorem 3.2 we refer to [15]. The iterative solution of (3.9,3.10) is done by a multigrid algoritm with respect to an adaptively generated hierarchy of simplicial triangulations with a distributed smoothing process featuring a defect correction in the subspaces of irrotational vector fields building the kernels of the discrete curl-operator (for details see [2, 15, 16, 17]). An efficient and reliable residual based a posteriori error estimator can be provided following the ideas in [3].

4. Numerical results

In case of the electrothermomechanical coupling effects in IHV-modules we are interested in the temperature distribution and the distribution of the mechanical stresses in the device. In particular, we focus on such regions where high equivalence stresses $\sigma_E := (\sum_{i,j=1}^{3}(\sigma_{ij}^{(D)})^2)^{1/2}$, with $\sigma^{(D)} := \sigma - \mathrm{tr}\,(\sigma)\mathbf{I}$ as the deviatoric stress tensor, are likely to occur such as the soldered joints between the copper blocks housing the IGBTs resp. power diodes and the first copper plate (soldered joint I) and between the copper ground plate and the copper plate above (soldered joint II) (cf. Figure 1). Figure 3 shows the simplicial triangulation of a realistic IHV-module as used as electrical drive in trams whereas Figure 4 diaplays the temperature distribution in the clipping plane corresponding to the soldered joint I. As expected, temperature peaks occur at the centers of the two parts of the joint located just below the copper housings. Figures 5 and 6 represent the equivalence stresses for the soldered joints I and II ranging from high stresses with a magnitude of several hundred MPa (red) to low stresses (blue). We see that for the soldered joint I high stresses appear at the centers whereas for the soldered joint II the maxima occur at the lateral boundary of the joint which complies well with experimental experience.

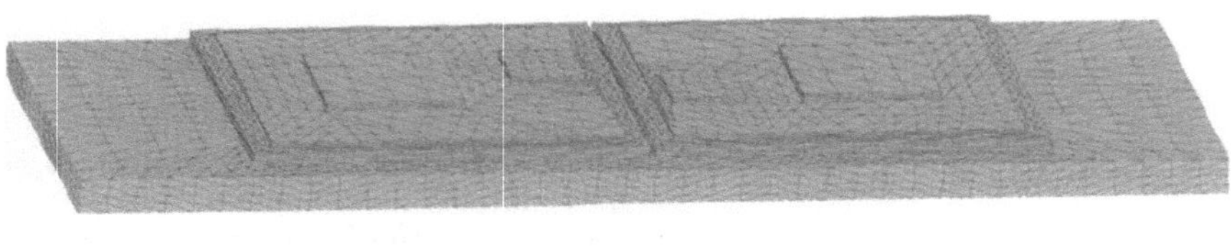

Fig. 3: Triangulation of the IHV-module

For individual bus bars of converter modules featuring several ports such as displayed in Figure 2b, the distributions of the scalar electric potential and the magnetic vector potential are of primary interst. Based on a simpilcial triangulation of the bus bar as shown in Figure 7, the corresponding results of numerical simulations are visualized in Figures 8 and 9. In particular, for a prespecified influx and outflux of currents at the individual ports, Figure 8 represents the distribution of the electric potential wheras Figure 9 displays the magnetic vector potential in a vicinity of two neighboring ports.

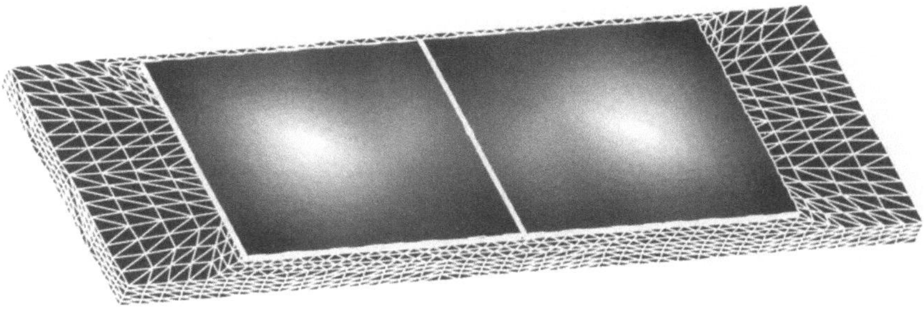

Fig. 4: Temperature distribution (cross section of soldered joint I)

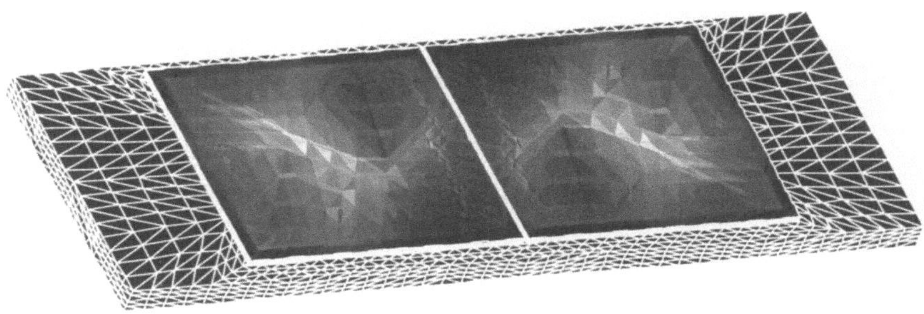

Fig. 5: Equivalence stress (cross section of soldered joint I)

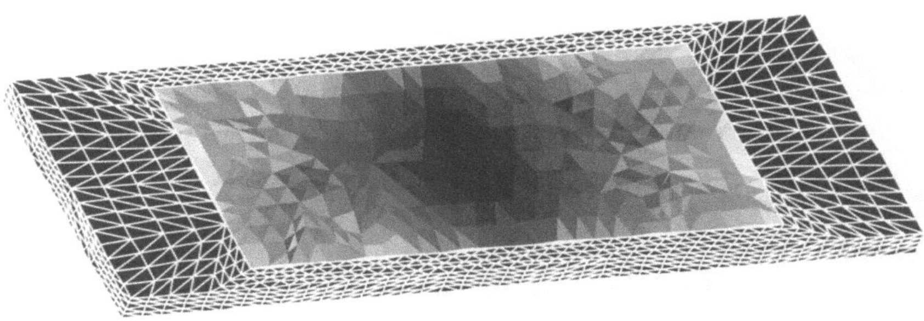

Fig. 6: Equivalence stress (cross section of soldered joint II)

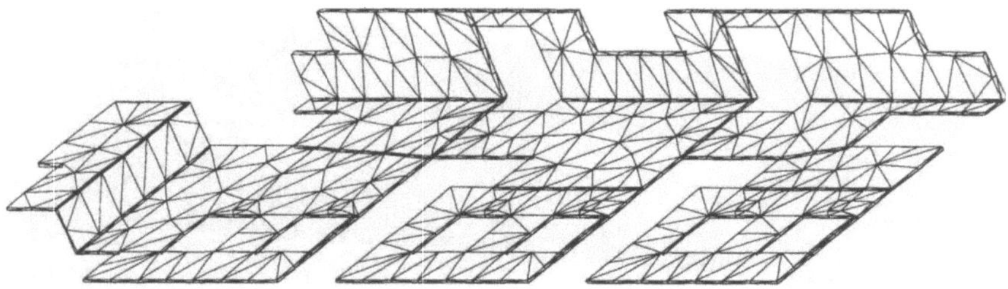

Fig. 7: Triangulation of an individual bus bar

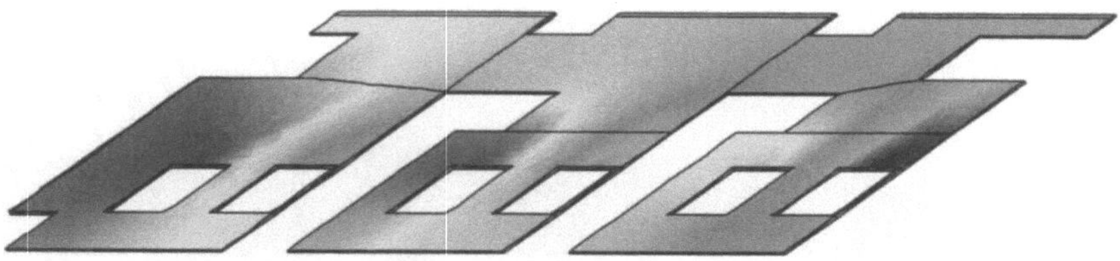

Fig. 8: Distribution of the electric potential

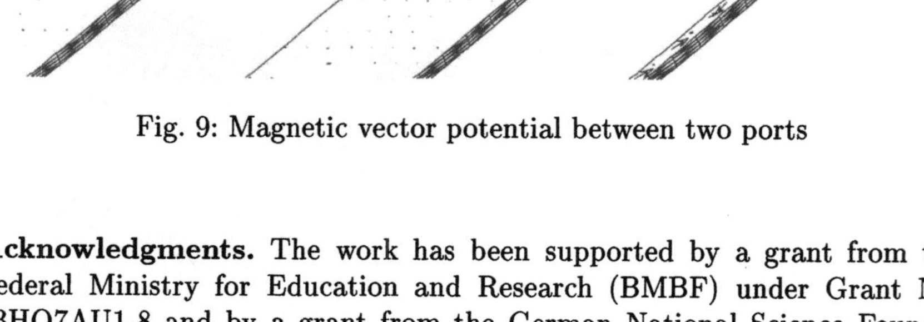

Fig. 9: Magnetic vector potential between two ports

Acknowledgments. The work has been supported by a grant from the Federal Ministry for Education and Research (BMBF) under Grant No. 03HO7AU1-8 and by a grant from the German National Science Foundation (DFG) under Grant No. HO877/4-1.

References

1. Agouzal, A. and Thomas, J.M.; Une méthode d'elements finis hybrides en decomposition de domains. Math. Modell. Numer. Anal. bf 29, 749-764 (1995)
2. Beck,R., Deuflhard,P., Hiptmair,R., Hoppe,R.H.W, and Wohlmuth,B.; Adaptive multilevel methods for edge element discretizations of Maxwell's equations. Surveys of Math. in Industry **8**, 271-312 (1999)
3. Beck, R., Hiptmair, R., Hoppe, R.H.W., and Wohlmuth, B.; Residual based a posteriori error estimators for eddy current computation. to appear in M^2AN Math. Modelling and Numer. Anal.
4. Ben Belgacem, F. and Maday, Y.; The mortar element method for three dimensional finite elements. *RAIRO* M^2AN **31** (1997), 289-302
5. Ben Belgacem, F. and Maday, Y.; The mortar and primal hybrid mortar finite element method of class $H(\text{curl})$. *Preprint*, Université Paul Sabatier, Toulouse, 1997
6. Ben Belgacem, F., Buffa, A., and Maday, Y.; The mortar finite element method for 3D Maxwell equations: First results. Report 99023, Laboratoire d'Analyse Numérique, Université Pierre et Marie Curie, Paris, 1999
7. Bernardi, C., Debit, N., and Maday, Y.; Coupling finite elements and spectral elements: First results. Math. Comp. **54**, 21-39 (1990)
8. Braess, D. and Dahmen, W.; Stability estimates of the mortar finite element method for 3-dimensional problems. East-West J. Numer. Math. **6**, 249-263 (1998)
9. Braess, D., Dahmen, W., and Wieners, C.; A multigrid algorithm for the mortar finite element method. Preprint, RWTH-Aachen, 1998 (submitted)
10. Dürndorfer, St., Gradinaru, V., Hoppe, R.H.W., König, E.-R., Schrag, G., and Wachutka, G.; Numerical simulation of microstructured semiconductor devices, transducers, and systems. In: High Performance Scientific and Engineering Computing. Proc. "Int. FORTWIHR-Symposium", Munich, March 1998 (Bungartz, H., Durst, F.,and Zenger, Chr.; eds.), pp. 309-323, Lecture Notes in Computational Science and Engineering Vol. **8**, Springer, Berlin-Heidelberg-New York, 1999
11. Engelmann,B., Hoppe,R.H.W., Iliash,Y., Kuznetsov,Y., Vassilevski, Y., and Wohlmuth,B.; Adaptive macro-hybrid finite element methods. In: Proc. 2nd European Conference on Numerical Methods (ENUMATH 97), Heidelberg, Sept. 29 - Oct. 3, 1997 (Bock,H.G.,Brezzi,F.,Glowinski,R.,Kanschat,G.Kuznetsov,Y., Périaux,J., and Rannacher,R.;eds.), pp. 294-302, World Scientific, Singapore, 1998
12. Engelmann,B., Hoppe,R.H.W., Iliash,Y., Kuznetsov,Y., Vassilevski, Y., and Wohlmuth,B.; Adaptive finite element methods for domain decomposition on nonmatching grids. to appear in: "Parallel Solution of PDEs", IMA Volume in Mathematics and its Applications, Springer, Berlin-Heidelberg-New York, 1999
13. Hiptmair,R. and Hoppe,R.H.W.; Mixed finite element discretization of continuity equations arising in semiconductor simulation. In: Proc. Conf. Math. Modeling and Simulation of Electrical Circuits and Devices, Oberwolfach, July 1992 (R.Bank, R.Bulirsch, H.Gajewski, and K.Merten; eds.), pp. 197-217, Birkhäuser, Basel, 1993
14. Hiptmair,R., Hoppe,R.H.W., and Wohlmuth,B.; Coupling problems in microelectronic device simulation. In: Notes on Numerical Fluid Mechanics. Vol. 51, Proc. Conf. 11th GAMM Seminar Kiel, (W.Hackbusch and G.Wittum; eds.), pp. 86-95, Vieweg, Braunschweig, 1995

15. Hoppe, R.H.W.; Mortar edge elements in R^3. East-West J. Numer. Anal. **7**, 159-173 (1999)
16. Hoppe, R.H.W.; Adaptive multilevel finite element methods in domain decomposition on nonmatching grids. to appear in Proc. "11th Int. Conf. on Domain Decomposition Methods", Greenwich, July 1998 (Cross, M. et al.; eds.), 1999
17. Hoppe, R.H.W.; Adaptive mortar edge elements in the computation of eddy currents. to appear in Proc. Conf. "Analysis and Approximation of Boundary Value Problems", Jyväskylä (Finland), October 1998 (Neittaanmäki, P. et al.; eds.), 1999
18. Hoppe, R.H.W., Iliash, Y., Kuznetsov, Y., Vassilevski, Y., and Wohlmuth, B.; Analysis and parallel implementation of adaptive mortar element methods. East-West J. Numer. Math. **6**, 223-248 (1998)
19. Markovich, P.A., Ringhofer, Chr.A., and Schmeiser, Chr.; Semiconductor Equations. Springer, Berlin-Heidelberg-New York, 1990
20. Nédélec, J.; Mixed finite elements in R^3. *Numer. Math.* **35** 1980, 315-341
21. Selberherr, S.; Analysis and Simulation of Semiconductor Devices. Springer, Berlin-Heidelberg-New York, 1984
22. Wachutka, G.; Rigorous thermodynamic treatment of heat generation and conduction in semiconductor device modeling. IEEE Trans. Computer-Aided Design of Integrated Circuits and Systems **T-CAD 9**, 1141-1149 (1990)
23. Wachutka, G.; Tailored modeling of miniaturized electrothermomechanical systems using thermodynamic methods. In: Micromechanical Systems, DSC - Vol. 40 (D. Cho et al.; eds.), pp. 183-198, American Society of Mechanical engineers, 1992

Matrix Compression for the Radiation Heat Transfer in Exhaust Pipes

M. Bebendorf and S. Rjasanow

Department of Mathematics, University of Saarland, Germany

Summary. A mathematical model for the heat transfer in a system of exhaust pipes of an automobile is described. The forced convection of the exhaust gas, the heat conduction and the heat transfer due to radiation are coupled. An effective numerical solution of the boundary integral equation for the radiation heat transfer is presented. The analytical form of the matrix entries for the collocation method is derived. Adaptive cross approximation is used for the compression of the system matrix. Some numerical examples for the matrix compression are presented.

1. Introduction

Catalytic converters play an important role in the reduction of the pollution gases in the exhaust of an automobile engine. Modern catalytic converters are very efficient, the reduction of the CO, NO_x and of the various hydrocarbons contained in the exhaust being more than 90%. However, during the warm-up cycle the temperature of the catalytic converter is too low and the efficiency of the converter can be further improved by shortening the warm-up cycle. There are different technical ideas for this purpose. One of the most popular is to use the insulated exhaust pipes first introduced in 1991 [1]. The insulating split between two pipes should prevent the energy produced by the engine from getting lost on the way to the catalytic converter. The three-dimensional geometry of an insulated exhaust pipe is presented in **Figure 1.1**. Because of current and future government regulations for emission control there is permanent urgency not only to improve the efficiency of exhaust systems but also to predict output during the development of new automobile models. Thus the mathematical modelling and numerical simulation of exhaust systems become more and more important. In the previous paper [6] we discussed the simplified situation of the straight insulated exhaust pipe. The special structure of the problem leads to special properties of the discrete system, i.e. the matrix is of Toeplitz structure. Using these properties we have designed a fast numerical algorithm for the numerical simulation of the warm-up cycle. In this paper we consider the most general situation of pipes having the geometry which is really used in industry.

The paper is organised as follows. In Section 2 we give a short description of the mathematical model for the heat transfer in an insulated exhaust pipe. In Section 3 we give a short review of the different approximation techniques for full dense matrices. In Section 4 we present the main ideas of our new approximation technique, called Adaptive Cross Approximation, and finally, in Section 5, we present the results of computations and draw some conclusions.

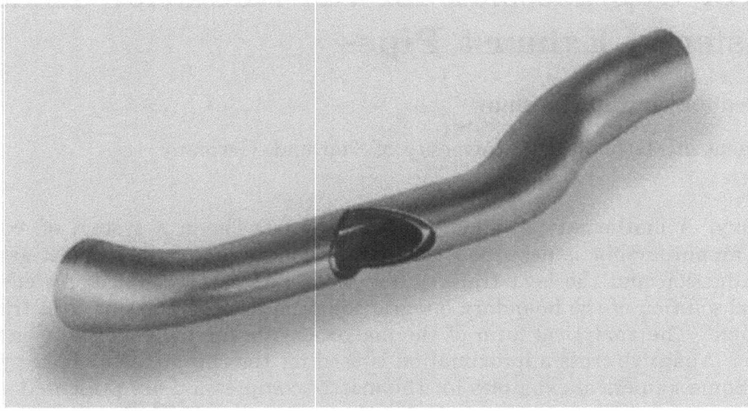

Fig. 1.1. The three-dimensional geometry of an insulated exhaust pipe

2. Mathematical model

We consider an insulated exhaust pipe of length L having the cross-section depicted in **Figure 2.1**. We consider the following physical processes in the

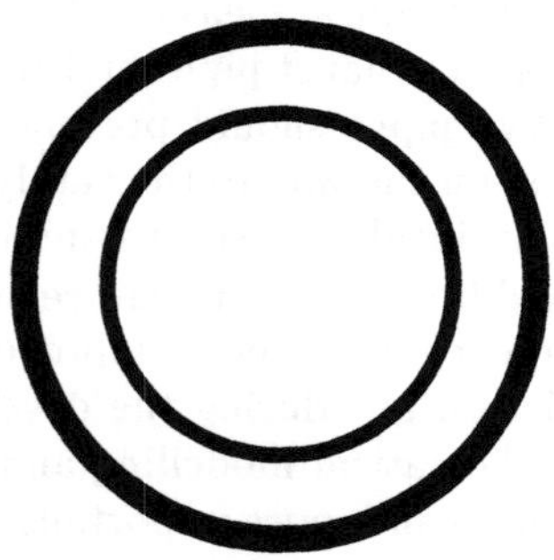

Fig. 2.1. Cross-section geometry of an insulated exhaust pipe

pipe:

1. Heat transfer due to the forced convection in the inner pipe;
2. Heat transfer due to conduction within and between both walls of the exhaust pipe;
3. Heat transfer due to radiation in the insulating split.

The exact description of the first two equations, the corresponding boundary and initial conditions as well as the standard numerical procedures can be found in our previous paper [6]. In this paper we concentrate on the numerical solution of the radiation heat transfer equation which can be written in the following form:

$$((\mathcal{I} - (1-\epsilon)\mathcal{B})\, R(x))\,(y) = (1-\epsilon)\left(\mathcal{B}^2 U(x)\right)\right)(y). \tag{2.1}$$

Here $R(x), x \in \Gamma$ denotes the reflected part of the radiation energy on the inner surfaces of the insulating split Γ the integral operator $\mathcal{B}$ is defined as

$$(\mathcal{B}R(x))\,(y) = \frac{1}{\pi}\int_{\Gamma} \kappa(x,y)\frac{(n_x, y-x)(n_y, x-y)}{|x-y|^4} R(x)\, ds_x\,, y \in \Gamma \tag{2.2}$$

whereby the function $U(x)$ is an abbreviation of

$$U(x) = \epsilon\,\sigma\, T^4(x) \;\; x \in \Gamma. \tag{2.3}$$

The constant ϵ is the emissivity of the steel and σ denotes the Boltzmann constant

$$\sigma = 5.669 \cdot 10^{-8}. \tag{2.4}$$

The numerical solution of equation (2.1) begins with the discretisation of the surface Γ using a system of plane, triangle panels

$$\Gamma \approx \Gamma_h = \bigcup_{j=1}^{N} \Gamma_j\,. \tag{2.5}$$

The centres of the mass of the panels Γ_i build the system of collocation points y_i

$$y_i = \frac{1}{3}\left(x_i^{(1)} + x_i^{(2)} + x_i^{(3)}\right)\,,\; i = 1, \ldots N\,. \tag{2.6}$$

The piecewise constant ansatz r for the unknown function $R(x)$ and u for the function $U(x)$ leads to the following system of collocation equations

$$A\,r = f\,,\; A \in \mathbb{R}^{N\times N}\,,\; r, f \in \mathbb{R}^N\,. \tag{2.7}$$

The matrix A and the right hand side of this system are of the following form:

$$A = I - (1-\epsilon)B\,,\; f = (1-\epsilon)B^2 u\,. \tag{2.8}$$

The elements of the matrix B are defined as

$$b_{i,j} = \frac{1}{\pi}\int_{\Gamma_j} \kappa(x, y_i)\frac{(n_x, y_i - x)(n_{y_i}, x - y_i)}{|x - y_i|^4}\, ds_x. \tag{2.9}$$

If the triangle Γ_j is completely visible from the collocation point y_i then the corresponding element of the matrix B can be computed analytically in the following way. We first consider the situation where the projection of the point y_i into the plane defined by the triangle Γ_j coincides with one of its edges. First we rotate and translate the system of coordinates in such a way

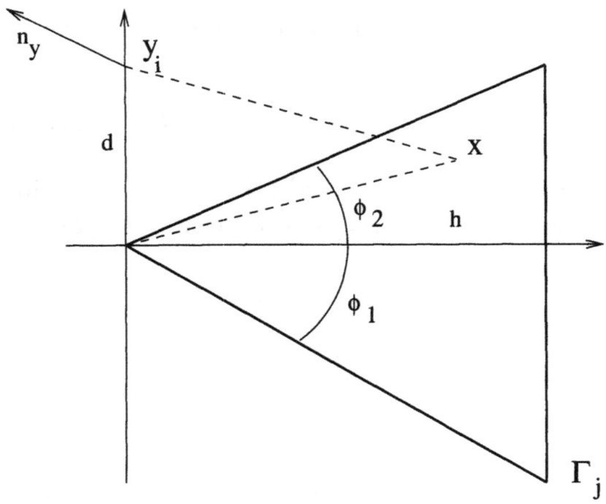

Fig. 2.2. Computation of the element of the matrix

that the origin coincides with a projection of the point y_i, the x_1-axis is directed along the height of the triangle and the x_3-axis is directed along the vector n_x. Let us denote the corresponding rotation matrix by V. In these new coordinates, the unit normal vectors n_x and n_y take the following form:

$$Vn_x = (0,0,1)^T\,,\; Vn_{y_i} = (\alpha_1, \alpha_2, \alpha_3)^T\,. \tag{2.10}$$

This situation is depicted in **Figure 2.2.** The integral will be evaluated in polar coordinates in the plane of the triangle Γ_j. Using the notation from **Figure 2.2** we obtain

$$b_{i,j} = \frac{d}{\pi} \int\limits_{\phi_1}^{\phi_2} \int\limits_{0}^{\frac{h}{\cos(\phi)}} \frac{\alpha_1 \rho \cos(\phi) + \alpha_2 \rho \sin(\phi) - \alpha_3 d}{(\rho^2 + d^2)^2}\, \rho \, d\rho d\phi. \tag{2.11}$$

After simple calculations we obtain the following expression for $b_{i,j}$.

$$\begin{aligned} b_{i,j} \;&=\; -\frac{1}{2\pi}\frac{\alpha_1 d + \alpha_3 h}{\sqrt{d^2+h^2}} \left[\arctan\left(h^{-1}\left(\sqrt{d^2+h^2}\tan(\frac{\phi}{2}) + d\right)\right)\right]_{\phi_1}^{\phi_2} \\ &-\; \frac{1}{2\pi}\frac{\alpha_1 d + \alpha_3 h}{\sqrt{d^2+h^2}} \left[\arctan\left(h^{-1}\left(\sqrt{d^2+h^2}\tan(\frac{\phi}{2}) - d\right)\right)\right]_{\phi_1}^{\phi_2} \\ &+\; \frac{1}{2\pi}\left[\arctan\left(\frac{h}{d\cos(\phi)}\right)(\alpha_1 \sin(\phi) - \alpha_2\cos(\phi))\right]_{\phi_1}^{\phi_2}. \end{aligned} \tag{2.12}$$

If the projection of the point y_i does not coincide with one of the edges of the triangle Γ_j then we can always reduce the calculation of the integral to the calculation of the three integrals of the previous type. Thus we obtain the analytical result in any case. If the triangle Γ_j is not completely visible from y_i then the visible part has a polygonal form and therefore the integral can be computed using some decomposition of the visible part into a set of triangles. This can be a very complicated and rather technical task.

3. Approximation of the full dense matrices

The system matrix A of the usual boundary element method has the following properties. The matrix is dense, i.e. the memory requirements are $Mem(A) = O(N^2)$. N denotes the size of the matrix here. The matrix can be asymptotically badly conditioned, i.e. in many applications the condition number is $\kappa(A) = O(N)$. The numerical solution of the resulting system of linear equations $Ay = b$ using classical direct solvers is extremely expensive, i.e. the numerical work is $Op(y = A^{-1}b) = O(N^3)$. The only possible increase in the efficiency of the boundary element methods is due to using iterative methods for the resulting system of linear equations. Two main ingredients are needed for the iterative methods. The first is an effective preconditioner B with $\kappa(B^{-1}A) = O(1)$. The second is an effective procedure for the fast matrix-vector multiplication with $Op(As) = O(N)$. These two problems have been of continuous interest in research over the last 15 years.

There are two main preconditioning ideas. The first is based on the spectral equivalence of the boundary element matrices constructed using the same discretisation of the same problem on different but topologically equivalent surfaces. Thus one can choose some structured surface (i.e. surface of revolution) to obtain a specially structured system matrix (i.e. multi-level Toeplitz-like matrix) (see [5],[7],[8]). Since such matrices can be handled very efficiently, the optimal preconditioner is constructed. There are clear disadvantages of this preconditioning. It is not easy to construct a well-structured surface if the given surface is complicated (i.e. multiple connected, etc.). The preconditioning will also require uniform discretisation on the structured domain and it can come into conflict with the necessity of the adaptive discretisation of the given domain. The second preconditioning strategy is based on using the operator of the opposite order on the same surface and using the same discretisation (see [12]). It is much more convenient for the majority of problems but the application of the preconditioning matrix is now as expensive as the matrix-vector multiplication itself. Thus research on fast matrix-vector multiplication becomes even more important.

The first fast matrix-vector multiplication tool, **the Fast Multipole Method**, was presented in 1985 by V. Rokhlin [13],[14]. The problem was to compute the force applied to each of the large number of charged particles in each time step. The main ideas here were to divide the whole space into a near and a far field and to cluster particles in the far field. The Fast Multipole Method is widely used for many applications (see also [15]). **The Panel**

Clustering algorithms for boundary integral equations, first published by Hackbusch and Novak [4], are based on very similar ideas.

4. Adaptive cross approximation

4.1 Matrix partitioning

Let $P = \{\Gamma_1, \dots, \Gamma_N\}$ be the set of panels. If we use a quadrature formula to approximate the integrals, the approximant's properties with respect to availability of low-rank approximants come from the kernel. Thus we concentrate on the investigation of the matrices

$$A \in \mathbb{R}^{N \times N}, \quad a_{ij} = \kappa(x_i, x_j).$$

We first subdivide the index set $I \times I$, where $I = \{1, \dots, N\}$, into subsets $t_1^i \times t_2^i$, $i = 1, \dots, n$ so that with $S_t = \{x_j,\, j \in t\}$ either the admissibility condition

$$\operatorname{diam} S_{t_2^i} \leq \eta \operatorname{dist}(S_{t_1^i}, S_{t_2^i})$$

is fulfilled or one of the index sets t_1^i and t_2^i has just one element.

In [4] a set of clusters $\mathcal{T}$ that possesses a tree structure is used to suitably subdivide the set P with respect to a fixed point. We will use a set of cluster pairs having a tree structure for the partitioning of $P \times P$ (the Cartesian product of the set of panels with itself). This set $\mathcal{T}'$ is constructed from the set $\mathcal{T}$ by applying the following recursion to (Γ_h, Γ_h). Take a cluster pair (t_1, t_2), $t_1, t_2 \in \mathcal{T}$. If t_1 and t_2 both have children t_{11}, t_{12} and t_{21}, t_{22} in $\mathcal{T}$ respectively, then assign the pairs (t_{11}, t_{21}), (t_{11}, t_{22}), (t_{12}, t_{21}) and (t_{12}, t_{22}) as children to the cluster pair (t_1, t_2) and add them to $\mathcal{T}'$. Now repeat these steps with each child. It is obvious that the covering of $\Gamma_h \times \Gamma_h$ with the smallest number of cluster pairs can be computed by the following algorithm.

Set $D = \emptyset$ and call *Partition*$((I, I), D)$, where *Partition* is the following recursive procedure.

procedure *Partition*$((t_1, t_2), D)$
  if (t_1, t_2) is admissible
  then $D := D \cup \{(t_1, t_2)\}$
  else apply the procedure to each of the children of (t_1, t_2) in $\mathcal{T}'$.

Each generated index pair $(t_1, t_2) \in D$ is admissible. Recently Hackbusch and Khoromskij [2], [3] have proven that the storage requirement for such matrices is $O(N \log N)$. However, they did not investigate the dependency on the parameter η which is essential for our algorithm since the approximation error will be controlled by this parameter.

In [9] it is shown that under the assumption of a quasi-uniform panel set P, i.e. there is a constant c_u so that

$$c_u \text{diam}\, \Gamma_j > h, \text{ for all } \Gamma_j \in P,$$

where $h = \max_{\Gamma_j \in P} \text{diam}\, \Gamma_j$, for the number of generated blocks $\mathcal{N}$ it holds that

$$\mathcal{N} = O(N^{1+\alpha}\eta^{-2(d-1)}), \ \forall \alpha > 0 .$$

The partitioning of the matrix can be done so that the number of operations for this purpose is of the same order as $\mathcal{N}$.

4.2 Low-rank approximation

If the generating function $\kappa : \mathbb{R}^d \times \mathbb{R}^d \to \mathbb{R}$ is asymptotically smooth (see [10]), i.e. there are constants c_1, $c_2 > 0$ and $g \in \mathbb{R}$ so that for any multi-index $\alpha \in \mathbb{N}_0^d$

$$|\partial_y^\alpha \kappa(x,y)| \leq c_1\, p!\, c_2^p\, |x-y|^{g-p}, \quad p = |\alpha|,$$

it can be shown using a Taylor expansion that an admissible block can be well approximated by low-rank matrices. It is known that the best low-rank approximant is obtained by using singular value decomposition. But for SVD it is necessary to generate the whole block, which finally leads to a complexity of order N^2.

In [11] a result on the existence of low-rank approximants using only a small part of the original matrix can be found. In this result the so-called maximum volume submatrices are used. But to find them within a block in general all entries have to be investigated. In [9] we proposed an algorithm that computes the same kind of approximant but uses only a small part of the original matrix.

Assume that we have an admissible block of dimension $m \times n$. For $k = 1, 2, \ldots$ calculate

$$\begin{aligned} (\widetilde{u}_k)_i &= \kappa(x_i, y_{j_k}) - \sum_{l=1}^{k-1} (v_l)_{j_k} (u_l)_i, \quad i = 1, \ldots, m \\ u_k &= (\widetilde{u}_k)_{i_k}^{-1}\, \widetilde{u}_k \\ (v_k)_j &= \kappa(x_{i_k}, y_j) - \sum_{l=1}^{k-1} (u_l)_{i_k} (v_l)_j, \quad j = 1, \ldots, n, \end{aligned}$$

where in each step j_k is chosen so that $\widetilde{u}_k \neq 0$ and i_k so that $|(\widetilde{u}_k)_{i_k}| \geq |(\widetilde{u}_k)_i|$. For the approximant S_k it holds that

$$S_k = \sum_{l=1}^{k} u_l v_l^T .$$

Thus for the whole approximation we need only the evaluation of $\kappa(x_i, y_{j_k})$, $i = 1, \ldots, m$, and $\kappa(x_{i_k}, y_j)$, $j = 1, \ldots, n$. The rest are algebraic transformations, which are easy to implement, whereby it should be remembered that the entries of u_k at the positions i_l and the entries of v_k at the positions j_l, $l < k$ are zero.

To obtain S_k we need $(m+n)k$ units of storage and $O((m+n)k^2)$ operations to generate the approximant S_k.

In [9] it has been shown that using this method to generate an approximant of accuracy ε, i.e. $\|A - \widetilde{A}\|_F \leq \varepsilon$, we need

$$\mathcal{N}_{st} = O(N^{1+\alpha}\varepsilon^{-\alpha})$$

storage and the number of operations for a matrix-vector multiplication and for the generation of the approximant must not exceed $O(N^{1+\alpha}\varepsilon^{-\alpha})$.

Although we will use a bounded number of iteration steps it is useful to stop the algorithm if a prescribed accuracy is reached. Since we do not want to calculate the whole original matrix we cannot compute the error exactly. Thus a good approximation for the error is the only way to control the algorithm.
Since

$$R_{n_{p+1}} = R_{n_p} - \sum_{l=n_p+1}^{n_{p+1}} u_l v_l^T$$

and $\|R_{n_{p+1}}\|_F$ is of one order smaller than $\|R_{n_p}\|_F$ the value

$$\| \sum_{l=n_p+1}^{n_{p+1}} u_l v_l^T \|_F$$

may be used as a good approximation to $\|R_{n_p}\|_F$.
When computing $\| \sum_{l=n_p+1}^{n_{p+1}} u_l v_l^T \|_F$ it should be noted that

$$\| \sum_{l=n_p+1}^{n_{p+1}} u_l v_l^T \|_F^2 = \sum_{l,k=n_p+1}^{n_{p+1}} \left(\sum_{i=1}^{m} (u_l)_i (u_k)_i \right) \left(\sum_{j=1}^{n} (v_l)_j (v_k)_j \right).$$

The evaluation of the last expression can be made in $(m+n)(r_p^d)^2$ operations.

5. Numerical examples

We begin this subsection with the following trivial situation. Let Γ be the inner surface of the unit sphere

$$\Gamma = \left\{ x \in \mathbb{R}^3 \, , \; |x| = 1 \right\}.$$

Then the kernel

$$\frac{1}{\pi} \frac{(n_x, y - x)(n_y, x - y)}{|x - y|^4}$$

of the radiation heat transfer equation can be simplified as follows:

$$n_x = -x \,, \; n_y = -y$$

and therefore

$$\frac{1}{\pi}\frac{(n_x, y-x)(n_y, x-y)}{|x-y|^4} = \frac{1}{\pi}\frac{(1-(x,y))(1-(y,x))}{(2-2(x,y))^2} = \frac{1}{4\pi}.$$

Thus, any appropriate discretisation of this problem will lead to the matrix B of the rank one.

For the numerical tests of the Adaptive Cross Approximation procedure we will consider a geometry of a simple pipe given by the following parametric representation

$$\Gamma = \left\{ x \in \mathbb{R}^3 \,, \; x = \begin{pmatrix} \cos(2\pi\, t) \\ \sin(2\pi\, t) \\ 4\, z \end{pmatrix} \,, \; (t,z) \in (0,1)^2 \right\} .$$

The panels are plane triangles appearing after the canonical discretisation of the parameter domain $(0,1)^2$ using n_t and n_z discretisation points. The whole number of panels is then $N = 2\, n_t\, n_z$. In the following table we present the results of the matrix compression with accuracy $\varepsilon = 10^{-6}$ in the Frobenius norm achieved by the Adaptive Cross Approximation procedure. The first column of this matrix contains the dimension of the matrix N, the second column gives the amount of memory in megaabytes which would be needed without the approximation, the third the corresponding value for the compressed matrix, and the last the percentage of memory saved.

Table 5.1. The results of the compression

Dimension	MBytes	MBytes	Percentage
512	2	2	100%
1024	8	5.0	63%
2048	32	12.6	39.5%
4096	128	27.4	21.4%
8192	512	63.6	12.4%
16384	2048	163.8	8.0%
32768	8192	380.9	4.6%

References

1. T. Nording. Neuartiges Konzept für Abgaskrümmer, Vorrohre und Katalysatoren. *Motortechnische Zeitschrift*, 52: 206–210, 1991.

2. W. Hackbusch. A sparse matrix arithmetic based on $\mathcal{H}$-matrices. I. Introduction to $\mathcal{H}$-matrices. *Computing*, 62(2): 89–108, 1999.
3. W. Hackbusch and B. N. Khoromskij. A Sparse $\mathcal{H}$-Matrix Arithmetic. Part II: Application to Multi-Dimensional Problems. Technical report, Max-Planck-Institut, Leipzig, 1999.
4. W. Hackbusch and Z. P. Nowak. On the fast matrix multiplication in the boundary element method by panel clustering. *Numer. Math.*, 54(4): 463–491, 1989.
5. S. Rjasanow. Effective Algorithms with Block Circulant Matrices. *Linear Algebra Appl.*, 202: 55-69, 1994.
6. S. Rjasanow. Heat Transfer in an Insulated Exhaust Pipe. *Journal Eng. Math.*, 29: 33–49, 1995.
7. S. Rjasanow. Optimal preconditioner for boundary element formulation of the Dirichlet problem in elasticity. *Math. Methods in Appl. Sciences*, 18: 603–613, 1995.
8. S. Rjasanow. The Structure of the Boundary Element Matrix for the Three-dimensional Dirichlet Problem in Elasticity. *Num. Linear Algebra Appl.*, 5(3): 203–217, 1998.
9. M. Bebendorf. Low-rank approximation of boundary element matrices. to appear in Numerische Mathematik.
10. A. Brandt. Multilevel computations of integral transforms and particle interactions with oscillatory kernels. *Comput. Phys. Comm.*, 65: 24–38, 1991.
11. S. A. Goreinov, E. E. Tyrtyshnikov and N. L. Zamarashkin. A theory of pseudoskeleton approximations. *Linear Algebra Appl.*, 261: 1–21, 1997.
12. Steinbach, O. and Wendland, W. L. The construction of some efficient preconditioners in the boundary element method. *Numerical treatment of boundary integral equations. Adv. Comput. Math.* 9(1-2): 191–216, 1998.
13. Rokhlin, V. Rapid solution of integral equations of classical potential theory. *J. Comput. Phys.*, 60: 187–207, 1985.
14. Greengard, L. and Rokhlin, V. A fast algorithm for particle simulations. *J. Comput. Phys.*, 73: 325–348, 1987.
15. Greengard, L. and Rokhlin, V. A new version of the fast multipole method for the Laplace equation in three dimensions. *Acta numerica, Cambridge* : 229–269, 1997.

An Iterative Substructuring Method with Lagrange Multipliers for Elasticity Problems Using Approximate Neumann Subdomain Solvers

A. Klawonn

SCAI - Institute for Algorithms and Scientific Computing
GMD - German National Research Center for Information Technology
Schloss Birlinghoven, D-53754 Sankt Augustin, Germany.
E-mail: klawonn@gmd.de, URL: http://www.gmd.de/SCAI/people/klawonn.

Summary. The FETI algorithms are among the best tested domain decomposition methods for elliptic partial differential equations with many applications in structural mechanics. A new algorithm which allows for approximate Neumann subdomain solves and is based on an efficient member of this class of preconditioners, is presented. Finally, numerical results are given for a plane stress cantilever model problem indicating the scalability of the new algorithm.

1. Introduction

The **F**inite **E**lement **T**earing and **I**nterconnecting (FETI) method is a computationally very efficient and well tested domain decomposition method with Lagrange multipliers, introduced by Farhat and Roux [3]. In a variant of the FETI method, presented in Farhat, Mandel, and Roux [2], a Neumann and a Dirichlet finite element problem is solved exactly on each subdomain, in each iteration. Reformulating the preconditioned system of the FETI method as a saddle point problem with both, primal and dual variables as unknowns, we present a new domain decomposition method based on a block–triangular preconditioner which gives the opportunity to use approximate subdomain solvers. For a related approach based on block–diagonal preconditioners, see Klawonn and Widlund [5].

The equations of linear elasticity model the displacement of a linear elastic material under the action of external and internal forces. We denote the elastic body by $\Omega \subset \mathbf{R}^d, d = 2, 3$, and its boundary by $\partial\Omega$ and assume that one part of the boundary, Γ_0, is clamped, i.e. with homogeneous Dirichlet boundary conditions, and that the rest, $\Gamma_1 := \partial\Omega \setminus \Gamma_0$, is subject to a surface force $\mathbf{g}$, i.e. a natural boundary condition. We can also introduce an internal volume force $\mathbf{f}$, e.g. gravity. The appropriate space for a variational formulation is the Sobolev space $H^1_{\Gamma_0}(\Omega) := \{\mathbf{v} \in H^1(\Omega)^d \ : \ \mathbf{v}_{|\Gamma_0} = 0\}$. The linear elasticity problem consists in finding the displacement $\mathbf{u} \in H^1_{\Gamma_0}(\Omega)$ of the elastic body Ω, such that

$$G \int_\Omega \varepsilon(\mathbf{u}) : \varepsilon(\mathbf{v}) dx \ + \ G\beta \int_\Omega \operatorname{div} \mathbf{u} \operatorname{div} \mathbf{v} \, dx = \langle \mathbf{F}, \mathbf{v} \rangle \quad \forall \mathbf{v} \in H^1_{\Gamma_0}(\Omega). \quad (1.1)$$

Here G and β are material parameters depending on the Poisson ratio ν and Young's modulus E. In the case of plane stress, we have $G = E/(1+\nu), \beta = 1/(1-\nu^2)$ and for plane strain and three dimensional elasticity, we have $G = E/(1+\nu), \beta = \nu/(1-2\nu)$. Furthermore, $\varepsilon_{ij}(\mathbf{u}) := \frac{1}{2}(\frac{\partial u_i}{\partial x_j} + \frac{\partial u_j}{\partial x_i})$ is the linearized strain tensor, and $\varepsilon(\mathbf{u}) : \varepsilon(\mathbf{v}) = \sum_{i,j=1}^{d} \varepsilon_{ij}(\mathbf{u})\varepsilon_{ij}(\mathbf{v})$, $\langle \mathbf{F}, \mathbf{v} \rangle := \sum_{i=1}^{d} \int_{\Omega} f_i v_i \, dx + \sum_{i=1}^{d} \int_{\Gamma_1} g_i v_i \, d\sigma$. The associated bilinear form $a(\mathbf{u}, \mathbf{v})$ is defined as the left hand side of equation (1.1). In this article, we only consider the case of compressible elasticity, i.e., the Poisson ratio ν is bounded away from 1/2. Thus, it also follows from Korn's first inequality that the bilinear form $a(\cdot,\cdot)$ is uniformly elliptic. We can therefore successfully discretize the system (1.1) with low–order, conforming finite elements, such as linear or bilinear elements. We assume that a shape regular triangulation τ^h of Ω is given with a typical element diameter of h. We denote by $\mathbf{W}^h(\Omega) \subset H^1_{\Gamma_0}(\Omega)$ the corresponding conforming space of finite element functions, e.g. piecewise linear or bilinear continuous functions.

2. The FETI method

In this section, we give a brief review of the original FETI method introduced in Farhat and Roux [3] and the variant with a Dirichlet preconditioner with multiplicity scaling discussed, e.g., in Rixen and Farhat [8]. The first Dirichlet preconditioner for FETI was defined in Mandel, Farhat, and Roux [2]. For more detailed descriptions and proofs, we refer to [1, 8, 4, 2, 7] and the references therein.

Let the domain $\Omega \subset \mathbf{R}^d$ be decomposed into N non-overlapping subdomains $\Omega_i, i = 1, \ldots, N$, each of which is the union of elements and such that the finite element nodes on the boundaries of neighboring subdomains match across the interface $\Gamma := \left(\bigcup_{i=1}^{N} \partial\Omega_i\right) \setminus \partial\Omega$. Let the corresponding conforming finite element spaces be $\mathbf{W}_i = \mathbf{W}^h(\Omega_i), i = 1, \ldots, N$, and let $\mathbf{W} := \prod_{i=1}^{N} \mathbf{W}_i$ be the associated product space. When it is necessary to use vectors of nodal values, which define the elements of a finite element space, we underline, e.g. $\underline{\mathbf{W}}$ is the product space that corresponds to $\mathbf{W}$. Analogously, we denote the vector of nodal values associated with the finite element function $\mathbf{u}$ by $\underline{\mathbf{u}}$.

For each subdomain $\Omega_i, i = 1, \ldots, N$, we assemble the local stiffness matrices K_i and local load vectors $\underline{\mathbf{f}}_i$ by integrating the appropriate expressions over individual subdomains. We denote the local vectors of nodal values by $\underline{\mathbf{u}}_i$. Let us also define a matrix $B = [B_1, \ldots, B_N]$ such that the values of a continuous vector $\underline{\mathbf{u}}$, that are associated with more than one subdomain, coincide when $B\underline{\mathbf{u}} = 0$; this enforces the intersubdomain continuity.

By introducing a vector of Lagrange multipliers $\underline{\lambda}$ to enforce the intersubdomain continuity constraint, we obtain the saddle point problem:

Find $(\underline{\mathbf{u}}, \underline{\lambda}) \in \underline{\mathbf{W}} \times \underline{\mathbf{U}}$, such that

$$\left.\begin{array}{rcl} K\underline{\mathbf{u}} + B^t\underline{\lambda} & = & \underline{\mathbf{f}} \\ B\underline{\mathbf{u}} & = & 0 \end{array}\right\}, \tag{2.1}$$

where $\underline{\mathbf{u}} = [\underline{\mathbf{u}}_1, \ldots, \underline{\mathbf{u}}_N]^t, \underline{\mathbf{f}} = [\underline{\mathbf{f}}_1, \ldots, \underline{\mathbf{f}}_N]^t$, and $K = diag_{i=1}^N K_i$ is a block–diagonal matrix. The local stiffness matrices K_i are positive semidefinite. The problem (2.1) is uniquely solvable if and only if $\ker(K) \cap \ker(B) = \{0\}$. This condition holds since the original finite element model is elliptic. Following Rixen and Farhat [8], we consider the case where a maximum number of redundant Lagrange multipliers is introduced, i.e., when all possible pairs of degrees of freedom of the primal variable $\underline{\mathbf{u}}$, that belong to the same nodal point x on the interface Γ, are connected by a Lagrange multiplier. We note that the solution $\underline{\lambda}$ of (2.1) is in general only unique up to an additive vector from $\ker(B^t)$. The space of Lagrange multipliers $\underline{\mathbf{U}}$ is therefore chosen as the $\mathrm{range}(B)$; see also discussion below.

We will also use a full rank matrix R, built from the rigid body motions on the interior subdomains, such that $\mathrm{range}(R) = \ker(K)$.

The solution of the first equation in (2.1) exists if and only if $\underline{\mathbf{f}} - B^t\underline{\lambda} \in \mathrm{range}(K)$; this constraint will lead to the introduction of a projection P. We obtain $\underline{\mathbf{u}} = K^\dagger(\underline{\mathbf{f}} - B^t\underline{\lambda}) + R\underline{\alpha}$ if $\underline{\mathbf{f}} - B^t\underline{\lambda} \perp \ker(K)$, where $K^\dagger$ is a pseudoinverse of K and $\underline{\alpha}$ has to be determined; see discussion below. Substituting $\underline{\mathbf{u}}$ into the second equation of (2.1) and considering the component orthogonal to BR, gives

$$\left.\begin{array}{rcl} PF\underline{\lambda} & = & P\underline{\mathbf{d}} \\ G^t\underline{\lambda} & = & \underline{\mathbf{e}} \end{array}\right\} \tag{2.2}$$

with $G := BR, F := BK^\dagger B^t, \underline{\mathbf{d}} := BK^\dagger\underline{\mathbf{f}}, P := I - G(G^tG)^{-1}G^t$, and $\underline{\mathbf{e}} := R^t\underline{\mathbf{f}}$. We note that P is an orthogonal projection from $\underline{\mathbf{U}}$ onto $\ker(G^t)$. Any solution $\underline{\lambda}$ of (2.1) and (2.2) yields the same solution $\underline{\mathbf{u}}$ of (2.1) if $\underline{\alpha} := -(G^tG)^{-1}G^t(\underline{\mathbf{d}} - F\lambda)$ is chosen; see Mandel, Tezaur, and Farhat [7, Theorem 2.4]. We define the space of admissible increments by $\underline{\mathbf{V}} := \ker(G^t)$. The original FETI method is a conjugate gradient method in the space $\underline{\mathbf{V}}$ applied to

$$PF\underline{\lambda} = P\underline{\mathbf{d}}, \qquad \underline{\lambda} \in \underline{\lambda}_0 + \underline{\mathbf{V}} \tag{2.3}$$

with an initial approximation $\underline{\lambda}_0$ chosen such that $G^t\underline{\lambda}_0 = \underline{\mathbf{e}}$.

We consider now a preconditioned variant, the Dirichlet preconditioner with multiplicity scaling, see, e.g., Rixen and Farhat [8] and Farhat, Mandel, and Roux [2] for an earlier variant. To define the preconditioner, we first introduce scaling matrices D_i, that operate on the Lagrange multiplier space, as follows: For a nodal point x on $\partial\Omega_i \cap \partial\Omega_j$, consider the Lagrange multiplier connecting the corresponding two degrees of freedom in $\underline{\mathbf{W}}_i$ and $\underline{\mathbf{W}}_j$, respectively. Then, the diagonal entry of D_i for that point is chosen as the reciprocal of its multiplicity, i.e. the number of subdomains it belongs to. We can now define a scaled jump operator $B_D := [D_1B_1, \ldots, D_NB_N]$, and the FETI preconditioner M^{-1} by

$$M^{-1} = B_D \begin{bmatrix} O & O \\ O & S \end{bmatrix} B_D^t$$

with the block–diagonal matrix $S = diag_{i=1}^N S_i$. The matrix S is the Schur complement of K obtained by eliminating the interior degrees of freedom of each of the subdomains. This computation can clearly be carried out in parallel and results in the block–diagonal matrix S which only operates on the degrees of freedom on the subdomain boundaries. In the application of M^{-1} to a vector, N independent Dirichlet problems have to be solved in each iteration step; it is therefore often called the Dirichlet preconditioner (with multiplicity scaling (MS)). This preconditioner was introduced by Rixen and Farhat in [8, section 5] in a framework of mechanically consistent preconditioners. A convergence analysis will be given in Klawonn and Widlund [6].

To keep the search directions of this preconditioned conjugate gradient method in the space $\underline{\mathbf{V}}$, the application of the preconditioner M^{-1} has to be followed by another application of the projection P. Hence, the Dirichlet variant of the FETI method is the conjugate gradient algorithm applied to

$$PM^{-1}PF\underline{\lambda} = PM^{-1}P\underline{\mathbf{d}}, \qquad \underline{\lambda} \in \underline{\lambda}_0 + \underline{\mathbf{V}}. \tag{2.4}$$

It is well known that a convergence bound for the conjugate gradient method is determined by a condition number estimate of the preconditioned system. The next theorem is proven in Klawonn and Widlund [6] for scalar elliptic equations of second order in three dimensions; analogous arguments can be used in two dimensions. Combining it with the techniques used in [5], extends the result to linear elasticity.

Theorem 2.1. *The spectral condition number κ of the FETI method defined by the Dirichlet preconditioner M satisfies*

$$\kappa(PM^{-1}PF) \leq C\,(1 + \log(H/h))^2.$$

Here, C is independent of h and H.

Let us remark that the algorithm considered in this section can be enhanced, such that its convergence rate is still optimal for problems with discontinuous coefficients, cf. [1] and, for a convergence analysis, [6]. Note that another interesting, closely related FETI algorithm for non–redundant Lagrange multipliers is presented and analyzed in [6].

3. The new domain decomposition method with approximate subdomain solves

Using the decomposition $\underline{\lambda} = \underline{\lambda}_0 + \underline{\mu}$, with $\underline{\mu} \in \underline{\mathbf{V}}$, we can rewrite (2.3) as

$$PBK^{\dagger}B^t\underline{\mu} = PBK^{\dagger}(\underline{\mathbf{f}} - B^t\underline{\lambda}_0). \tag{3.1}$$

Since $\underline{\mathbf{u}} = K^\dagger(\underline{\mathbf{f}} - B^t\underline{\lambda}) + R\underline{\alpha}$, we see, by using $\underline{\mu} \in \underline{\mathbf{V}}$ i.e. $P\underline{\mu} = \underline{\mu}$, that the solution of (3.1), can also be obtained by solving

$$\begin{bmatrix} K & (PB)^t \\ PB & O \end{bmatrix} \begin{bmatrix} \underline{\mathbf{u}} \\ \underline{\mu} \end{bmatrix} = \begin{bmatrix} \underline{\mathbf{f}} - B^t\underline{\lambda}_0 \\ 0 \end{bmatrix}. \tag{3.2}$$

We note that in this formulation we are not enforcing $B\underline{\mathbf{u}} = 0$ but only its projected version $PB\underline{\mathbf{u}} = 0$. The addition of an element of $\ker(K)$ does not change the solution $\underline{\mathbf{u}}$ of the first equation in (3.2). Since $B^t\underline{\mu} \perp \ker(K)$ for $\underline{\mu} \in \underline{\mathbf{V}}$, this is also true for the second equation in (3.2). We use this fact to post-process $\underline{\mathbf{u}}$, such that $B\underline{\mathbf{u}} = 0$ is finally satisfied. This can be done by setting $\underline{\mathbf{u}}_{cor} := \underline{\mathbf{u}} - R(G^tG)^{-1}G^tB\underline{\mathbf{u}}$; it easily follows that $B\underline{\mathbf{u}}_{cor} = PB\underline{\mathbf{u}} = 0$. Thus, we first compute the component of the solution in $\text{range}(K)$ and then add the correct null space component, such that the solution has no jumps across the interface.

To solve the saddle point problem (3.2), we suggest GMRES with a block–triangular preconditioner. Let us point out that other Krylov space methods, which are suitable for nonsymmetric systems, e.g., BICGSTAB, QMR, TFQMR, etc., can also be applied. We refer to section 4. for numerical experiments with other Krylov space methods. We note that our algorithm will be designed such that the first component of the iterates belongs to $\text{range}(K)$.

Our preconditioner has the form

$$\mathcal{B}_U = \begin{bmatrix} \widehat{K} & (PB)^t \\ O & -\widehat{M} \end{bmatrix} \text{ or } \mathcal{B}_L = \begin{bmatrix} \widehat{K} & O \\ PB & -\widehat{M} \end{bmatrix}.$$

The upper triangular preconditioner will be used for right preconditioning and the lower for preconditioning from the left. For simplicity, we describe our algorithm only for left preconditioning and drop the subscript L from now on. All the results easily carry over to the case of right preconditioning.

It is assumed that $\widehat{K}$ is symmetric and a good preconditioner for $K + D_H Q$, where $Q = diag_{i=1}^N Q_i$ is a block–diagonal matrix with Q_i the mass matrices associated with the mesh on Ω_i and $D_H = diag_{i=1}^N(H_i^{-2}I_i)$ a diagonal matrix. Here H_i denotes the diameter of the subdomain Ω_i. We further assume that $\widehat{M}$ is symmetric and a good preconditioner for M, i.e. we assume there exists constants m_0, k_0, k_1, m_0, m_1 with $0 < c \le m_0 \le m_1 \le C < \infty$ and $0 < c \le k_0 \le k_1 \le C < \infty$, such that

$$\begin{array}{lllllll} k_0\, \underline{\mathbf{u}}^t(K + D_HQ)\underline{\mathbf{u}} & \le & \underline{\mathbf{u}}^t\widehat{K}\underline{\mathbf{u}} & \le & k_1\, \underline{\mathbf{u}}^t(K + D_HQ)\underline{\mathbf{u}} & \forall\underline{\mathbf{u}} \in \underline{\mathbf{W}}, \\ m_0\, \underline{\lambda}^t M\underline{\lambda} & \le & \underline{\lambda}^t\widehat{M}\underline{\lambda} & \le & m_1\, \underline{\lambda}^t M\underline{\lambda} & \forall\underline{\lambda} \in \underline{\mathbf{V}}. \end{array} \tag{3.3}$$

Here c and C are generic constants independent of, or only weakly dependent on, the mesh size. Let us note that the quality of the preconditioners $\widehat{K}, \widehat{M}$ is reflected in the convergence behavior of the new algorithm, e.g., when $\widehat{K}$ is build using an ILU decomposition, the overall convergence rate of our new algorithm can depend on H/h which is a measure for the size of the subproblems; see section 4.. This can be improved by choosing a $\widehat{K}$ based on optimal preconditioners on each subdomain, e.g., multigrid or algebraic multigrid. We also note that all that is required here are good preconditioners for quite benign positive definite problems on individual subregions. In order to have a computationally efficient preconditioner, we must also assume that $\widehat{K}^{-1}$ and $\widehat{M}^{-1}$ can be applied to a vector at a low cost.

To guarantee that the iterates belong to range(K), we introduce the projection P_R onto range(K) by $P_R := I - R(R^tR)^{-1}R^t$. We recall that range$(R) = \ker(K)$ and note that R^tR is a block matrix with a 3×3 block for each interior subdomain; the expense of applying P_R to a vector is therefore very modest.

The resulting domain decomposition method is GMRES (or any other Krylov space method suitable for nonsymmetric systems, e.g. BICGSTAB, QMR, etc.), applied to the preconditioned system

$$\mathcal{B}^{-1}\mathcal{A}\underline{\mathbf{x}} = \mathcal{B}^{-1}\underline{\mathbf{F}}$$

with

$$\mathcal{A} = \begin{bmatrix} K & (PB)^t \\ PB & O \end{bmatrix}, \mathcal{B}^{-1} = \begin{bmatrix} P_R\widehat{K}^{-1}P_R^t & O \\ P\widehat{M}^{-1}P^tBP_R\widehat{K}^{-1}P_R^t & -P\widehat{M}^{-1}P^t \end{bmatrix}, \tag{3.4}$$

$\underline{\mathbf{x}} = [\underline{\mathbf{u}}, \underline{\mu}]^t, \underline{\mathbf{F}} = [\underline{\mathbf{f}} - B^t\underline{\lambda}_0, 0]^t$. We note that it is easy to see that only two matrix-vector products with the projection P and one with the projection P_R are required in each step. Let us also point out that the matrix–vector product of $\mathcal{B}^{-1}$ only needs one application each of $\widehat{K}^{-1}$ and $\widehat{M}^{-1}$ to a vector. The iterates of our algorithm obviously belong to $\underline{\mathbf{W}}_R \times \underline{\mathbf{V}}$ with $\underline{\mathbf{W}}_R :=$ range(K).

A theoretical analysis of the numerical scalability of this algorithm will be given elsewhere.

4. Numerical results

We applied our domain decomposition method to a plane stress problem described in section 1.. The Poisson ratio is $\nu = 0.3$ and the elasticity modulus $E = 2.1 \cdot 10^{11} N/m^2$, which models steel. The domain Ω is the unit square fixed on the left hand side and free on the other three edges except for the upper right corner, where we impose a non-homogeneous natural boundary condition in the form of a point force that has components in the positive x and y directions both equal to $10^5 N$. All our computations have been performed in Matlab 5.3. As Krylov space methods we use GMRES and BICGSTAB with a zero initial guess. The stopping criterion is $||\underline{\mathbf{r}}_n||_2/||\underline{\mathbf{r}}_0||_2 < 10^{-6}$, where $\underline{\mathbf{r}}_n$ and $\underline{\mathbf{r}}_0$ are the $n-$th and initial residual, respectively. Our domain Ω is decomposed into $N \times N$ square subdomains with $H := 1/N$. We carried out two different types of experiments for different combinations of preconditioners

$\widehat{K}$ and $\widehat{M}$, in order to analyze the numerical scalability of the method. In our first set of runs, we keep the dimension of the subproblems, and H/h, fixed and increase the number of subdomains and thus the overall problem size. Our second set of experiments is carried out with a fixed number of subdomains and increasing values of H/h resulting in an increased $1/h$. For both cases, we present results for $\widehat{M}$ constructed using the multiplicity scaling; cf. section 2.. To gain insight into the convergence behavior with inexact blocks $\widehat{K}$, we use preconditioners based on an incomplete Cholesky factorization (ILU). In the following, ILU(*tol*) is a threshold ILU factorization, with a threshold of *tol*, as provided in Matlab 5.3; any entry in a column of the Cholesky factor L is dropped if its magnitude is smaller than the drop tolerance *tol* times the norm of its column. Two different combinations are considered: 1) $\widehat{K} = K + \frac{1}{H^2}Q$ and $\widehat{M}^{-1} = M^{-1}$; 2) $\widehat{K}$ is built with ILU(10^{-3}) applied to $K + \frac{1}{H^2}Q$ and $\widehat{M}^{-1} = M^{-1}$. To obtain a first insight how the iteration counts of our new method compare to the original FETI algorithm, we also ran the conjugate gradient method to solve the preconditioned FETI system using the Dirichlet preconditioner with multiplicity scaling. We have stopped the iterations when the norm of the $k-$th projected residual was lower or equal than 10^{-6} times the norm of $\underline{\mathbf{d}}$.For the discussion of a stopping criterion for FETI often used in practice; cf. Farhat and Roux [4]. Let us point out that each iteration of BICGSTAB requires two matrix–vector products with $\mathcal{A}, \hat{\mathcal{B}}^{-1}$, respectively. Thus each BICGSTAB iteration is twice as expensive as one with GMRES. On the other hand, BICGSTAB only requires a three–term recursion, while GMRES in its present form uses full reorthogonalization. Other alternatives such as restarted or truncated GMRES are of course possible. Let us note that our results do not provide a thorough comparison, which should also include timings or flop counts, but primarily a first test of the numerical scalability of the new algorithm. All computational results are given in Tables 4.1 and 4.2.

Table 4.1. (I) : $\widehat{K} = K + 1/H^2 M_Q$ and $\widehat{M} = M$, (II) : $\widehat{K}$ =ILU(10^{-3}) of $K + 1/H^2 M_Q$ and $\widehat{M} = M$, and (III) : FETI with $\widehat{M} = M$.

H/h = 8		GMRES		BICGSTAB		CG
$1/h$	$1/H$	I	II	I	II	III
16	2	8	10	6	7	8
32	4	13	15	9	11	12
64	8	17	19	13	15	16
96	12	19	21	14	16	19
128	16	22	24	15	18	21

Table 4.2. (I) : $\widehat{K} = K + 1/H^2 M_Q$ and $\widehat{M} = M$, (II) : $\widehat{K}$ =ILU(10^{-3}) of $K + 1/H^2 M_Q$ and $\widehat{M} = M$, and (III) : FETI with $\widehat{M} = M$.

H = 1/4		GMRES		BICGSTAB		CG
$1/h$	H/h	I	II	I	II	III
16	4	11	14	8	10	11
32	8	13	15	9	11	12
64	16	15	23	11	15	14
96	24	16	28	12	17	15
128	32	17	35	14	24	16

Acknowledgments

The author would like to thank Olof Widlund for many helpful and interesting discussions and Daniel Rixen for parts of his Matlab routines.

References

1. Manoj Bhardwaj, David Day, Charbel Farhat, Michel Lesoinne, Kendall Pierson, and Daniel Rixen Application of the FETI method to ASCI problems: Scalability results on one thousand processors and discussion of highly heterogeneous problems University of Colorado at Boulder, Department of Aerospace Engineering, TR CU-CAS-99-05, March, 1999.
2. Charbel Farhat, Jan Mandel, and Francois-Xavier Roux. Optimal Convergence Properties of the FETI Domain Decomposition Method. *Comput. Methods Appl. Mech. Engng.*, 115:367–388, 1994.
3. Charbel Farhat and Francois-Xavier Roux. A Method of Finite Element Tearing and Interconnecting and its Parallel Solution Algorithm. *Int. J. Numer. Meth. Engng.*, 32:1205–1227, 1991.
4. Charbel Farhat and Francois-Xavier Roux. Implicit parallel processing in structural mechanics. *Computational Mechanics Advances*, 2(1):1–124, 1994.
5. Axel Klawonn and Olof B. Widlund. A domain decomposition method with Lagrange multipliers for linear elasticity. TR 780, Computer Science Department, Courant Institute, USA, February 1999. URL: `file://cs.nyu.edu/pub/tech-reports/tr780.ps.gz`.
6. Axel Klawonn and Olof B. Widlund. FETI and Neumann–Neumann Iterative Substructuring Methods: Connections and New Results. TR 796, Computer Science Department, Courant Institute, December 1999.
7. Jan Mandel, Radek Tezaur, and Charbel Farhat. A Scalable Substructuring Method by Lagrange Multipliers for Plate Bending Problems. *SIAM J. Numer. Anal.*, 36(5):1370–1391, 1999.
8. Daniel Rixen and Charbel Farhat. A simple and efficient extension of a class of substructure based preconditioners to heterogeneous structural mechanics problems. *Inter. J. Numer. Meth. Engrg.*, 44:489–516, 1999.

A New a Posteriori Error Estimator in Adaptive Direct Boundary Element Methods. The Neumann Problem

H. Schulz and O. Steinbach

Mathematisches Institut A, Universität Stuttgart, Germany

Summary. A new a posteriori error estimator for the Neumann problem with a second order elliptic partial differential operator solved by boundary element methods is proposed. Following the main ideas in [10] the method is based on an approximate solution of a second kind boundary integral equation by a Neumann series [11] in appropriate factor spaces to estimate the error of a previously computed solution of an arbitrary boundary element method with high accuracy.

1. Introduction

Let $\Omega \subset \mathbb{R}^n$ $(n = 2, 3)$ be a bounded domain with a Lipschitz continuous boundary Γ. For an elliptic and self–adjoint second order partial differential operator L and a corresponding conormal derivative operator $T_x, x \in \Gamma$, we consider the homogeneous Neumann boundary value problem

$$Lu(x) = 0 \quad \text{for } x \in \Omega\,, \quad (T_x u)(x) = g(x) \quad \text{for } x \in \Gamma\,. \tag{1.1}$$

Note that g has to satisfy the compatibility condition $\langle g, v\rangle_{L_2(\Gamma)} = 0$ for all $v \in \mathcal{R}$ where $\mathcal{R}$ is the solution space of the homogeneous Neumann boundary value problem, i.e.,

$$\mathcal{R} := \left\{ v \in H^{1/2}(\Gamma), v = \widetilde{v}_{|\Gamma}, \widetilde{v} \in H^1(\Omega) : L\widetilde{v} = 0 \text{ in } \Omega,\ T_x\widetilde{v} = 0 \text{ on } \Gamma \right\},$$

and that the solution u of (1.1) is only unique modulo $\mathcal{R}$.

We assume that there is given a fundamental solution $U^*(x, y)$ of L. Therefore we can write any solution of the boundary value problem (1.1) by using the representation formula as

$$u(\tilde{x}) = \int_\Gamma U^*(\tilde{x}, y) g(y) ds_y - \int_\Gamma u(y) T^*(\tilde{x}, y) ds_y \quad \text{for } \tilde{x} \in \Omega \tag{1.2}$$

with $T^*(\tilde{x}, y) = T_y U^*(\tilde{x}, y)$. Let us define for $x \in \Gamma$ the standard boundary integral operators as

$$(Vt)(x) = \int_\Gamma U^*(x, y) t(y) ds_y, \qquad (Du)(x) = -T_x \int_\Gamma T_y U^*(x, y) u(y) ds_y,$$

$$(Ku)(x) = \int_\Gamma T_y U^*(x, y) u(y) ds_y, \quad (K't)(x) = \int_\Gamma T_x U^*(x, y) t(y) ds_y.$$

The mapping properties of all of these boundary integral operators are well known, see [3]. We assume that the single layer potential V is $H^{-1/2}(\Gamma)$–elliptic and that the hypersingular integral operator D is $H^{1/2}(\Gamma)$–semi–elliptic, i.e.,

$$\langle V\tau,\tau\rangle_{L^2(\Gamma)} \geq c_1^V \cdot \|\tau\|^2_{H^{-1/2}(\Gamma)} \quad \text{for all } \tau \in H^{-1/2}(\Gamma)\,, \tag{1.3}$$

$$\langle Dv,v\rangle_{L^2(\Gamma)} \geq c_1^D \cdot \|v\|^2_{H^{1/2}(\Gamma)} \quad \text{for all } v \in H_0^{1/2}(\Gamma) \tag{1.4}$$

where

$$H_0^{1/2}(\Gamma) := \left\{ v \in H^{1/2}(\Gamma) \,:\, \langle v,w\rangle_{L^2(\Gamma)} = 0 \text{ for all } w \in \mathcal{R} \right\}. \tag{1.5}$$

Therefore,

$$\|v\|_D := \sqrt{\langle Dv,v\rangle_{L^2(\Gamma)}} \quad \text{and} \quad \|v\|_{V^{-1}} := \sqrt{\langle V^{-1}v,v\rangle_{L^2(\Gamma)}} \tag{1.6}$$

define equivalent norms in $H_0^{1/2}(\Gamma)$ and $H^{1/2}(\Gamma)$, respectively. Note that for $u_{|\Gamma} \in H_0^{1/2}(\Gamma)$, which is unknown, the solution $u(\tilde{x}), \tilde{x} \in \Omega$, defined by (1.2) is unique.

Let us assume that some boundary element approximation

$$u_h \in W_h := \operatorname{span}\{\varphi_k^\mu\}_{k=1}^M \subset H_0^{1/2}(\Gamma) \tag{1.7}$$

can be computed by an arbitrary boundary element method. In (1.7), W_h is defined with respect to a family of regular boundary element meshes with boundary elements Γ_ℓ of local mesh size h_ℓ, a global mesh size $h = \max_\ell h_\ell$, and the φ_k^μ are basis functions (for example splines of polynomial order μ). Moreover, we assume that an approximation property in W_h holds, i.e.,

$$\inf_{v_h \in W_h} \|v - v_h\|_{H^{1/2}(\Gamma)} \leq c \cdot h^{s-\frac{1}{2}} \cdot \|v\|_{H^s(\Gamma)} \quad \text{for all } v \in H^s(\Gamma) \tag{1.8}$$

with $s \leq \mu+1$. The error of the boundary element approximation u_h is defined by $e_h(x) := (u-u_h)(x)$ for $x \in \Gamma$. We are interested in the construction of an a posteriori error estimator η for estimating the error e_h in a suitable norm, such that

$$c_1 \cdot \eta \leq \|e_h\| \leq c_2 \cdot \eta \tag{1.9}$$

with positive constants c_1, c_2 independent of h. Moreover, we want to find local error indicators η_ℓ defined with respect to a boundary element Γ_ℓ which will be used to refine all boundary elements Γ_k where

$$\eta_k \geq \theta \cdot \max_\ell \eta_\ell \tag{1.10}$$

holds with some appropriate chosen refinement parameter $\theta \in [0,1]$.

A possible boundary element approach is to find $u_h \in W_h$ defined by the Galerkin variational formulation

$$\langle Du_h, v_h\rangle_{L^2(\Gamma)} = \langle((1-\sigma)I - K')g, v_h\rangle_{L^2(\Gamma)} \quad \text{for all } v_h \in W_h\,, \qquad (1.11)$$

which is uniquely solvable due to (1.4) [3]. By Cea's lemma and (1.8) we have

$$||u - u_h||_{H^{1/2}(\Gamma)} \leq c \cdot \inf_{v_h \in W_h} ||u - v_h||_{H^{1/2}(\Gamma)} \leq c \cdot h^{s-\frac{1}{2}} \cdot ||u||_{H^s(\Gamma)} \quad (1.12)$$

with $s \leq \min\{\rho, \mu + 1\}$ if $u \in H^\rho(\Gamma)$.

If the boundary element approximation u_h is given as the Galerkin solution of (1.11), one may use residual–based error estimators [1, 2, 5, 8, 9, 12, 13], which require the computation of the residual

$$r_h(x) := (De_h)(x) = ((1-\sigma(x))I - K')g(x) - (Du_h)(x) \quad \text{for } x \in \Gamma \quad (1.13)$$

as well as the evaluation and localization of $||r_h||_{H^{-1/2}(\Gamma)}$.

Following the main ideas in [10], we propose a new a posteriori error estimator η for boundary element methods related to the Neumann boundary value problem (1.1). This technique can be used for an arbitrarily given boundary element approximation $u_h \in W_h \subset H_0^{1/2}(\Gamma)$, i.e., we may use some other boundary integral equation and some other discretization scheme instead of the Galerkin method (1.11), for example, a collocation or qualocation scheme. As in [10], our method is based on the solution of an appropriate error equation, which is different from the original solution method and can be solved by using a (truncated) Neumann series where we have to take care of the fact that the Neumann problem is uniquely solvable in appropriate factor spaces only.

2. An a posteriori error estimator

Let us assume that an arbitrary boundary element approximation $u_h \in W_h \subset H_0^{1/2}(\Gamma)$ is given. Related to the representation formula (1.2), we define an approximate solution of the Neumann boundary value problem (1.1) by

$$\tilde{u}(\tilde{x}) := \int_\Gamma U^*(\tilde{x}, y)g(y)ds_y - \int_\Gamma u_h(y)T^*(\tilde{x}, y)ds_y \quad \text{for } \tilde{x} \in \Omega\,. \qquad (2.1)$$

Taking the limit $\Omega \ni \tilde{x} \to x \in \Gamma$ and using the jump relation of the double layer potential K we get

$$\tilde{u}(x) = (Vg)(x) + ((1-\sigma(x))I - K)u_h(x) \quad \text{for } x \in \Gamma\,. \qquad (2.2)$$

Note that from the representation formula (1.2) we obtain simultaneously

$$u(x) = (Vg)(x) + ((1-\sigma(x))I - K)u(x) \quad \text{for } x \in \Gamma\,. \qquad (2.3)$$

Subtracting (2.3) from (2.2) we can prove the following result:

Lemma 2.1. *The error $e_h := u - u_h$ of an arbitrarily given boundary element approximation u_h is a solution of the boundary integral equation*

$$(\sigma(x)I + K)(u - u_h)(x) = (\tilde{u} - u_h)(x) \quad \text{for } x \in \Gamma. \tag{2.4}$$

To compute the error e_h from (2.4) we require the solvability of (2.4) for a given right hand side. To this end we introduce the projection operator

$$P_{\mathcal{R}}: \; H^{1/2}(\Gamma) \longrightarrow H_0^{1/2}(\Gamma)$$

defined by

$$\langle P_{\mathcal{R}} v, w \rangle_{L^2(\Gamma)} = 0 \quad \text{for all } w \in \mathcal{R}.$$

Theorem 2.1 ([11]). *The operator $((1 - \sigma(x))I - K)$ acting on $H_0^{1/2}(\Gamma)$ is bounded by*

$$||[(1 - \sigma)I - K]v||_{V^{-1}} \leq c_K \, ||v||_{V^{-1}} \quad \text{for all } v \in H_0^{1/2}(\Gamma) \tag{2.5}$$

with

$$c_K = \frac{1}{2} + \sqrt{\frac{1}{4} - c_0} \qquad \text{where} \quad c_0 := c_1^V c_1^D \quad \text{and} \quad 0 < c_0 \leq \frac{1}{4}. \tag{2.6}$$

To construct a solution of (2.4) in $H_0^{1/2}(\Gamma)$ we consider the standard Neumann iteration [7]

$$\hat{e}_h^{(k+1)} := \hat{e}_h^{(k)} - P_{\mathcal{R}}[(\sigma(x)I + K)\hat{e}_h^{(k)} - (\tilde{u} - u_h)] \quad \text{for } k = 0, 1, 2, \ldots. \tag{2.7}$$

Assuming $\hat{e}_h^{(0)} \in H_0^{1/2}(\Gamma)$ we get $\hat{e}_h^{(k+1)} \in H_0^{1/2}(\Gamma)$ for all $k \geq 0$. Since for $v \in H_0^{1/2}(\Gamma)$ there holds $P_{\mathcal{R}} v = v$, we can write (2.7) as

$$\hat{e}_h^{(k+1)} := P_{\mathcal{R}}[((1 - \sigma(x))I - K)\hat{e}_h^{(k)} + (\tilde{u} - u_h)]$$

and using the initial guess $\hat{e}_h^{(0)} = 0$ we get

$$\hat{e}_h^{(k)} = \sum_{\ell=0}^{k-1} [P_{\mathcal{R}}((1 - \sigma(x))I - K)]^\ell P_{\mathcal{R}}(\tilde{u} - u_h) \quad \text{for } k = 1, 2, \ldots. \tag{2.8}$$

Due to Theorem 2.1, the Neumann series (2.8) converges in $H_0^{1/2}(\Gamma)$ yielding

$$e_h = \lim_{k \to \infty} \hat{e}_h^{(k)} \in H_0^{1/2}(\Gamma). \tag{2.9}$$

To get a computable error estimator we define

$$e_h^{(q)} := \sum_{\ell=0}^{q} [P_{\mathcal{R}}((1 - \sigma(x))I - K)]^\ell P_{\mathcal{R}}(\tilde{u} - u_h) \quad \text{for some } q \geq 0 \tag{2.10}$$

and

$$\eta^{(q)} := ||e_h^{(q)}||_D. \tag{2.11}$$

As in [10, Theorem 2.3] we can show the equivalence inequalities (1.9) for the a posteriori error estimator defined by (2.11):

Theorem 2.2. *The equivalence inequalities*

$$\frac{1}{1+c_K^{q+1}} \cdot \eta^{(q)} \le ||e_h||_D \le \frac{1}{1-c_K^{q+1}} \cdot \eta^{(q)} \tag{2.12}$$

hold with the positive constant $c_K < 1$ as given in (2.6).

3. Implementation

For the practical implementation of the proposed error estimator (2.11) we need to compute $e_h^{(0)} := \tilde{u} - u_h$ first. Using (2.2), we get

$$e_h^{(0)}(x) = (Vg)(x) - (\sigma I + K)u_h(x) \quad \text{for } x \in \Gamma\,. \tag{3.1}$$

Due to Theorem 2.2 ($q = 0$) we can use $e_h^{(0)}$ to compute the global error estimator (2.11) as well as some appropriate local error indicators which will be described later in this section. To use the proposed error estimator for some $q > 0$ we have to apply the truncated Neumann series (2.10). For the numerical evaluation of (2.10) we need to introduce some finite–dimensional approximation. Let $W_{\tilde{h}} \subset H_0^{1/2}(\Gamma)$ be a trial space of splines of polynomial order μ with respect to a refined triangulation $\Gamma_{\tilde{h}}$ with $\tilde{h} \le ch$ sufficiently small. Let the approximation property (1.8) be satisfied for $W_{\tilde{h}}$. For $v \in H^1(\Gamma)$ we define the L^2–Galerkin projection $Q_{\tilde{h}}v = v_{\tilde{h}} \in W_{\tilde{h}}$ by

$$\langle v_{\tilde{h}}, w_{\tilde{h}}\rangle_{L^2(\Gamma)} = \langle v, w_{\tilde{h}}\rangle_{L^2(\Gamma)} \quad \text{for all } w_{\tilde{h}} \in W_{\tilde{h}}\,. \tag{3.2}$$

For what follows we assume that $Q_{\tilde{h}} : H^1(\Gamma) \to H^1(\Gamma)$ is bounded, i.e.,

$$||Q_{\tilde{h}}v||_{H^1(\Gamma)} \le c \cdot ||v||_{H^1(\Gamma)}\,. \tag{3.3}$$

Note that (3.3) is satisfied under some additional mesh restrictions [4] which still allow strong local mesh refinements as in adaptive algorithms. For example, for $n = 2$ and piecewise linear trial functions, a local mesh ratio $h_k/h_{k+1} \le 4$ is allowed [4]. Based on (3.3) we are able to prove the following error estimate:

Lemma 3.1. *Let the L^2–Galerkin projection $Q_{\tilde{h}} : H^1(\Gamma) \to W_{\tilde{h}} \subset H^1(\Gamma)$ be stable. Then,*

$$||(I - Q_{\tilde{h}})v||_D \le c \cdot \tilde{h}^{1/2} \cdot ||v||_{H^1(\Gamma)} \quad \textit{for all } v \in H^1(\Gamma)\,. \tag{3.4}$$

Proof. For any $w \in H^{-1}(\Gamma)$ we define $w_{\tilde{h}} \in W_{\tilde{h}}$ by

$$\langle w_{\tilde{h}}, \tau_{\tilde{h}}\rangle_{H^{-1}(\Gamma)} = \langle w, \tau_{\tilde{h}}\rangle_{H^{-1}(\Gamma)} \quad \text{for all } \tau_{\tilde{h}} \in W_{\tilde{h}}\,.$$

If $w \in H^{-1/2}(\Gamma)$, we obtain from the approximation property of $W_{\tilde{h}}$ the error estimate

$$||w - w_{\tilde{h}}||_{H^{-1}(\Gamma)} \leq c \cdot \tilde{h}^{1/2} \cdot ||w||_{H^{-1/2}(\Gamma)} . \tag{3.5}$$

Then, using duality, the Galerkin orthogonality (3.2), the stability estimate (3.3) and the error estimate (3.5), this gives

$$\begin{aligned} ||(I - Q_{\tilde{h}})v||_{H^{1/2}(\Gamma)} &= \sup_{0 \neq w \in H^{-1/2}(\Gamma)} \frac{|\langle (I - Q_{\tilde{h}})v, w\rangle_{L^2(\Gamma)}|}{||w||_{H^{-1/2}(\Gamma)}} \\ &= \sup_{0 \neq w \in H^{-1/2}(\Gamma)} \frac{|\langle (I - Q_{\tilde{h}})v, w - w_{\tilde{h}}\rangle_{L^2(\Gamma)}|}{||w||_{H^{-1/2}(\Gamma)}} \leq c \cdot \tilde{h}^{1/2} \cdot ||v||_{H^1(\Gamma)} . \end{aligned}$$

Since $W_{\tilde{h}} \subset H_0^{1/2}(\Gamma)$ and since (1.6) defines an equivalent norm in $H_0^{1/2}(\Gamma)$ the assertion follows. □

Now we can define an approximate sequence of (2.10) as

$$\tilde{e}_h^{(\ell)} := \tilde{e}_h^{(\ell-1)} + z_{\tilde{h}}^{(\ell)}, \quad z_{\tilde{h}}^{(\ell)} = Q_{\tilde{h}}((1-\sigma)I - K)z_{\tilde{h}}^{(\ell-1)} \quad \text{for } \ell = 1, \ldots, q \tag{3.6}$$

and with $z_{\tilde{h}}^{(0)} = \tilde{e}_h^{(0)} = Q_{\tilde{h}} e_h^{(0)}$. Note that due to $Q_{\tilde{h}} : L^2(\Gamma) \to W_{\tilde{h}} \subset H_0^{1/2}(\Gamma)$ we do not need to apply $P_{\mathcal{R}}$ in the discrete case. Hence, the error estimator (2.11) takes the form

$$\tilde{\eta}^{(q)} := ||\tilde{e}_h^{(q)}||_D . \tag{3.7}$$

Consequently, we can formulate the equivalence inequalities (1.9) for the computable error estimator (3.7), see also [10, Theorem 4].

Theorem 3.1. *Let $\tilde{e}_h^{(q)}$ be defined as in (3.6) and let (3.7) be the corresponding a posteriori error estimator. Then,*

$$\begin{aligned} \frac{1}{1 + c_K^{q+1}} \cdot \left\{ \tilde{\eta}^{(q)} - c \cdot q \cdot \tilde{h}^{1/2} \cdot ||e_h^{(0)}||_{H^1(\Gamma)} \right\} &\leq ||e_h||_D \leq \\ &\leq \frac{1}{1 - c_K^{q+1}} \cdot \left\{ \tilde{\eta}^{(q)} + c \cdot q \cdot \tilde{h}^{1/2} \cdot ||e_h^{(0)}||_{H^1(\Gamma)} \right\} . \end{aligned}$$

Therefore, if $\tilde{h} \leq ch$ is sufficiently small, $\tilde{\eta}^{(q)}$ defined in (3.7) is a global error estimator satisfying (1.9).

To derive local error indicators we may define

$$\tilde{\eta}_k^{(q)} := \sqrt{|\langle D\tilde{e}_h^{(q)}, \tilde{e}_h^{(q)}\rangle_{L^2(\Gamma_k)}|} \quad \text{for } k = 1, \ldots, N. \tag{3.8}$$

The final adaptive boundary element algorithm reads as follows:

1. For a given boundary element mesh Γ_h compute a boundary element approximation $u_h \in W_h$, e.g., by solving (1.11).
2. Compute $\tilde{e}_h^{(0)}$ by solving

$$\langle \tilde{e}_h^{(0)}, v_{\tilde{h}}\rangle_{L^2(\Gamma)} = \langle e_h^{(0)}, v_{\tilde{h}}\rangle_{L^2(\Gamma)} \quad \text{for all } v_{\tilde{h}} \in W_{\tilde{h}}$$

using (3.1) for the computation of the right hand.

3. Compute the error function $\tilde{e}_h^{(q)}$ for some fixed $q > 0$ via (3.6), i.e., by solving
$$\langle z_{\tilde{h}}^{(k)}, v_{\tilde{h}} \rangle_{L^2(\Gamma)} = \langle [(1-\sigma)I - K] z_{\tilde{h}}^{(k-1)}, v_{\tilde{h}} \rangle_{L^2(\Gamma)} \quad \text{for all } v_{\tilde{h}} \in W_{\tilde{h}}$$
and $k = 1, \ldots, q$. Since $W_{\tilde{h}} \subset H_0^{1/2}(\Gamma)$ we already have $z_{\tilde{h}}^{(k)} \in H_0^{1/2}(\Gamma)$.
4. Compute all local error indicators (3.8) and the global error estimator (3.7). Stop, if a required accuracy is reached.
5. Refine all boundary elements Γ_k where (1.10) is satisfied and go to 1.

4. Numerical example

We consider a Neumann boundary value problem for the two–dimensional Laplace equation where Ω is the L shaped domain with the nodes $(0,0)$, $(0.25,0)$, $(0.25,0.25)$, $(-0.25,0.25)$, $(-0.25,-0.25)$ and $(0,-0.25)$. The Neumann data g are given in such a way that the solution of (1.1) is

$$u(x) = u(r,\varphi) = r^{2/3} \cos(\frac{2}{3}\varphi) \tag{4.1}$$

when using polar coordinates. Hence the regularity result $u \in H^\sigma(\Gamma)$ for all $\sigma < \frac{5}{3}$ holds, see [6]. To compute a boundary element approximation $u_h \in W_h$ we used the Galerkin variational formulation (1.11) discretized by using piecewise linear continuous trial functions ($\mu = 1$). Therefore, in the case of an uniform refinement we get the order of convergence when measuring the error in the energy norm (1.6) to be about $\frac{2}{3}$. Using the adaptive algorithm as described in the previous section with $q = 0$ and $\theta = 0.25$, the error is improved significantly as shown in Fig. 4.1, even when comparing the computing times needed, see Fig. 4.2.

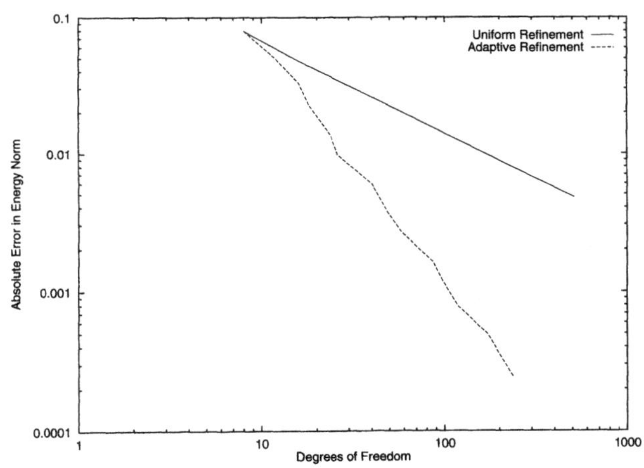

Fig. 4.1. Adaptive versus uniform refinement: Degrees of freedom

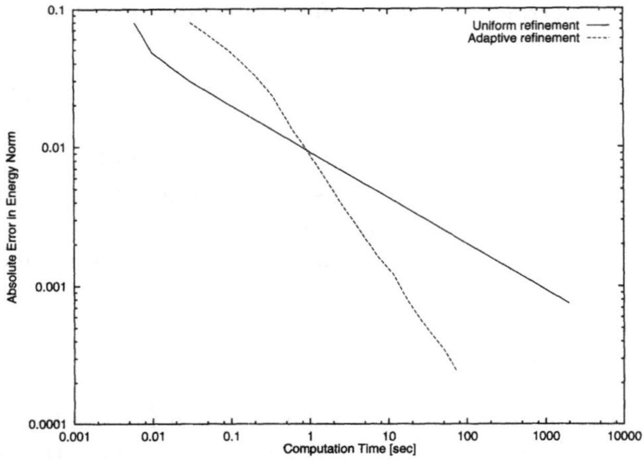

Fig. 4.2. Adaptive versus uniform refinement: Computing times.

References

1. C. Carstensen, An a posteriori error estimate for a first–kind integral equation. Math. Comp. 66 (1997) 139–156.
2. C. Carstensen, E. P. Stephan, Adaptive boundary element methods for some first–kind integral equations. SIAM J. Numer. Anal. 33 (1996) 2166–2183.
3. M. Costabel, Boundary integral operators on Lipschitz domains: Elementary results. SIAM J. Numer. Anal. 19 (1988) 613–626.
4. M. Crouzeix, V. Thomeé, The stability in L_p and W_p^1 of the L_2–projection onto finite element function spaces. Math. Comp. 48 (1987) 521–532.
5. B. Faermann, Local a posteriori error estimators for the Galerkin discretization of boundary integral equations. Numer. Math. 79 (1998) 43–76.
6. P. Grisvard, Elliptic Problems in Nonsmooth Domains. Pitman, Boston, 1985.
7. G. C. Hsiao, E. Schnack, W. L. Wendland, Hybrid coupled finite–boundary element methods for elliptic systems of second order. Comput. Methods Appl. Engrg., to appear.
8. J. Saranen, W. L. Wendland, Local residual-type error estimates for adaptive boundary element methods on closed curves. Appl. Anal. 48 (1993) 37–50.
9. H. Schulz, Über lokale und globale Fehlerabschätzungen für adaptive Randelementmethoden. Doctoral Thesis, University of Stuttgart, 1997.
10. H. Schulz, O. Steinbach, A new a posteriori error estimator in adaptive direct boundary element methods. The Dirichlet problem. Calcolo (1999) accepted.
11. O. Steinbach, W. L. Wendland, On C. Neumann's method for second order elliptic systems in domains with non–smooth boundaries. In preparation.
12. T. Tran, Local error estimates for the Galerkin method applied to strongly elliptic integral equations on open curves. SIAM J. Numer. Anal. 33 (1996) 1484–1493.
13. W. L. Wendland, De–Hao Yu, Adaptive boundary element methods for strongly elliptic integral equations. Numer. Math. 53 (1988) 539–558.

Efficient Elasto-Plastic Simulation

C. Wieners

ICA, University of Stuttgart, Germany

Summary. In this paper we describe a method for the construction of radial return algorithms to the plasticity models discussed in ALBER [1]. For the classical examples this shows that the algorithms in SIMO-HUGHES [5] can be derived by this method in a systematic way. We apply the method to viscoplasticity with nonlinear isotropic and kinematic hardening. The combination of parallel multigrid methods with the radial return algorithm results in a very efficient algorithm. The performance of the method is illustrated by a numerical example.

1. Introduction

We consider the quasi-static process of the deformation of materials with memory, where the history of the deformation is described by internal variables. For a wide range of models in solid mechanics the evolution of the displacement field u and of the internal variables z are determined by the principle of maximal dissipation. In addition, in case of linearized strains and associated plasticity, the constitutive equations are determined by the free energy functional and a monotone function describing the material history evolution.

In particular, we will consider here the class of *generalized standard materials* which are of gradient-monotone type as described in [1]. There, the material history is determined by a monotone function of the form $\partial\chi$. The classical plasticity models can be described in this framework.

In computational plasticity, the time discretization by the implicit Euler method leads to a radial return formulation, which is a commonly used numerical method in plasticity, cf. [5]. This method combines the local solution of a nonlinear system at every material point describing the material history evolution with the global linear variational problem corresponding to the equilibrium condition. The linearization of the nonlinear problem in every time step leads to the consistent tangent operator introduced in [6]; here we give a general definition of the tangent operator for our abstract model.

The paper is organized as follows. In Sect. 2. we state a general model for quasi-static plasticity involving internal history parameters. For this model we introduce radial return algorithms in Sect. 3. In Sect. 4. we show how this formalism applies to generalized standard materials. As an example we consider in Sect. 5. viscoplasticity with nonlinear isotropic and kinematic hardening, and the implementation is discussed in Sect. 6. Finally, in Sect. 7. we demonstrate the performance of the method by a numerical experiment realized with the software package *UG* [2], which combines a flexible finite element library with parallel multigrid methods.

2. A general framework for quasi-static plasticity

Let $\Omega \subset \mathbf{R}^3$ be a domain, let $\Gamma \subset \partial\Omega$ be a subset of the boundary, and let $[0, t_{\max}]$ be a time interval. Furthermore, we prescribe body forces

$$b \colon \Omega \times [0, t_{\max}] \longrightarrow \mathbf{R}^3,$$

surface tractions

$$s \colon \partial\Omega \setminus \Gamma \times [0, t_{\max}] \longrightarrow \mathbf{R}^3$$

and boundary displacements

$$d \colon \Gamma \times [0, t_{\max}] \longrightarrow \mathbf{R}^3.$$

We consider the following problem: find a displacement field

$$u \colon \Omega \times [0, t_{\max}] \longrightarrow \mathbf{R}^3$$

with $u(x,t) = d(x,t)$ a. e. in $\Gamma \times [0, t_{\max}]$ and internal variables

$$z \colon \Omega \times [0, t_{\max}] \longrightarrow \mathbf{R}^m,$$

such that the equilibrium condition

$$\int_\Omega T(F(x,t), z(x,t)) \cdot \nabla v(x)\, dx = L(t)(v), \qquad F = 1 + \nabla u \tag{2.1}$$

holds for all v such that $v_{|\Gamma} = 0$ and a. a. $t \in [0, t_{\max}]$, and such that the constitutive equation for the internal variables

$$\dot{z}(x,t) \in f(F(x,t), z(x,t))$$

is fulfilled a. e. in $\Omega \times [0, t_{\max}]$, subject to initial data

$$(u(x,0), z(x,0)) = (u_0(x), z_0(x)) \qquad \text{a. e. in } \Omega.$$

Here,

$$T \colon \mathbf{R}^9 \times \mathbf{R}^m \longrightarrow \mathbf{R}^9$$

describes the strain-stress relation,

$$f \colon \mathbf{R}^9 \times \mathbf{R}^m \longrightarrow \mathcal{P}(\mathbf{R}^m)$$

describes the material history, and L denotes the load functional combining the body forces and the surface traction

$$L(t)(v) = \int_\Omega b(x,t) v(x)\, dx + \int_{\partial\Omega\setminus\Gamma} s(x,t) v(x)\, da.$$

Finally, we assume that the initial data are admissible, i. e., $(u_0, z_0) \in \mathrm{dom}(f)$ and (u_0, z_0) satisfy the equilibrium condition (2.1) for $t = 0$.

3. Return mapping algorithms for incremental plasticity

We consider the incremental approximation of the quasi-static model, where

$$0 = t_0 < t_1 < \cdots < t_J = t_{\max}$$

is a time series and $\tau_j = t_j - t_{j-1}$.

Now, the implicit Euler method reads as follows: find

$$(u_j, z_j)\colon \Omega \longrightarrow \mathbf{R}^3 \times \mathbf{R}^m, \qquad j = 1, \ldots, J$$

such that $u_j(x) = d(x, t_j)$ a. e. in Γ, satisfying

$$\int_\Omega T(F_j(x), z_j(x)) \cdot \nabla v(x)\, dx = L(t_j)(v)$$

for all v with $v_{|\Gamma} = 0$ and

$$\frac{1}{\tau_j}\Big(z_j(x) - z_{j-1}(x)\Big) \in f(F_j(x), z_j(x)), \qquad F_j = 1 + \nabla u_j. \tag{3.1}$$

Basic assumption in computational plasticity

If the nonlinear equation (3.1) has a solution, and if this solution can be described by a function

$$Z\colon (0, \infty) \times \mathbf{R}^9 \times \mathbf{R}^m \longrightarrow \mathbf{R}^m \tag{3.2}$$

in the form

$$z_j(x) = Z(\tau_j, F_j(x), z_{j-1}(x)),$$

the time increment is determined by the response function

$$R\colon (0, \infty) \times \mathbf{R}^9 \times \mathbf{R}^m \longrightarrow \mathbf{R}^9$$

defined by

$$R(\tau_j, F_j(x), z_{j-1}) := T(F_j(x), Z(\tau_j, F_j(x), z_{j-1})). \tag{3.3}$$

The consistent tangent operator

$$C\colon (0, \infty) \times \mathbf{R}^9 \times \mathbf{R}^m \longrightarrow \mathbf{R}^{81}$$

introduced in [6] is defined by

$$C(\tau_j, F_j, z_{j-1}) := \frac{\partial}{\partial F} T(F_j, Z(\tau_j, F_j, z_{j-1})).$$

The displacement increment is determined by the following nonlinear variational problem: find u_j with $u_j(x) = d(x, t_j)$ on Γ such that

$$\int_\Omega R(\tau_j, F_j(x), z_{j-1}(x)) \cdot \nabla v(x)\, dx = L(t_j)(v) \tag{3.4}$$

for all v with $v_{|\Gamma} = 0$.

The nonlinear problem (3.4) is solved by a Newton method:

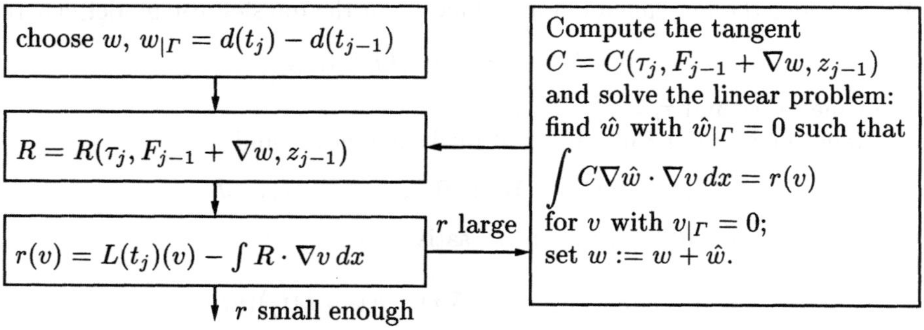

This gives $u_j = u_{j-1} + w$; then, z_j is defined by (3.2).

4. Standard materials

Standard materials are described by the free energy functional

$$\psi \colon \mathbf{R}^9 \times \mathbf{R}^m \longrightarrow \mathbf{R}$$

and the convex dissipation potential

$$\chi \colon \mathbf{R}^m \longrightarrow \mathbf{R}$$

by the hyperelastic stress response

$$T(F, z) = \frac{\partial}{\partial F} \psi(F, z)$$

and the material history evolution

$$f(F, z) = \partial \chi(q), \tag{4.1}$$

where the dissipation depends only on

$$q = -\frac{\partial}{\partial z} \psi(F, z). \tag{4.2}$$

Then, the solution of (3.1) is characterized by the inequality

$$\tfrac{1}{\tau_j}(z_j - z_{j-1}) \cdot (q_j - q) + \chi(q) \geq \chi(q_j) \qquad \text{for all } q \in \mathbf{R}^m,$$

where the internal variable z_j is related with q_j by

$$q_j = -\frac{\partial}{\partial z} \psi(F_j, z_j).$$

Standard materials can be generalized by replacing $\partial\chi$ in (4.1) by a (maximal) monotone operator g such that

$$f(F, z) = g(q) = g\Big(-\frac{\partial}{\partial z} \psi(F, z)\Big). \tag{4.3}$$

5. An example with nonlinear hardening

We consider small strain cyclic viscoplasticity with nonlinear isotropic and kinematic hardening under isothermal conditions, cf. LEMAITRE-CHABOCHE [4, Sect. 6.4.3]. In this model we have $z = (\varepsilon^p, \varepsilon^n, p) \in \mathbf{R}^6 \times \mathbf{R}^6 \times \mathbf{R}$ for the internal variables, where ε^p is the plastic strain, ε^n is the kinematic hardening variable, and p describes the isotropic hardening. The free energy is given by

$$\psi(1 + \nabla u, \varepsilon^p, \varepsilon^n, p) = \mu\,|\mathrm{dev}(\varepsilon(u)) - \varepsilon^p|^2 + \frac{\kappa}{2}\,\mathrm{div}(u)^2 + \frac{C}{3}\,|\varepsilon^p - \varepsilon^n|^2 + h(p)$$

with $\mathrm{dev}(\varepsilon) = \varepsilon - \frac{1}{3}\mathrm{trace}(\varepsilon)\,\mathbf{I}$, the linearized strain $\varepsilon(u) = \frac{1}{2}(\nabla u + \nabla u^T)$ and $h(p) = Qp - \frac{Q}{b}(1 - \exp(-bp))$. The linear stress response is given by

$$\sigma := T(1 + \nabla u, \varepsilon^p) = 2\mu\ \mathrm{dev}(\varepsilon(u) - \varepsilon^p) + \kappa\ \mathrm{div}(u)\,\mathbf{I}\,.$$

The relation (4.2) has the form

$$\Big(\mathrm{dev}(\sigma - X), X, R\Big) = \Big(2\mu\,\mathrm{dev}(\varepsilon(u) - \varepsilon^p) - \tfrac{2C}{3}(\varepsilon^p - \varepsilon^n), \tfrac{2C}{3}(\varepsilon^p - \varepsilon^n), h'(p)\Big),$$

and the material evolution is described by

$$(\dot{\varepsilon}^p, \dot{\varepsilon}^n, \dot{p}) = f(1 + \nabla u, \varepsilon^p, \varepsilon^n, p) := \gamma\ \left(\frac{\mathrm{dev}(\sigma - X)}{|\mathrm{dev}(\sigma - X)|}, \sqrt{\frac{3}{2}\frac{G}{C}}\,X, \sqrt{\frac{2}{3}}\right) \quad (5.1)$$

with the return parameter

$$\gamma = \chi_n'\Big(\sqrt{\tfrac{3}{2}}\ |\mathrm{dev}(\sigma - X)| - R\Big), \quad (5.2)$$

where

$$\chi_n(s) = \frac{K}{n+1}\left\langle\frac{s-k}{K}\right\rangle^{n+1}, \qquad \langle s\rangle = \max\{0, s\}. \quad (5.3)$$

The model depends on the material parameters $\mu, \kappa, K > 0$, $C, Q, G, b, k \geq 0$. It describes a material of monotone type, where the operator g is defined by (5.1), (5.2) and (5.3). For the derivation of the evolution equation from a suitable dissipation functional we refer to [4, Sect. 6.4.3].

The limit $n \longrightarrow \infty$ describes the classical Prandtl-Reuß plasticity, where

$$\gamma \in \partial\chi_\infty\Big(\sqrt{\tfrac{3}{2}}\ |\mathrm{dev}(\sigma - X)| - R\Big), \qquad \chi_\infty(s) = \begin{cases} 0 & \text{for } s - k \leq K \\ \infty & \text{for } s - k > K. \end{cases}$$

Here,

$$\{\sigma \in \mathbf{R}^6 \mid \sqrt{\tfrac{3}{2}}\ |\mathrm{dev}(\sigma - X)| - R - k \leq K\}, \qquad R = Q(1 - \exp(-bp))$$

describes the convex domain of admissible generalized stresses. The convergence properties of return mapping algorithms for the model of Prandtl-Reuß with nonlinear hardening are analyzed in [3, 7].

6. The computation of the return parameter

The evaluation of the radial return (3.3) requires the solution of the nonlinear problem (3.1). In this model, the nonlinear problem (3.1) has the form

$$(\varepsilon_j^p, \varepsilon_j^n, p_j) - (\varepsilon_{j-1}^p, \varepsilon_{j-1}^n, p_{j-1}) = \gamma_j \left(\frac{d_j}{|d_j|}, \sqrt{\frac{3}{2}\frac{G}{C}} X_j, \sqrt{\frac{2}{3}} \right)$$

depending on the return parameter $\gamma_j = \tau_j \left\langle \frac{\sqrt{\frac{3}{2}}|d_j| - R_j - k}{K} \right\rangle^n$ and

$$(d_j, X_j, R_j) = \Big(2\mu \operatorname{dev}(\varepsilon(u_j) - \varepsilon_j^p) - \tfrac{2C}{3}(\varepsilon_j^p - \varepsilon_j^n), \tfrac{2C}{3}(\varepsilon_j^p - \varepsilon_j^n), h'(p_j) \Big).$$

It is possible to reduce this problem to a nonlinear equation in $\mathbf{R}$

$$\gamma_j = \tau_j \Gamma(\gamma_j; 2\mu \operatorname{dev}(\varepsilon(u_j) - \varepsilon_{j-1}^p) - X_{j-1}, X_{j-1}, p_{j-1}) \tag{6.1}$$

for the return parameter γ_j. Therefore, Γ can be evaluated from fixed values for $F_j, \tilde{d}_j = 2\mu \operatorname{dev}(\varepsilon(u_j) - \varepsilon_{j-1}^p) - X_{j-1}, X_{j-1}, p_{j-1}$, the hardening response

$$R_j = h'\left(p_{j-1} + \sqrt{\tfrac{2}{3}}\, \gamma_j \right),$$

and $|d_j|$; this can be computed as follows:

$$X_j - X_{j-1} = \tfrac{2C}{3}(\varepsilon_j^p - \varepsilon_{j-1}^p) - \tfrac{2C}{3}(\varepsilon_j^n - \varepsilon_{j-1}^n) = \tfrac{2C}{3}(\varepsilon_j^p - \varepsilon_{j-1}^p) - \sqrt{\tfrac{2}{3}} G \gamma_j X_j$$

gives for the backstress increment

$$X_j - X_{j-1} = \frac{\gamma_j}{1 + \sqrt{\frac{2}{3}} G \gamma_j} \left(\frac{2C}{3} \frac{d_j}{|d_j|} - \sqrt{\frac{2}{3}} G X_{j-1} \right).$$

Inserting the backstress increment into

$$\begin{aligned} d_j &= \tilde{d}_j - 2\mu(\varepsilon_j^p - \varepsilon_{j-1}^p) - (X_j - X_{j-1}) \\ &= \tilde{d}_j - \gamma_j \left(2\mu + \frac{\frac{2}{3} C}{1 + \sqrt{\frac{2}{3}} G \gamma_j} \right) \frac{d_j}{|d_j|} + \frac{\sqrt{\frac{2}{3}} G \gamma_j}{1 + \sqrt{\frac{2}{3}} G \gamma_j} X_{j-1} \end{aligned}$$

yields

$$|d_j| = \left| \tilde{d}_j + \frac{\sqrt{\frac{2}{3}} G \gamma_j}{1 + \sqrt{\frac{2}{3}} G \gamma_j} X_{j-1} \right| - 2\mu \gamma_j - \frac{\frac{2}{3} C \gamma_j}{1 + \sqrt{\frac{2}{3}} G \gamma_j}$$

and

$$\frac{d_j}{|d_j|} = \frac{\tilde{d}_j + \frac{\sqrt{\frac{2}{3}} G \gamma_j}{1 + \sqrt{\frac{2}{3}} G \gamma_j} X_{j-1}}{\left| \tilde{d}_j + \frac{\sqrt{\frac{2}{3}} G \gamma_j}{1 + \sqrt{\frac{2}{3}} G \gamma_j} X_{j-1} \right|}.$$

Note that this reduces to the method in [5, Sect. 3.7] for linear hardening, where $n = 1$, $G = 0$, and $h'(p) = Qp$.

7. A numerical experiment

The complete algorithm is realized in *UG*. For a comprehensive overview of the modular design of *UG* we refer to [2]. It supports parallel multigrid methods. The application to plasticity problems is described in detail in [8].

In the numerical algorithm the return parameter is determined at the Gauß points of the numerical quadrature. It is solved by a local Newton method for (6.1), where the derivative is computed numerically.

In the example we use the material parameters $E = 216000$, $\nu = 0.29$, $n = 24$, $K = 151$, $k = 82$, $C = 162400$, $G = 2800$, and $b = 8$, cf. [4, Tab. 6.3].

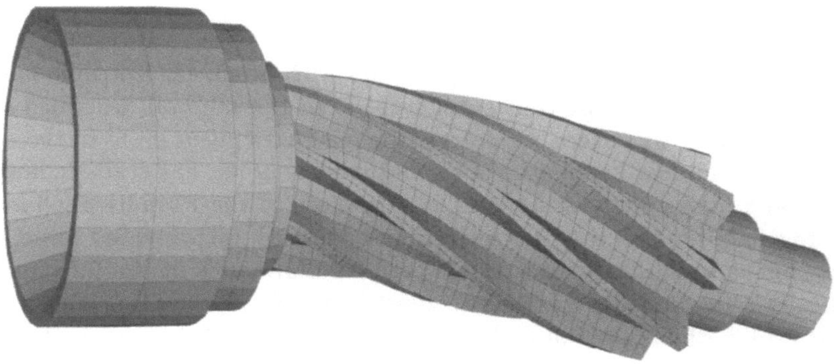

Fig. 7.1. Deformation of the gear geometry (scaled by a factor of 1000).

We compute the cyclic deformation of a gear, cf. Fig. 7.1. The resulting stress evolution at a suitable test point is presented in Fig. 7.2.

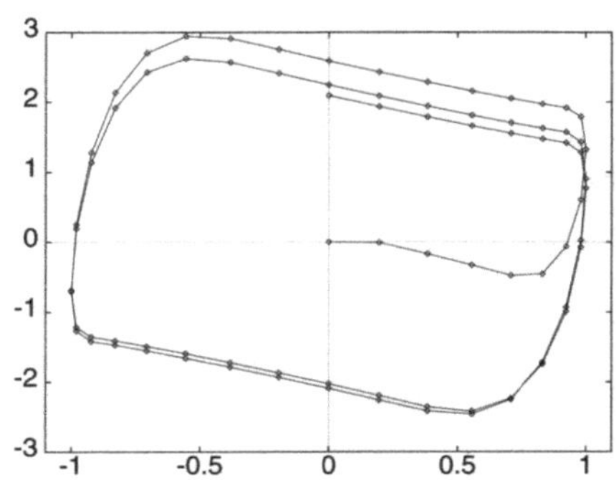

Fig. 7.2. Stress evolution for $L(t) = \sin(t)\, L_0$, $t \in [0, 5\pi]$. The computation with 80 loading steps and 256 Newton steps can be executed with 11456 elements (34887 unknowns) on 96 processors in 48 minutes.

8. Conclusion

We defined a method for the construction of radial return algorithms for an abstract model for materials with memory. In the special case, where the dissipation potential is convex and the free energy functional can be transformed by a transformation of the internal variables into a quadratic functional, the local material update is equivalent to a convex minimization problem; for more general models it is an open question to define sufficient criteria which guarantees that the local problems can be solved.

Algorithmically there is no difference in the formulation for linearized strains and for finite strains in the geometrical nonlinear case (see [5, Chap. 9]), but our analysis of the radial return in [7] cannot be extended to nonconvex free energy functionals.

For a special example of viscoplasticity with nonlinear isotropic hardening we showed that this construction leads to the method described in [5] in a systematic way. Here, we extend the examples given in [5] to the application to nonlinear kinematic hardening models in [4]. For realistic material parameters the viscoplastic model is very close to the Prandtl-Reuß model, where a fully implicit time discretization is required; we presented a numerical experiment which underlines the superior performance of the fully implicit time discretization also for the viscoplastic case. In combination with parallel multigrid methods the full method results in a very efficient algorithm.

References

1. H.-D. Alber, *Materials with memory*, vol. 1682 of Lecture Notes in Mathematics, Springer, 1998.
2. P. Bastian, K. Birken, K. Johannsen, S. Lang, N. Neuss, H. Rentz-Reichert, and C. Wieners, *UG – a flexible software toolbox for solving partial differential equations*, Computing and Visualization in Science, 1 (1997), pp. 27–40.
3. R. Blaheta, *Convergence of Netwon-type methods in incremental return mapping analysis of elasto-plastic problems*, Comput. Meth. Appl. Mech. Engrg., 147 (1997), pp. 167–185.
4. J. Lemaitre and J. L. Chaboche, *Mechanics of solid materials*, Cambridge University press, 1994.
5. J. C. Simo and T. J. R. Hughes, *Computational inelasticity*, Springer, 1998.
6. J. C. Simo and R. L. Taylor, *Consistent tangent operators for rate-independent elastoplasticity*, Comput. Meth. Appl. Mech. Engrg., 48 (1985), pp. 101–118.
7. C. Wieners, *Orthogonal projections onto convex sets and the application to problems in plasticity*, tech. rep., Universität Stuttgart, SFB 404 Preprint 99/15, 1999.
8. C. Wieners, *Theorie und Numerik der Prandtl-Reuß Plastizität*, Universität Heidelberg, 1999. Habilitationsschrift, submitted.

Section 5

Contact and Fracture

Edited by W. L. Wendland

Sensitivity and Optimal Control in Contact Mechanics

G. Szefer

Institut of Structural Mechanics, Cracow University of Technology

Summary. In this paper sensitivity and optimal control problems with bodies undergoing contact are formulated and discussed. Starting with a variational formulation of the contact boundary value problem, a sensitivity and optimal control problem will be stated. Then, using the method of the adjoint system, the sensitivity operators for a given functional are constructed. Special attention is paid to variation of integrals defined over the contact area. A physical interpretation of the adjoint system in the presence of contact is given.

1. Introduction

Many dynamical systems considered in mechanical engineering, robotics – and in recent years also in civil engineering – lead to problems of controllable structures experiencing contact. Buildings with friction–controlled sliding bearings, structures with active dampers, elastic folding of shells in deployable structures etc. are examples of this type. Such classes of problems lead to questions of control and sensitivity of deformation processes in terms of contact phenomena.
The aim of the present paper is to formulate sensitivity and optimal control problems with bodies undergoing contact. Our main goal is to show how sensitivity operators of the system are influenced by the presence of contact and how optimal control problems in terms of unilateral contact can be solved.
The content of the paper is as follows. We start with the formulation of the optimal control problem. Then we pass to sensitivity of functionals arising due to contact. Parameter and shape sensitivity are distinguished. The adjoint–system approach with a clear physical explanation of the contact terms influence is given. We include a list of remarks and comments to emphasize the features of the considered problem.

2. Setting of the optimal control and sensitivity problem

Denoting by $\mathbf{u}$ a state function which describes the behaviour of a given dynamical system, by $\mathbf{s}$ the controller function (abbreviated as "control"), one can formulate any optimal control problem in an abstract schematic form as follows.
Find the control $\mathbf{s} \in U$ and the state $\mathbf{u} \in V$ (where V, U are topological linear

spaces) subject to
the state equation

$$A(\mathbf{s},\mathbf{u}) = f(s) \tag{2.1}$$

and constraints

$$\mathbf{s} \in U_{ad} \subset U \quad , \quad \mathbf{u} \in V_{ad} \subset V \tag{2.2}$$

such that a given objective functional

$$J(\mathbf{s},\mathbf{u}) = \int_V F_0(s,u)dV + \int_S f_0(s)d\Gamma \tag{2.3}$$

attains its minimum. Here A is a given, obviously differential operator (with corresponding boundary and initial conditions), f is the prescribed external action, the admissible sets U_{ad}, V_{ad} are defined obviously by a system of equalities and inequalities, and the performance (cost) functions (2.3) describe a measure of system response.
Sensitivity consists in determining the variation δJ_i of any functional $J_i(s,u)$ provided that the control s changes on variation $\delta s, (s+\delta s) \in U_{ad}$. The construction of operators $\delta J_i = A_i \delta s$ which show explicitly the relationship between the variations δJ_i and δs constitutes the main goal of sensitivity analysis. This enables one to formulate necessary optimality conditions or at least to analyze the modification of the state due to variation in choosen design variables. In contact mechanics where an elastic body B^+ is in dynamic unilateral, frictional contact with another target body B^-, the state equation (2.1) is obviously expressed by means of the virtual power principle, which takes the form

$$\begin{aligned}
&\int_V \sigma : \varepsilon(\mathbf{v}) = \int_V \mathbf{b}\mathbf{v}dV + \int_S \mathbf{p}\mathbf{v}d\Gamma - \int_V \rho\ddot{\mathbf{u}}\mathbf{v}dV + \int_{\Gamma_c} \mathbf{t}_n\mathbf{v}_n d\Gamma + \int_{\Gamma_c} \mathbf{t}_T\mathbf{v}_T d\Gamma \\
&\forall \mathbf{v} \equiv \delta\mathbf{u} \in V_0 \\
&\mathbf{u}(\mathbf{x},t_0) = \mathbf{u_0}(x), \dot{\mathbf{u}}(\mathbf{x},t_0) = \mathbf{v_0}(x)
\end{aligned} \tag{2.4}$$

with the Signorini unilateral contact conditions

$$\begin{aligned}
&\Gamma_c : d_n \geq 0, t_n \leq 0 \qquad d_n > 0 \Rightarrow t_n = 0 \\
&d_n = 0 \Rightarrow t_n \leq 0 \quad t_n d_n = 0
\end{aligned} \tag{2.5}$$

(which expresses the normal force interaction between the bodies).
Here

$\mathbf{u}, \varepsilon, \sigma$	displacement vector, strain and stress tensor fields
$t_n = (\sigma\mathbf{n})\mathbf{n}$,	
$\mathbf{t}_n = t_n\mathbf{n}$	normal contact stresses
$\mathbf{t}_T$	tangential contact stresses described by a friction law
Γ_c	contact zone
d_n	distance between bodies undergoing contact

$\mathbf{v}$	virtual velocity (or displacement in the case of statics)
$\mathbf{n}$	unit normal vector
ρ	mass density
$\mathbf{b}, \mathbf{p}$	prescribed external body and surface forces
$(:)$	inner product of tensors
V_0	space of kinematical admissible velocities
$V = V^+ \cup V^-$	domain occupied by the bodies B^+ and B^-
$S = S^+ \cup S^-$	part of the boundary with prescribed surface tractions

Setting now the parameter sensitivity contact problem, assume that the external loads $\mathbf{p}$ and $\mathbf{b}$ or internal stress distribution σ depend on the control $\mathbf{s}$. So all terms appearing in (2.4) are control dependent.
Provided $\mathbf{s}$ varies on $\delta\mathbf{s}$ all the above mentioned integrals change too. But in contact problems not only the stress tensor σ but also the contact stresses $\mathbf{t}_n$ and $\mathbf{t}_T$ change due to dependence on the state $\mathbf{u}$. Variation of $\mathbf{u}$ on $\delta\mathbf{u}^s$ induces variation of the contact area $\Gamma_c(\mathbf{u}(\mathbf{s}))$. Thus dependent shape variations $\delta\varphi$ of the contact zone occur. This result must be taken into account in the computation of sensitivity operators. Therefore, we need a formula for the variation of the shape integral

$$J = \int_{\Gamma_c} \Psi(\mathbf{u}) d\Gamma, \tag{2.6}$$

namely ([3],[6])

$$\delta J = \int_{\Gamma_c} [\frac{\partial \Psi}{\partial \mathbf{u}} \delta\mathbf{u}^s + (\frac{\partial \Psi}{\partial n} - 2k\Psi)\delta\varphi_n] d\Gamma + \int_{\partial\Gamma_c} \Psi \nu \delta\varphi ds, \tag{2.7}$$

where
k - is the mean curvature of the surface Γ_c
$\delta\varphi_n = \delta\varphi \cdot \mathbf{n}$, $\delta\varphi$ - shape variation
ν - unit vector normal to $\partial\Gamma_c$ and tangential to Γ_c, pointing toward the outside of Γ_c.
The following sensitivity problem can be now stated.
Find δJ of the given functional (2.3) subject to the boundary–value contact problem (2.4) - (2.5).
It will be

$$\delta J = \int_V (\frac{\partial F}{\partial \mathbf{u}} \delta\mathbf{u}^s + \frac{\partial F}{\partial \mathbf{s}} \delta\mathbf{s}) dV + \int_S \frac{\partial f}{\partial s} \delta s d\Gamma = \delta J^s + \delta J^u, \tag{2.8}$$

where for clarity the parts depending on $\delta\mathbf{u}^{\mathbf{s}}$ and $\delta\mathbf{s}$ are denoted separately by

$$\delta J^u = \int_V \frac{\partial F}{\partial \mathbf{u}} \delta\mathbf{u}^s dV \quad , \quad \delta J^s = \int_V \frac{\partial F}{\partial \mathbf{s}} \delta\mathbf{s} dV + \int_S \frac{\partial f}{\partial \mathbf{s}} \delta\mathbf{s} d\Gamma. \tag{2.9}$$

In order to find $\delta\mathbf{u}^s$, consider the variational equation

$$\int\limits_V \sigma(\delta\mathbf{u}^s) : \varepsilon(\mathbf{v})dV + \int\limits_V \frac{\partial\sigma}{\partial\mathbf{s}}\delta\mathbf{s} : \varepsilon(\mathbf{v})dV = \int\limits_V \frac{\partial\mathbf{b}}{\partial\mathbf{s}}\delta\mathbf{s}\mathbf{v}dV + \int\limits_S \frac{\partial\mathbf{p}}{\partial\mathbf{s}}\delta\mathbf{s}\mathbf{v}d\Gamma$$

$$- \int\limits_V \rho\delta\ddot{\mathbf{u}}^s\mathbf{v}dV + \delta L_{cN} + \delta L_{cT}, \tag{2.10}$$

which is deduced from (2.4) due to variation with respect to varying $\mathbf{s}$. For simplicity, the integrals over Γ_c (the last two terms in equation (2.4)), which express the virtual power of the contact normal and frictional stresses, are denoted by L_{cN} and L_{cT} respectively. These terms will be discussed in detail below.

Remark 1 – The form of the first integral in (2.10) is valid for linear elasticity. Thus the form $\sigma(\mathbf{s},\mathbf{u}) : \varepsilon(\mathbf{v})$ is bilinear with respect to $\mathbf{u}$ and $\mathbf{v}$.

Remark 2 – In the nonlinear case (large deformations) the above mentioned integral takes the form $\int\limits_{V_R} \mathbf{S}\mathbf{F}^T : \nabla\mathbf{v}dV_R$, where $\mathbf{S}$ - II Piola-Kirchhoff stress tensor, $\mathbf{F}$ - deformation gradient, ∇ - gradient operator and thus the variation yield $\int\limits_{V_R} (\frac{\partial\mathbf{S}}{\partial\mathbf{E}}\delta\mathbf{E}^s\mathbf{F}^T + \mathbf{S}\delta\nabla^T\mathbf{u}^s) : \nabla\mathbf{v}dV$ (here $\mathbf{E}$ - is the Green strain tensor).

Remark 3 – The variational equation (2.10) results under the additional assumption that all integrals in (2.4) are Gateaux differentiable.
Return now to the contact terms L_{cN} and L_{cT}. Their variation needs some additional interpretation. The nondifferentiable and multivalued Signorini conditions (2.5) are replaced by a regularization term

$$t_n^\varepsilon = \begin{cases} 0 & d_n > 0 \\ \frac{d_n}{\varepsilon} & d_n \leq 0 \end{cases} \tag{2.11}$$

(for small deformation $d_n = u_n - g$, where g is the gap between the contacting bodies). Hence,

$$L_{cN}^\varepsilon = \int\limits_{\Gamma_c} t_n^\varepsilon v_n d\Gamma = \int\limits_{\Gamma_c} \frac{d_n}{\varepsilon} v_n d\Gamma = \int\limits_{\Gamma_c} \frac{u_n - g}{\varepsilon} v_n d\Gamma. \tag{2.12}$$

The regularization of the frictional stress vector $\mathbf{t}_T = -\mu t_n \frac{\dot{\mathbf{u}}_T}{|\dot{\mathbf{u}}_T|}$ has the form

$$t_{T\alpha}^\varepsilon = -\mu(t_n, |\dot{u}_T|)t_n\Psi_\alpha^\varepsilon \quad , \quad \Psi_\alpha^\varepsilon = \begin{cases} \frac{\dot{u}_{T\alpha}}{\varepsilon} & |\dot{\mathbf{u}}_T| \leq \varepsilon \\ \mathrm{sign}\dot{\mathbf{u}}_T\mathbf{e}_T & |\dot{\mathbf{u}}_T| > \varepsilon, \end{cases} \tag{2.13}$$

where
μ - friction coefficient
$\dot{\mathbf{u}}_T = \dot{\mathbf{u}}^+ - \dot{\mathbf{u}}^-$ - sliding velocity
ε - sufficient small number, $\alpha = 1, 2$.

Thus

$$L^{\varepsilon}_{cT} = \int_{\Gamma_c} t^{\varepsilon}_{T\alpha} v_{T\alpha} d\Gamma = - \int_{\Gamma_c} \mu t_n \Psi^{\varepsilon}_{\alpha} v_{T\alpha} d\Gamma. \tag{2.14}$$

According to formula (2.7) the variations of the contact functionals (2.12) and (2.14) take the form

$$\begin{aligned} \delta L^{\varepsilon}_{cN} &= \int_{\Gamma_c} \frac{\delta u^s_n}{\varepsilon} v_n d\Gamma + \delta L^{\varphi}_{cN}, \\ \delta L^{\varepsilon}_{cT} &= - \int_{\Gamma_c} [(\frac{\partial \mu}{\partial t_n} \delta t^s_n + \frac{\partial \mu}{\partial |\dot{u}_T|} \delta \dot{u}^s_T) t^{\varepsilon}_n + \mu \delta t^s_n] \Psi^{\varepsilon}_{\alpha} v_{T\alpha} d\Gamma \\ &- \int_{\Gamma_c} \mu t_n \delta \Psi^{\varepsilon}_{\alpha} v_{T\alpha} d\Gamma + \delta L^{\varphi}_{cT} \end{aligned} \tag{2.15}$$

with the substitutions

$$\begin{aligned} \delta L^{\varphi}_{cN} &= \int_{\Gamma_c} (\frac{\partial t^{\varepsilon}_n}{\partial n} - 2kt_n) \delta \varphi_n v_n d\Gamma + \int_{\partial \Gamma_c} t_n v_n \nu \delta \varphi ds, \\ \delta L^{\varphi}_{cT} &= \int_{\Gamma_c} (\frac{\partial t_{T\alpha}}{\partial n} - 2kt_{T\alpha}) \delta \varphi_n v_{T\alpha} d\Gamma + \int_{\partial \Gamma_c} t_{T\alpha} v_{T\alpha} \nu \delta \varphi ds. \end{aligned} \tag{2.16}$$

The integrals (2.16) describe an increment of the contact power which results from the changes of the contact zone due to deformation. This means that the shape variation $\delta\varphi$ depends on $\delta \mathbf{u}^s$ and thus it will be

$$\begin{aligned} &\delta \varphi = \delta \mathbf{u}^s \qquad \text{on } \Gamma_c, \\ &t_n = 0 \Rightarrow \frac{\partial t_n}{\partial \mathbf{u}} \delta \mathbf{u}^s + \frac{\partial t_n}{\partial \mathbf{x}} \delta \varphi = 0 \quad \text{for} \quad \mathbf{x} \in \partial \Gamma_c, \\ &\delta \varphi = 0 \qquad \text{when} \qquad \partial \Gamma_c \quad \text{is fixed.} \end{aligned} \tag{2.17}$$

So the variation of the contact terms represented in equation (2.10) by the sum $\delta L_c = \delta L_{cN} + \delta L_{cT}$ are fully expressed by means of $\delta \mathbf{u}^s$ (formulae (2.15)-(2.17)).

The above derived formulae are rather complicated, therefore before doing the next step of analysis some comments are necessary.

Comments: In addition to purely mathematical problems (like existence of derivatives, regularity of the problem etc.) the fundamental question arises: Is it practically (numerically) possible to perform a sensitivity analysis for such complex and sensitive (in the physical meaning) phenomena like frictional contact? The following remarks should bring some explanation to this question.

1. The majority of nonlinear boundary value problems (unilateral contact and friction are nonlinear even when small deformations are considered) are solved by means of an incremental approach (iteratively). As we know the standard procedure consists in execution of mapping $(\mathbf{b}, \mathbf{p}) : \lambda \in [0,1] \to (\lambda \mathbf{b}, \lambda \mathbf{p})$ with the partition of the interval $0 = \lambda_0 < \lambda_1 < ... < \lambda_N = 1$. Hence $\mathbf{u} : [\mathbf{0}, \mathbf{1}] \to \mathbf{u}(\lambda), \mathbf{u}(\lambda_\mathbf{i} + \triangle\lambda) = \mathbf{u}^{\mathbf{i+1}} = \mathbf{u}^{\mathbf{i}} + \triangle\mathbf{u}, \triangle\mathbf{u} = \mathbf{u}'\triangle\lambda, \mathbf{i} = \mathbf{0}, \mathbf{1}...\mathbf{N}$. Increments $\triangle\lambda$ of the load parameter λ evoke increments of all state quantities $\triangle\sigma, \triangle\varepsilon, \triangle t_n, \triangle t_{T\alpha}$ etc. and lead to mappings $\Gamma_c(\lambda^i) \to \Gamma_c(\lambda^{i+1})$, too (see [5]).
So we see, that in fact any incrementally realized calculation of unilateral (and or frictional) contact includes terms considered by the derivation the formula (2.15).
One can interpret the incremental approach as a sensitivity procedure with the control parameter $s \equiv \lambda \in \Re^1$.
A proper parameter sensitivity analysis differs in the treatment of the independent variables $\lambda \in [0,1]$ and $\mathbf{s}(\mathbf{x}, \mathbf{t}) \in \mathbf{U_{ad}}$. Partition of λ results from the method of solution of the boundary value problem whereas incrementation (variation) of $\mathbf{s}$ means redesign of the dynamic system under consideration. By using an additional optimization procedure the sequence of $\delta\mathbf{s}$ lead to an optimal solution of the problem (2.1)-(2.3). It is however worth noting that the sensitivity procedure repeats in fact the same calculational algorithm (with the same difficulties) like this which is obviously used to solve the contact problem. So, the same computational programs can be used for this reason.
2. The analysis of the stick-slip process needs especially careful treatment. Many sliding rules are quoted in the literature ([5]). Replacing the discontinuous Coulomb friction law by its regularization (2.13), one sees that the increment $\delta\Psi_\alpha^\varepsilon$ in the expression (2.15) differs from 0 for $|\dot{\mathbf{u}}_T| \leq \varepsilon$ only (otherwise this term vanishes). Thus it will be $\int_{\Gamma_c} \mu t_n \delta\Psi_\alpha^\varepsilon v_{T\alpha} d\Gamma = \int_{\Gamma_c} \mu t_n \frac{\delta \dot{u}_{T\alpha}}{\varepsilon} v_{T\alpha} d\Gamma \neq 0$.
Bearing in mind the conclusion of the previous remark, one should distinguish between the calculation of all terms in (2.15) (and hence in equation (2.10)) to find the solution of (2.4) (2.5) for any $\mathbf{b}, \mathbf{p}$ by fixed $\mathbf{s}$ and that for varying $\mathbf{s}$ (which means redesign for given $\mathbf{b}, \mathbf{p}$).
3. The numerical treatment of the obviously high-frequent oscillatory character of the contact phenomena (see [5]) is the most time-consuming part of the analysis. Sensitivity analysis of another objective functional of type (2.3) is much simpler. The main difficulty is then connected with the determination of the state variation $\delta\mathbf{u}^s$. For this reason the variational equation (2.10) (with the substitutions (2.15),(2.16)) must be solved. This method will be called the direct sensitivity approach

3. Adjoint system

In order to avoid the computationally expensive calculation of $\delta\mathbf{u}^{\mathrm{s}}$ needed in (2.8), let us consider on augmented functional (instead of (2.3))

$$\begin{aligned} J(s,\mathbf{u})^* &= J(s,\mathbf{u}) \\ &-[B(s,\mathbf{u},\mathbf{v}) - L(s,\mathbf{v}) - L_{CN}(s,\mathbf{uv}) - L_{CT}(\mathbf{u},\mathbf{v}) + b(\ddot{\mathbf{u}},\mathbf{v})] \end{aligned} \tag{3.1}$$

where
$B(s,\mathbf{u},\mathbf{v}), L(s,\mathbf{v}), \mathbf{b}(\ddot{\mathbf{u}},\mathbf{v})$ stand for virtual power of strains, external forces and inertial force respectively and
$L_{CN}(\mathbf{u},\mathbf{v}), L_{CT}(\mathbf{u},\mathbf{v})$ are the virtual power of the contact forces mentioned in (2.10).
Performing variation of (3.1), one obtains

$$\begin{aligned} &\delta J^* = \delta J_u + \delta J_s - \delta B_u - \delta B_s + \delta L_s + \delta L_{CN} + \delta L_{cT} - \delta b_u = \\ &(\delta J_u - \delta B_u + \delta L_{cN} + \delta L_{cT} - \delta b_u) + \delta J_s - \delta B_s + \delta L_s, \end{aligned} \tag{3.2}$$

where the denotations (2.9), (2.15) shoud be used and the subscripts in forms B, L, b mean variations with respect to $\mathbf{u}$ and $\mathbf{s}$, respectively.
Notice that the equation (2.10) has the form

$$\delta B_u + \delta B_s = \delta L_s + \delta L_{cN} + \delta L_{cT} - \delta b_u. \tag{3.3}$$

Specifying now an adjoint problem
$\mathbf{v}:\qquad \delta B_u = \delta J_u + \delta L_{cN} + \delta L_{cT} - \delta b_u,$
that is

$$\begin{aligned} &\int_V \sigma(\mathbf{s},\mathbf{v}) : \varepsilon(\delta\mathbf{u}^s) dV = \int_V \frac{\partial F_0}{\partial u} \delta\mathbf{u}^s dV + \int_S \frac{\partial f_0}{\partial u} \delta\mathbf{u}^s dS + \int_{\Gamma_c} \frac{\delta u_n^s}{\varepsilon} v_n d\Gamma \\ &- \int_{\Gamma_c} \{[(\frac{\partial\mu}{\partial t_n}\delta t_n^s + \frac{\partial\mu}{\partial\dot{u}_T}\delta\dot{u}_T^s) t_n^\varepsilon + \mu\delta t_n^s]\Psi_\alpha^\varepsilon + \mu t_n \frac{\delta\dot{u}_{T\alpha}}{\varepsilon}\} v_{T\alpha} d\Gamma \\ &-\delta L_{cN}^\varphi(\delta\mathbf{u}^s) + \delta L_{cT}^\varphi(\delta\mathbf{u}^s) - \int_V \rho\ddot{\mathbf{v}}\delta\mathbf{u}^s dV, \\ &\mathbf{v}(t_k) = \dot{\mathbf{v}}(t_k) = 0 \qquad \forall \quad \delta\mathbf{u}^s \in V_0, \end{aligned} \tag{3.4}$$

one sees that the first term in (3.2) (the sum in brackets) vanishes. So in view of (3.3) and (3.4), the first-order sensitivity of the functional (2.9) is expressed as follows

$$\begin{aligned} &\delta J = \delta J_s + \delta L_s - \delta B_s = \int_V \frac{\partial F_0}{\partial \mathbf{s}} \delta\mathbf{s} dV + \int_S \frac{\partial f_0}{\partial \mathbf{s}} \delta\mathbf{s} dV + \\ &+ \int_V \frac{\partial \mathbf{b}}{\partial \mathbf{s}} \delta\mathbf{s}\mathbf{v} dV + \int_{\Gamma_\sigma} \frac{\partial \mathbf{p}}{\partial \mathbf{s}} \delta\mathbf{s}\mathbf{v} d\Gamma - \int_V \frac{\partial \sigma}{\partial \mathbf{s}} \delta\mathbf{s} : \varepsilon(\mathbf{v}) dV, \end{aligned} \tag{3.5}$$

that is in terms of the primary and adjoint fields $\mathbf{u}$ and $\mathbf{v}$ and explicitly by $\delta\mathbf{s}$ only.

It is of great qualitative as well as of quantitative importance to emphasize that the adjoint variable $\mathbf{v}$ has a clear physical meaning. It is namely the virtual velocity field which is equal to displacements of the system resting on a conjugate elastic foundation with fictitious reactions $t_n^* = k_N^* v_n$ and $t_{T\alpha}^* = k_{T\alpha}^* v_{t\alpha}$ loaded by generalized forces $\mathbf{b}^* = \frac{\partial F_0}{\partial \mathbf{u}}$, $\mathbf{p}^* = \frac{\partial f_0}{\partial \mathbf{u}}$ (Fig.1)
Here $k_n^*, k_{T\alpha}^*$ denote the expression in (3.4) which stand in the integrals over Γ_c as multiplier by $\mathbf{v}$. In order to perform the parameter sensitivity analysis of the system the following iterative procedure can be applied.

i) For given loads $\mathbf{b}, \mathbf{p}$ and presumed control $\mathbf{s}^0$ find the solution of (2.4),(2.5). For this reason the standard incremental approach is obviously used. A finite element technique in the space domain and a suitable method of time integration are mostly applied.

ii) Compute the variation δJ and variations $\delta L_{cN}, \delta L_{cT}$ from (2.15), (2.16), taking into account (2.17).

iii) Solve the adjoint equation (3.4) (always linear with respect to $\mathbf{v}$).

iv) Use formula (3.5) to calculate the sensitivity of J.

Remark 4 Looking for the optimal control, the steepest descent method can be used. The required descent direction $\mathbf{w} = -\nabla\mathbf{J}$ (the gradient of the considered objective function) can be found for different increments $\delta\mathbf{s}$.

Remark 5 A wide class of optimal control problems (e.g. intelligent structures etc.) lead after space discretization to the matrix equation

$\mathbf{M}\ddot{\mathbf{q}} + \mathbf{K}_T\dot{\mathbf{q}} + (\mathbf{K} + \mathbf{K}_N + \mathbf{K}_{TN})\mathbf{q} = \mathbf{F} + \mathbf{F}^s$,

where $\mathbf{M}, \mathbf{K}, \mathbf{K_T}, \mathbf{K_N}, \mathbf{K_{TN}}$ stand for mass, stiffness, friction damping and contact matrices

$\mathbf{q}, \mathbf{F}, \mathbf{F^s}$ are state, external load and control vectors respectively.

Thus, Pontryagin's maximum principle has to be used. In this case it is worth making a note that for autonomous systems (and otherwise by suitable substitutions one can convert the system into such a form) the corresponding system Hamiltonian $H(\mathbf{s}, \mathbf{q}, \mathbf{v}, t)$ attains for the optimal solution $\mathbf{s}_{opt}, \mathbf{q}_{opt}$ a constant value with respect time t. This feature (calculation of $H(s(t), q(t), v(t), t)$) is a good verification of the computation (is the optimum achieved or not).

Remark 6 Shape sensitivity (when the shape-transformation process is specified by providing the rule changes of the boundary of the contactor body) can be stated and analyzed in a similar manner. Shape derivatives (or the so called material derivative approach) are then used to determine the sensitivity of a given functional (see [1], [3], [6]). For an objective function of the form (2.3), one obtains then the expression

$$\delta J = \int\limits_{V_\varphi} \frac{\partial F_0}{\partial u} \delta\mathbf{u} dV + \int\limits_{S_\varphi} F_0 \delta\varphi \cdot \mathbf{n} d\Gamma + \int\limits_{S_\varphi} [\frac{\partial f_0}{\partial u}\delta\mathbf{u} + (\frac{\partial f_0}{\partial n} + 2k f_0)\delta\varphi \cdot \mathbf{n}] d\Gamma +$$

$$\int\limits_{\partial S_\varphi} f_0 \nu \cdot \varphi ds,$$

where the shape variations $\delta\varphi$ are now independent. Suitable adjoint fields can be introduced similarly. Sometimes it is convenient to formulate the state equation in a dual form (with the advantage that integrals which describe the behaviour of the system are defined over the contact area Γ_c, only - see eg [6]).

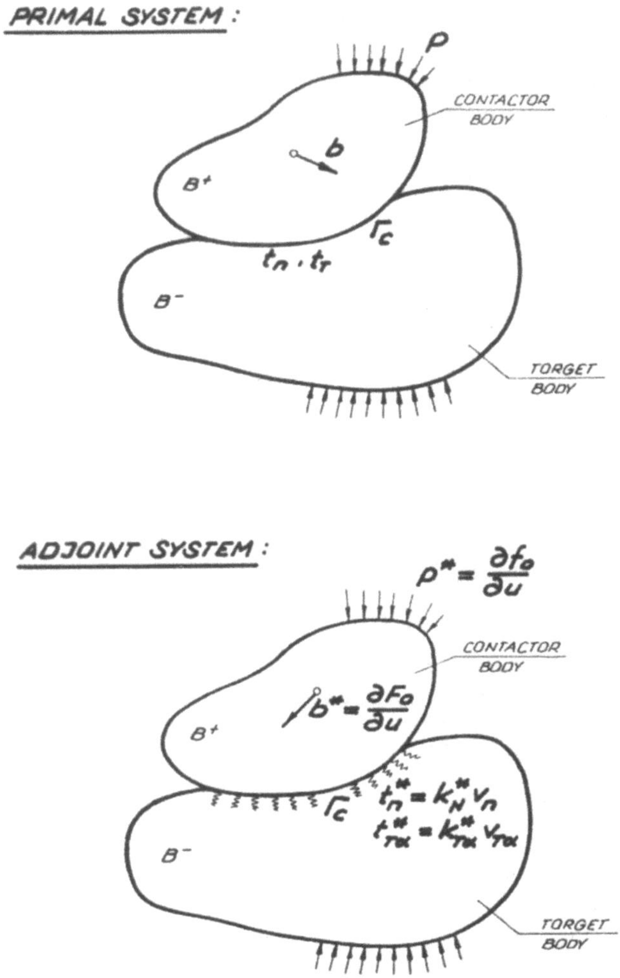

Fig. 1

4. Conclusions

In this paper the statement and features of sensitivity of continuous dynamical systems in terms of unilateral contact has been presented. Attention was paid to parameter sensitivity. As mentioned in remark 6, shape sensitivity, which is evoked by treating the boundary shape of the body undergoing contact as a design variable, can be performed by using the same adjoint–system approach. Sensitivity expressions of contact terms provide an analytical tool to analyze contact phenomena and their influence on the behaviour of the dynamical system. Together with other optimization techniques (like Pontryagin's maximum principle, the Milyutin-Dobovitsky theorem etc) they constitute the basis for finding optimal solutions for controllable structures.

References

1. Haslinger, J., Neittaanmaki, P. *Finite Element Approximation for Optimal Shape Design*, J.Wiley & Sons, 1988
2. Kleiber, M., Antunez, H., Hin, T., Kowalczyk, P. *Parameter Sensitivity in Nonlinear Mechanics*, J.Wiley & Sons, 1997
3. Mróz, Z. *Variational methods in sensitivity analysis and optimal design*, Eur.J.Mech. A/Solides, **13**, 4-suppl.
4. Paczelt, I., Szabo, T. *Optimal shape design for contact problems*, Struct. Optim. **7**, 66-75, 1994.
5. Szefer, G. *Dynamic contact of bodies experiencing large deformations*, Acta Mechanica, 125, 217-233, 1997
6. Szefer, G. *Shape sensitivity in contact problems of elastic bodies*, in First World Congr. Multidisc. Optim. 335-340, Pergamon 1995
7. Szefer, G. *Konzeption der adjungierten Gleichung in der optimalen Steuerung elastischer Systeme*, TH Leipzig, Wissenschaftl. Zeitschr. 8/1984, 3, 129-143

On the Treatment of Contact Problems in Elasto-Plasticity

M. Bach, C. Eck and M. Schulz

Mathematical Institute A, University of Stuttgart, Germany

Summary. We present a boundary integral formulation for a quasistatic elasto-plastic contact problem with Coulomb friction. The used model is valid for material satisfying a general flow law including hardening in the framework of the theory of small deformations. Our approach is based on a penalization and smoothing of the Coulomb friction law and a symmetric representation of the Steklov–Poincaré operator. For the solution of the non–linear evolution problem a return mapping algorithm is formulated.

1. Introduction

Contact problems have important applications in different fields of mechanics, for example in earthquake engineering, civil engineering or machine dynamics. Since the contact areas in many applications are very small, the transmitted force densities are usually rather big and lead to plastic deformations. Therefore it is suitable to use an elasto-plastic model for the material. For the solution of contact problems, the boundary element method is particularly suitable, because the nonlinear contact and friction boundary conditions are formulated on the boundary of the bodies only.

In this contribution we consider a contact problem for an elasto-plastic body and a rigid obstacle. The material law is given by a general flow law including hardening. The friction is modeled by the classical Coulomb law.

Following an approach proposed by Korneev and Langer [4], the elasto-plastic material law is formulated as an evolution problem for the displacement field, the plastic stresses and the hardening parameters. The formulation involves a nonlinear monotone differential operator acting on the displacements. Using the classical penalty method and smoothing of the norm in the friction term, the problem can be formulated as a nonlinear variational equation. A boundary integral formulation of this variational equation is presented. It is discretized by a symmetric boundary element method described in [3] for the corresponding elastic contact problem. For the solution of the resulting nonlinear boundary integral equation we employ a return mapping algorithm.

2. Problem setting

For simplicity of the presentation we consider the quasistatic contact of an elasto-plastic body $\Omega \subset \mathbb{R}^d, d \in \{2,3\}$, with a rigid foundation. The approach can be extended to the contact of several elasto-plastic bodies.

First, we present the model for the elasto-plastic deformation following the classical plasticity theory of D.C. Drucker [2] and the concept of V.G. Korneev and U. Langer [4]. If there occure only small deformations, the strain tensor can be decomposed in an elastic part and a plastic part

$$\varepsilon(u) = \left(\frac{1}{2}\left(\frac{\partial u_i}{\partial x_j} + \frac{\partial u_j}{\partial x_i}\right)\right)_{i,j=1}^{d} = \varepsilon^{(e)} + \varepsilon^{(p)}. \tag{2.1}$$

The stress–strain relation in the elastic regime is given by the Hooke law

$$\sigma = E\varepsilon^{(e)}.$$

For the plastic deformation $\varepsilon^{(p)}$ an elastoplastic evolution law including hardening is used. Let $\kappa \in \mathbb{R}^l$ be a vector of hardening parameters. Then the yield surface is characterized by a yield function $F = F(\sigma, \kappa)$ via $F(\sigma, \kappa) = 0$. The evolution of the internal variables is given by the hardening rule

$$\dot{\kappa} = \Lambda\, r(\sigma, \kappa), \qquad \kappa(0) = \kappa_0, \tag{2.2}$$

with a continuous function $r : \mathbb{R}^6 \times \mathbb{R}^l \to \mathbb{R}^l$ and the associated flow law

$$\dot{\varepsilon}^{(p)} = \Lambda \frac{\partial F}{\partial \sigma}(\sigma, \kappa). \tag{2.3}$$

The proportionality factor $\Lambda \geq 0$ in both relations is the same.

The loading and unloading criteria are characterized by the compatibility condition

$$\frac{d}{dt} F(\sigma, \kappa) = 0.$$

This condition ensures that the state of stress which leads to plastic deformation lies always on the yield surface. If $F(\sigma, \kappa) < 0$, we have purely elastic deformations. If the stresses reach the yield surface $F(\sigma, \kappa) = 0$ we have to distinguish the following three cases:

- If $\frac{\partial F}{\partial \sigma}(\sigma, \kappa)\dot{\sigma} < 0$ the process is called unloading, because the previous strain increment was plastic and the actual is elastic.
- If $\frac{\partial F}{\partial \sigma}(\sigma, \kappa)\dot{\sigma} = 0$ the process is termed neutral loading, there is no change in the hardening parameters and in the plastic strains.
- If $\frac{\partial F}{\partial \sigma}(\sigma, \kappa)\dot{\sigma} > 0$ we have the case of active loading, a change of the stress state is followed by plastic deformations.

For the proportional factor Λ we have

$$\Lambda = 0 \quad \text{for} \quad F(\sigma,\kappa) < 0 \vee \left(F(\sigma,\kappa) = 0 \wedge \frac{\partial F}{\partial \sigma}\dot{\sigma} \le 0 \right),$$

$$\Lambda = \frac{\frac{\partial F}{\partial \sigma} E}{\frac{\partial F}{\partial \sigma} E \frac{\partial F}{\partial \sigma} - \frac{\partial F}{\partial \kappa} r(\sigma,\kappa)} \dot{\varepsilon} \quad \text{for} \quad F(\sigma,\kappa) = 0 \wedge \frac{\partial F}{\partial \sigma}\dot{\sigma} > 0 \,.$$

Our goal is the formulation of the stress–strain relations in a unique form for all possible processes of loading. For simplification we use the relation

$$\operatorname{sign}\left(\frac{\partial F}{\partial \sigma}\dot{\sigma} \right) = \operatorname{sign}\left(\frac{\partial F}{\partial \sigma} E\dot{\varepsilon} \right), \tag{2.4}$$

which holds for all regular points σ on the yield surface satisfying $\dot{\sigma} = E_{ep}\dot{\varepsilon}$ with $\dot{\varepsilon} = \varepsilon(\dot{u})$ and E_{ep} as follows (see [4, 5])

$$E_{ep} := E - \frac{E \frac{\partial F}{\partial \sigma} \frac{\partial F}{\partial \sigma}^T E}{\frac{\partial F}{\partial \sigma}^T E \frac{\partial F}{\partial \sigma} - \frac{\partial F}{\partial \kappa}^T r(\sigma,\kappa)} \,.$$

Now the possible states of stresses can be reformulated in terms of the strains and, introducing the abbreviation $\psi(\sigma,\kappa) := \left(\frac{\partial F}{\partial \sigma}^T E \frac{\partial F}{\partial \sigma} - \frac{\partial F}{\partial \kappa}^T r(\sigma,\kappa) \right)^{-1}$, the Heaviside function $\Theta_+(y)$ and the function $\Phi(y) := \Theta_+(y)\, y$, we get the following evolution equations

$$\dot{\varepsilon}^p = e(\dot{\varepsilon},\sigma,\kappa) \quad := \quad \frac{\partial F}{\partial \sigma} \Phi \left(\frac{\partial F}{\partial \sigma}^T E\varepsilon(\dot{u}) \right) \Theta_+(F(\sigma,\kappa))\psi(\sigma,\kappa), \tag{2.5}$$

$$\dot{\kappa} = k(\dot{\varepsilon},\sigma,\kappa), \quad := \quad r(\sigma,\kappa)\, \Phi \left(\frac{\partial F}{\partial \sigma}^T E\varepsilon(\dot{u}) \right) \Theta_+(F(\sigma,\kappa))\psi(\sigma,\kappa). \tag{2.6}$$

Altogether this results in a compact formulation of the stress–strain relation describing the elasto–plastic deformation of material with hardening:

$$\dot{\sigma} = E\,\dot{\varepsilon} - E\, e(\dot{\varepsilon},\sigma,\kappa), \tag{2.7}$$

$$\dot{\kappa} = k(\dot{\varepsilon},\sigma,\kappa). \tag{2.8}$$

Using a decomposition of the stress tensor into fictitious elastic and plastic parts, $\sigma = \sigma^{(e)}(u) - \sigma^{(p)} = E\varepsilon(u) - E\varepsilon^{(p)}$, the equilibrium equations of the quasistatic elasto-plastic problem reads as

$$-\operatorname{div}\sigma^{(e)}(u) = f - \operatorname{div}\sigma^{(p)} \,. \tag{2.9}$$

To formulate the corresponding boundary condition, we assume that the boundary $\Gamma := \partial\Omega$ is composed of the three disjoint, measurable parts $\Gamma_U, \Gamma_F, \Gamma_C$, with Dirichlet conditions

$$u = g \text{ on } \Gamma_U, \tag{2.10}$$

Neumann conditions

$$\tau(u) = b \text{ on } \Gamma_F, \tag{2.11}$$

where $\tau_i = \sigma_{ij}(u)n_j$ denotes the components of the boundary traction and $n = \{n_j\}$ is the outer normal vector of the boundary, and contact and friction conditions on Γ_C. The contact is described by the Signorini condition

$$u_N \leq g_N,\ \tau_N \leq 0,\ \tau_N(u_N - g_N) = 0, \tag{2.12}$$

where $u_N := u \cdot N, \tau_N := \tau \cdot N$, and g_N measures the distance of Γ_C to the rigid foundation in the direction of the given vector N [1, 3].

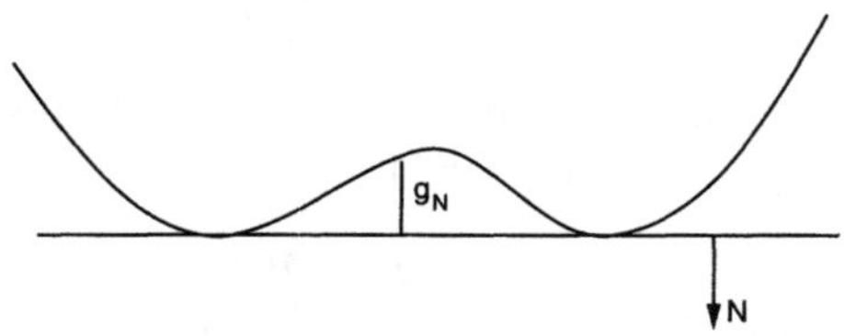

Fig. 2.1. Contact condition on Γ_C

The friction is modeled by the Coulomb law

$$\begin{aligned} \dot{u}_T = 0 \quad &\Rightarrow \quad |\tau_T| \leq \mathcal{F}(0)\,|\tau_N|\,, \\ \dot{u}_T \neq 0 \quad &\Rightarrow \quad \tau_T = -\mathcal{F}(\dot{u}_T)\,|\tau_N|\,\frac{\dot{u}_T}{|\dot{u}_T|}, \end{aligned} \tag{2.13}$$

with $u_T := u - u_N N, \tau_T := \tau - \tau_N N$ and time derivative $\dot{u}$ of u. Due to the dependence of the friction coefficient $\mathcal{F}$ on the velocity, we can model both sticking and slipping at least approximately.

Neglecting the inertia forces in the equilibrium equations, the boundary value problem (2.9), (2.10)–(2.13) describes a quasistatic contact problem with friction of an elasto-plastic body and a rigid foundation. The formulation of the problem is completed by initial conditions $u(x)|_{t=0} = u_0(x), \sigma(x)|_{t=0} = \sigma_0(x), \kappa(x)|_{t=0} = \kappa_0(x)$.

Now we use a time discretization

$$\dot{u}(t^\ell) \approx \frac{u(t^\ell) - u(t^{\ell-1})}{t^\ell - t^{\ell-1}} = \frac{\delta u}{\delta t},$$

with δu as increment of the displacement, δt as increment of the time, where t^ℓ, $\ell = 0, 1, 2, \ldots$ denote some fixed time steps. In what follows we denote $\delta\sigma^{(p)}$ as increment of the plastic stresses, $\delta\kappa$ as increment of the hardening parameters, and $\delta v := v - u(t^{\ell-1})$ for a general function v. The quasistatic problem is thus approximated by a sequence of static problems.

Taking into account the weak formulations

$$u_N \leq g_N, \quad \tau_N(v_N - u_N) \geq 0\,, \qquad \text{for all } v_N \leq g_N\,,$$

$$\tau_T \cdot (v_T - u_T) + \mathcal{F}(\delta u)|\tau_N|(|\delta v_T| - |\delta u_T|) \geq 0 , \qquad \text{for all } v_T ,$$

of the Signorini contact condition (2.12) and the (time-discretized) Coulomb friction law (2.13) (with modified coefficient of friction $\mathcal{F}$), the weak formulation of the static problem is given with the cone

$$\mathcal{K} := \left\{ v \in H^1(\Omega) \mid v = g \text{ on } \Gamma_U \text{ and } v_N \leq g_N \text{ on } \Gamma_C \right\}$$

of admissible functions by

Find a displacement field $u \in \mathcal{K}$ such that for all $v \in \mathcal{K}$

$$\begin{aligned} \int_\Omega \left(\sigma^{(e)}(\delta u) - \delta\sigma^{(p)}\right) \varepsilon(v-u)\, dy + \int_{\Gamma_C} \mathcal{F}(\delta u_T)\, |\tau_N(u)|\, (|\delta v_T| - |\delta u_T|)\, ds_y \\ \geq \int_\Omega f \cdot (v-u)\, dy + \int_{\Gamma_F} b \cdot (v-u)\, ds_y , \end{aligned} \qquad (2.14)$$

$$\delta\sigma^{(p)} = Ee(\varepsilon(\delta u), \sigma, \kappa) ,$$
$$\delta\kappa = k(\varepsilon(\delta u), \sigma, \kappa) .$$

Our aim is the approximation of the variational inequality by a variational equation and the derivation of a boundary element formulation. In order to reach the first aim we use the penalty method and replace the Signorini condition by the nonlinear boundary condition

$$\tau_N(u) = -\Phi_\lambda(u_N - g_N) , \qquad (2.15)$$

with, for example, $\Phi_\lambda(x) := \frac{1}{\lambda}[x]_+$, $\lambda > 0$, $[\cdot]_+ := \max\{0, \cdot\}$. In addition we replace the non-differentiable terms $|\delta v_T|, |\delta u_T|$ by differentiable approximations with a convex, continuously differentiable function Ψ_η satisfying $|\Psi_\eta(v) - |v|| \leq \eta$. Then we obtain the following variational equation having the set of admissable functions $g + \mathcal{V}$ with

$$\mathcal{V} := \left\{ v \in H^1(\Omega) \mid v = 0 \text{ on } \Gamma_U \right\} .$$

Find a displacement field $u \in g + \mathcal{V}$ such that for all $v \in \mathcal{V}$

$$\begin{aligned} \int_\Omega \left(\sigma^{(e)}(\delta u) - \delta\sigma^{(p)}\right) \varepsilon(v)\, dy + \int_{\Gamma_C} \Phi_\lambda(\delta u_N - \delta g_N)\, v_N\, ds_y \\ + \int_{\Gamma_C} \mathcal{F}(\delta u_T)\, \Phi_\lambda(\delta u_N - \delta g_N)\, \nabla\Psi_\eta(\delta u_T) \cdot v_T\, ds_y \\ = \int_\Omega f \cdot v\, dy + \int_{\Gamma_F} b \cdot v\, ds_y , \end{aligned} \qquad (2.16)$$

$$\delta\sigma^{(p)} = Ee(\varepsilon(\delta u), \sigma, \kappa) ,$$
$$\delta\kappa = k(\varepsilon(\delta u), \sigma, \kappa) .$$

3. Boundary element method

We employ the symmetric boundary element method proposed in [3]. Let us introduce the well-known boundary integral operators

$$V\psi(x) := \int_\Gamma U(x,y)\psi(y)\,ds_y \qquad \text{(single layer operator)},$$

$$K\varphi(x) := \int_\Gamma (T_yU)(x,y)\varphi(y)\,ds_y \qquad \text{(double layer potential)},$$

$$K'\psi(x) := \int_\Gamma (T_xU)(x,y)\psi(y)\,ds_y \qquad \text{(adjoint double layer potential)},$$

$$D\varphi(x) := -T_x \int_\Gamma (T_yU)(x,y)\varphi(y)\,ds_y \qquad \text{(hypersingular operator)},$$

where U denotes the fundamental solution of the equilibrium equations and $T_{x/y}$ the boundary stress operator with respect to variable x/y. Using the symmetric realization of the Steklov-Poincaré operator, i.e.

$$S := \left(\frac{1}{2}I + K'\right) V^{-1} \left(\frac{1}{2}I + K\right) + D \tag{3.1}$$

in order to ensure a symmetric matrix for the Dirichlet-Neumann map in a Galerkin discretization, we get the following boundary integral formulation having the set of admissable functions $g + \tilde{H}^{\frac{1}{2}}(\Gamma)$ with

$$\tilde{H}^{\frac{1}{2}}(\Gamma) := \left\{ v \in H^{\frac{1}{2}}(\Gamma) \,\middle|\, v = 0 \text{ on } \Gamma_U \right\}.$$

Find a displacement field $u \in g + \tilde{H}^{\frac{1}{2}}(\Gamma)$ *such that for all* $v \in \tilde{H}^{\frac{1}{2}}(\Gamma)$

$$\begin{aligned}
&\int_\Gamma S(\delta u)\cdot v\,ds_y + \int_{\Gamma_C} \Phi_\lambda\,(\delta u_N - \delta g_N)\,v_N\,ds_y \\
&\qquad + \int_{\Gamma_C} \mathcal{F}\,(\delta u)\,\Phi_\lambda\,(\delta u_N - \delta g_N)\,\nabla\Psi_\eta\,(\delta u_T)\cdot v_T\,ds_y \\
&\quad = \int_\Gamma V^{-1}N_1(f)\cdot v\,ds_y + \int_\Gamma V^{-1}N_2(\delta\sigma^{(p)})\cdot v\,ds_y + \int_{\Gamma_F} b\cdot v\,ds_y\,, \qquad (3.2)
\end{aligned}$$

$$\delta\sigma^{(p)} = Ee(\varepsilon(\delta u),\sigma,\kappa)\,,$$
$$\delta\kappa = k(\varepsilon(\delta u),\sigma,\kappa)\,,$$

where N_1, N_2 *denote the Newton potentials*

$$N_1(f)(x) := \int_\Omega U(x,y)f(y)\,dy\,, \quad N_2(g)(x) := \int_\Omega \nabla U(x,y)g(y)\,dy\,.$$

For the discrete Galerkin equations we refer to [3]. The Newton potentials can be evaluated efficiently by using the ficticious domain method, see e.g. [6]. This method is based on an effective calculation of a particular solution of the inhomogeneous problem in an artificial domain (including the original domain) via fast spectral or multigrid methods on uniform meshes.

Solving the problem (3.2) with an iterative solution scheme, it is necessary to ensure that the stresses are always on or inside the yield surface. To do this we use a return-mapping-algorithm, where, after calculation of an elastic predictor of the stresses and checking of the yield condition, a plastic corrector is calculated if the stresses lie outside the yield surface. Using for example the Von Mises flow condition with kinematic and isotropic hardening parameters $\kappa^{kin}, \kappa^{iso}$ and the constant hardening modules V^{kin}, V^{iso}, given as

$$F(\sigma, \kappa) = \sqrt{(s - V^{kin}\kappa^{kin})^T (s - V^{kin}\kappa^{kin})} - \sqrt{\frac{2}{3}}\,(\sigma_0 + V^{iso}\kappa^{iso})$$

with $s = (s_{ij})_{i,j=1,2,3}$ defined by $s_{ij} = \sigma_{ij} - \frac{1}{3}\sigma_{ij}\delta_{ij}$ for $i, j = 1, 2, 3$, and $\kappa = (\kappa^{kin}, \kappa^{iso})$, we get for the associated flow law

$$\delta\varepsilon^p_{j+1} = \varepsilon^p_{j+1} - \varepsilon^p_j = \Lambda_{j+1}\frac{\partial F}{\partial \sigma}\Big|_{j+1},$$

where $\frac{\partial F}{\partial \sigma} = \frac{\partial F}{\partial S} : \frac{\partial S}{\partial \sigma} = \frac{S}{\sqrt{S^T : S}}$ and $S := s - V^{kin}\kappa^{kin}$. For the increment of the kinematic and isotropic hardening it holds

$$\delta\kappa^{kin}_{j+1} = \frac{2}{3}\Lambda_{j+1}\frac{\partial F}{\partial \sigma}\Big|_{j+1} \qquad \text{and} \qquad \delta\kappa^{iso}_{j+1} = \sqrt{\frac{2}{3}}\Lambda_{j+1}.$$

Under the condition

$$\sqrt{S^T_{j+1} : S_{j+1}} - \sqrt{\frac{2}{3}}\,(\sigma_0 + V^{iso}\kappa^{iso}_{j+1}) = 0,$$

the plastic corrector s^p_{j+1} can be calculated as

$$s^p_{j+1} = s_{j+1} - s^{tst}_{j+1} = -2G\varepsilon_{j+1} = -2G\Lambda_{j+1}\frac{\partial F}{\partial \sigma}\Big|_{j+1},$$

with $G = \frac{E}{2(1+\nu)}$ and

$$\Lambda_{j+1} = \frac{F(s^{tst}, \kappa)|_{j+1}}{2G + \frac{2}{3}(V^{kin} + V^{iso})}$$

as proportionality factor.

4. Numerical example

As an example we present calculations for the plane contact problem of a purely elastic half disc (radius $r = 0.4$) with a rigid flat foundation. The non-dimensionalised material data are $E = 50$, $\nu = 0.3$ and $\mathcal{F} = 0.3$. In addition to the displacements $d = (0, -0.02)$ given on the upper part of the half disc, a volume force $f = (0, -0.15)$ acts on it. As penalty parameter we choose $\lambda = 10^{-4}$, as smoothing parameter we take $\eta = 10^{-5}$. We use a uniform grid with 256 nodes. This problem presents a preliminary step in an iteration procedure where the plastic stress part is considered like an additional volume force. In Fig. 4.1, both the normal and tangential tractions are depicted in the case of zero volume forces on the contact part of the boundary. Fig. 4.2 shows the corresponding boundary stresses in case of the given volume force f.

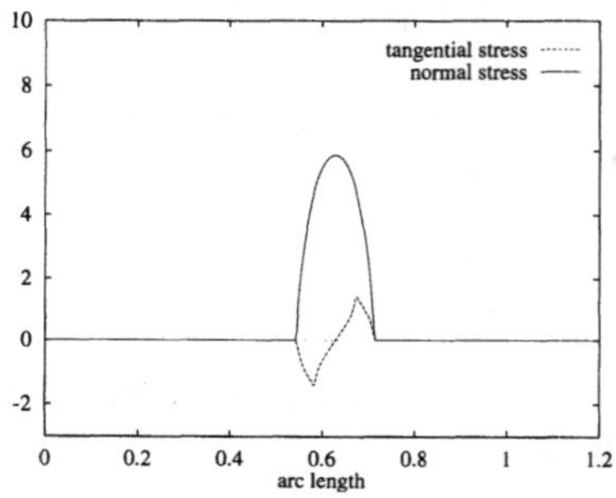

Fig. 4.1. Contact with friction and without volume force

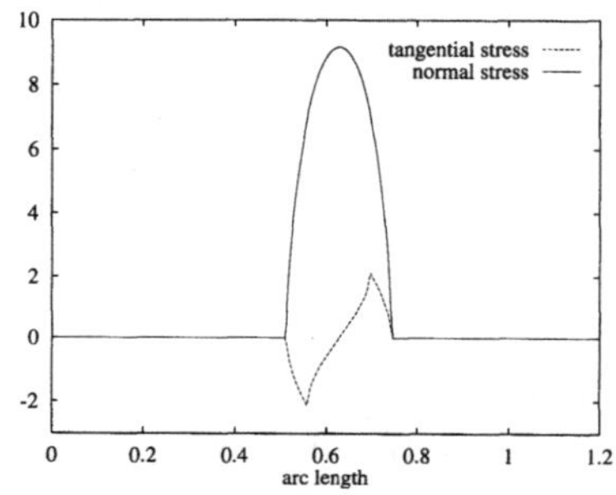

Fig. 4.2. Contact with friction and volume force $f = (0, -0.15)$

References

1. M. Bach, A.M. Khludnev, V.A. Kovtunenko: Derivatives of the energy functional for 2d-problems with a crack under Signorini and friction conditions. *Math. Meth. Appl. Sci.*, to appear.
2. D. C. Drucker, W. Prager: Soil Mechanics and Plastic Analysis of Limit Design. *Quart. Appl. Math.* **10**, 157–165, 1952.
3. Ch. Eck, O. Steinbach, W.L. Wendland. A symmetric boundary element method for contact problems with friction. *Mathematics and Computers in Simulation* **50**, 43-61, 1999.
4. V. G. Korneev, U. Langer: *Approximate solution of plastic flow theory problems.* Teubner–Texte zur Mathematik, Leipzig, 1984.
5. M. Schulz: A–Posteriori Fehlerschätzer für mittels Finiter Elemente modellierte elasto–plastische Verformungsvorgänge. Doctoral thesis, Universität Stuttgart, 1997.
6. O. Steinbach: Fast evaluation of Newton potentials in boundary element methods. *East-West J. Numer. Math.* **7**, 211-222, 1999.
7. M. Schulz, W. L. Wendland: On an adaptive finite element method for elasto–plastic deformations with hardening. Preprint **1997/39**, SFB 259, Universität Stuttgart, 1997.

On Contact Problems for Linear and Nonlinear Elastic Plates: Existence of Solutions and Application of Augmented Lagrangian Method

W. Bielski[2], A. Gałka[1], and J.J. Telega[1]

[1] Institute of Fundamental Technological Research, Pol. Acad. Sci., 00-049 Warsaw, ul. Swiętokrzyska 21
[2] Institute of Geophysics, Pol. Acad. Sci., 01-452 Warsaw, ul. Księcia Janusza 64

Abstract. The aim of this contribution is to study a class of unilateral contact problems for three models of elastic plates. Contact conditions are formulated on the face being in contact with an obstacle and not on the mid-plane of the plate. Such rigorous approach results in nonconvex minimization problems also in the case of thin, geometrically linear plates. Existence theorems are formulated for each model considered. Ito - Kunisch [2, 3] augmented Lagrangians methods have been extended to nonconvex problems. Specific example has also been provided.

1. Geometrically linear plates

1.1 Thin plates

Let $\Omega \subset \mathbb{R}^2$ be a sufficiently smooth domain and $\Gamma = \partial\Omega$ its boundary. Ω denotes the mid-plane of the undeformed plate. The plate occupies the region $\Omega \times (-h, h) \subset \mathbb{R}^3$. The boundary Γ is decomposed into two parts: Γ_0 and Γ_1 such that $\Gamma = \bar{\Gamma}_0 \cup \bar{\Gamma}_1, \Gamma_0 \cap \Gamma_1 = \emptyset$. Let $\boldsymbol{v} = (v_i)$, $v_i = v_i(x_\alpha, z)$ be a displacement vector of a point $(x_\alpha, z) \in \Omega \times (-h, h)$, $\alpha = 1, 2$; $i = 1, 2, 3$. The axis z is directed downwards. We assume the classical Kirchhoff-Love kinematical hypothesis:

$$\begin{aligned} v_\alpha(x_\beta, z) &= u_\alpha(x_\beta) - z w,_\alpha(x_\beta), \\ v_3(x_\beta, z) &\equiv w(x_\beta). \end{aligned} \tag{1.1}$$

Here $\boldsymbol{u} = (u_\alpha)$ stands for the in-plane displacement vector whilst w denotes the transverse displacement. Let $\boldsymbol{N} = (N_{\alpha\beta})$ and $\boldsymbol{M} = (M_{\alpha\beta})$ be the membrane force tensor and moment tensor, respectively, defined by

$$N_{\alpha\beta} = \int_{-h}^{h} \sigma_{\alpha\beta}\, dz, \quad M_{\alpha\beta} = \int_{-h}^{h} z\sigma_{\alpha\beta}\, dz. \tag{1.2}$$

Constitutive relations are given by:

$$N_{\alpha\beta} = A_{\alpha\beta\lambda\mu}\varepsilon_{\lambda\mu}(\boldsymbol{u}) \,, \qquad M_{\alpha\beta} = B_{\alpha\beta\lambda\mu}\kappa_{\lambda\mu}(w) \,.$$

Here $\varepsilon_{\alpha\beta}(\boldsymbol{u})$ and $\kappa_{\alpha\beta}(w)$ are the strain measures:

$$\varepsilon_{\alpha\beta}(\boldsymbol{u}) = (u_{\alpha,\beta} + u_{\beta,\alpha})/2 \,, \;\; \kappa_{\alpha\beta}(w) = -w,_{\alpha\beta} \,,$$

and

$$A_{\alpha\beta\lambda\mu} = \int_{-h}^{h} C_{\alpha\beta\lambda\mu} dz \ , \qquad B_{\alpha\beta\lambda\mu} = \int_{-h}^{h} z C_{\alpha\beta\lambda\mu} dz, \quad H_{\alpha\beta} = \int_{-h}^{h} c_{\alpha 3\beta 3} dz \ .$$

Here c_{ijkl} denotes the elasticity tensor of the material of the plate, and $C_{\alpha\beta\lambda\mu} = c_{\alpha\beta\lambda\mu} - c_{\alpha\beta 33} c_{33\lambda\mu} c_{3333}^{-1}$.

The equilibrium equations of the plate (in the absence of the obstacle) are:

$$N_{\alpha\beta,\beta} + p_\alpha = 0 \ , \qquad M_{\alpha\beta,\beta\alpha} + p = 0 \quad \text{in} \quad \Omega \ .$$

Let a continuous function

$$f : \Omega_1 \to \mathbb{R}, \ z = f(\boldsymbol{x}) \ , \ \Omega \subset \Omega_1,$$

determines a rigid obstacle. The unilateral condition is specified by [1]

$$w(\boldsymbol{x}) + h \leq f(x_\alpha + u_\alpha(\boldsymbol{x}) - h w_{,\alpha}(\boldsymbol{x})) \ . \tag{1.3}$$

The lower face of the plate may come into contact with the rigid obstacle. We introduce the set

$$K = \{(\boldsymbol{u}, w) \in H^1(\Omega)^2 \times H^2(\Omega) \mid (1.3) \text{ is satisfied for } \boldsymbol{x} \in \Omega\}.$$

Remark 1.1. If K is non-empty, then in general it is a non-convex set. K is a convex set provided that f is a concave function.

The boundary conditions are assumed in the form:

$$w = 0 \ , \ \frac{\partial w}{\partial \boldsymbol{n}} = 0 \ , \ \boldsymbol{u} = 0 \ \text{ on } \Gamma_0, \text{ meas } \Gamma_0 > 0.$$

Here $\boldsymbol{n}$ denotes the outer unit normal vector to Γ. We set

$$V = \{(\boldsymbol{u}, w) \in H^1(\Omega)^2 \times H^2(\Omega) | \ \boldsymbol{u} = 0, w = \tfrac{\partial w}{\partial \boldsymbol{n}} = 0 \text{ on } \Gamma_0\},$$

$$a(\boldsymbol{u}, \boldsymbol{v}) = \int_\Omega A_{\alpha\beta\lambda\mu}(\boldsymbol{x}) \varepsilon_{\alpha\beta}(\boldsymbol{u}) \varepsilon_{\lambda\mu}(\boldsymbol{v}) dx \ ,$$

$$b(w, t) = \int_\Omega B_{\alpha\beta\lambda\mu}(\boldsymbol{x}) \kappa_{\alpha\beta}(w) \kappa_{\lambda\mu}(t) dx \ ,$$

where $\boldsymbol{u}, \boldsymbol{v} \in H^1(\Omega)^2$ and $t, w \in H^2(\Omega)$.

The functional of the external loading is assumed in the form

$$L(\boldsymbol{u}, w) = \int\limits_\Omega (p_\alpha u_\alpha + p w) dx + \int\limits_{\Gamma_1} \left(r_\alpha u_\alpha + q w - \bar{M} \frac{\partial w}{\partial \boldsymbol{n}} \right) d\Gamma \ ,$$

where $r_\alpha, q, \bar{M} \in L^2(\Gamma_1)$. The functional of the total potential energy is given by

$$J(\boldsymbol{u}, w) = \frac{1}{2} a(\boldsymbol{u}, \boldsymbol{u}) + \frac{1}{2} b(w, w) - L(\boldsymbol{u}, w) \ .$$

Now we are in a position to formulate the first (nonconvex) minimization problem.

Problem ($\boldsymbol{P}$)

Find

$$\inf\{J(\boldsymbol{u}, w) \mid (\boldsymbol{u}, w) \in K \cap V\}.$$

We observe that on account of the unilateral condition (1.3) the in-plane and transverse displacements are interconnected. Consequently, problem (P) cannot be decomposed into membrane and plate problems. We recall that if the obstacle condition is imposed on the mid-plane of the plate then both problems are independent and only the bending problem is of a unilateral type. The strong form of the above minimization problem is now formulated provided that its solution is sufficiently regular.

Theorem 1.2. Problem (P) possesses at least one solution $(\tilde{\boldsymbol{u}}, \tilde{w}) \in K \cap V$. For the proof the reader is referred to [1].

Remark 1.3. The linearization of the r.h.s. of (1.3) was considered in [1].

1.2 Reissner's plate model

In a simple model of moderately thick plates accounting for transverse shear deformations it is assumed that

$$v_\alpha(\boldsymbol{x}, z) = u_\alpha(\boldsymbol{x}) + z\varphi_\alpha(\boldsymbol{x}) \ , \ v_3(x_\beta, z) \equiv w(x_\beta), \quad (\boldsymbol{x}, z) \in \Omega \times [-h, h] \ .$$

Here φ_α $(\alpha = 1, 2)$ denote the rotations of the plate transverse cross-sections. The strain measures are given by:

$$\varepsilon_{\alpha\beta}(\boldsymbol{u}) = u_{(\alpha,\beta)} = \frac{1}{2}\left(\frac{\partial u_\alpha}{\partial x_\beta} + \frac{\partial u_\beta}{\partial x_\alpha}\right), \ \rho_{\alpha\beta}(\boldsymbol{\varphi}) = \varphi_{(\alpha,\beta)}, \ d_\alpha(w, \boldsymbol{\varphi}) = w_{,\alpha} + \varphi_\alpha .$$

Let us denote by $\boldsymbol{T} = (T_\alpha)$ the transverse shear force vector. The constitutive relationships are given by

$$N_{\alpha\beta} = A_{\alpha\beta\lambda\mu}\varrho_{\lambda\mu}(\boldsymbol{\varphi}), \quad M_{\alpha\beta} = B_{\alpha\beta\lambda\mu}\rho_{\lambda\mu}(\boldsymbol{\varphi}) \ , \quad T_\alpha = H_{\alpha\beta}d_\beta(w, \boldsymbol{\varphi}).$$

The equilibrium equations have the form:

$$N_{\alpha\beta,\beta} + p_\alpha = 0, \quad M_{\alpha\beta,\beta} - T_\alpha + m_\alpha = 0 \ , \qquad T_{\alpha,\alpha} + p = 0 \ ,$$

provided that the obstacle is absent. The boundary conditions are:

$$\boldsymbol{u} = \boldsymbol{0} \ , \quad \boldsymbol{\varphi} = \boldsymbol{0} \ , \qquad w = 0 \text{ on } \Gamma_0 \ ,$$

where meas $\Gamma_0 > 0$.

We set

$$V_1 = \{(\boldsymbol{u}, w, \boldsymbol{\varphi}) \in H^1(\Omega)^2 \times H^1(\Omega) \times H^2(\Omega)^2 \mid \boldsymbol{u} = 0 \, , \boldsymbol{\varphi} = 0 \, , \ w = 0 \text{ on } \Gamma_0\} \, .$$

If a continuous function

$$f : \Omega_1 \to \mathbb{R}, \ \Omega \subset \Omega_1 \ , \ z = f(\boldsymbol{x})$$

defines the obstacle, then the condition of noninterpenetration is given by

$$w(\boldsymbol{x}) + h \leq f(x_\alpha + u_\alpha(\boldsymbol{x}) + h\varphi_\alpha(\boldsymbol{x})) \ , \ \boldsymbol{x} \in \Omega \ .$$

Consequently, the set of kinematically admissible displacements is defined by

$$K_1 = \{(\boldsymbol{u}, w, \boldsymbol{\varphi}) \in V_1 \mid w(\boldsymbol{x}) + h \leq f(x_\alpha + u_\alpha(\boldsymbol{x}) + h\varphi_\alpha(\boldsymbol{x})), \ \boldsymbol{x} \in \Omega\}.$$

The functional of the total potential energy is expressed by

$$J_1(\boldsymbol{u}, w, \boldsymbol{\varphi}) = \frac{1}{2}\int_\Omega [A_{\alpha\beta\lambda\mu}\varepsilon_{\alpha\beta}(\boldsymbol{u})\varepsilon_{\lambda\mu}(\boldsymbol{u}) + B_{\alpha\beta\lambda\mu}\rho_{\alpha\beta}(\boldsymbol{\varphi})\rho_{\lambda\mu}(\boldsymbol{\varphi}) + \\ + H_{\alpha\beta}d_\alpha(w, \boldsymbol{\varphi})d_\beta(w, \boldsymbol{\varphi})]d\boldsymbol{x} - L_1(\boldsymbol{u}, w, \boldsymbol{\varphi}),$$

where

$$L_1(\boldsymbol{u}, w, \boldsymbol{\varphi}) = \int_\Omega (p_\alpha u_\alpha + pw + m_\alpha\varphi_\alpha)dx + \int_{\Gamma_1} (r_\alpha u_\alpha + qw + \bar{M}_\alpha\varphi_\alpha)d\Gamma.$$

We formulate the second minimization problem.

Problem ($\boldsymbol{P_1}$)

Find

$$\inf\{J_1(\boldsymbol{u}, w, \boldsymbol{\varphi}) \mid (\boldsymbol{u}, w, \boldsymbol{\varphi}) \in K_1\}.$$

In general, this problem is also *nonconvex.* The next existence results is formulated as follows

Theorem 1.4. Problem (P_2) possesses at least one minimizer $(\tilde{\boldsymbol{u}}, \tilde{w}, \tilde{\boldsymbol{\varphi}}) \in K_1$.

2. Von Kármán plates

This model is still based on Kirchhoff-Love kinematical hypothesis. The strain measures are:

$$e_{\alpha\beta}(\boldsymbol{u}, w) = \varepsilon_{\alpha\beta}(\boldsymbol{u}) + \frac{1}{2}w_{,\alpha}\, w_{,\beta} \, ; \quad \kappa_{\alpha\beta}(w) = -w_{,\alpha\beta} \ ,$$

where $\varepsilon_{\alpha\beta}(\boldsymbol{u}) = u_{(\alpha,\beta)}$. We note that only the first strain measure is *nonlinear.*

The constitutive equations have the form

$$N_{\alpha\beta} = A_{\alpha\beta\lambda\mu}e_{\lambda\mu}(\boldsymbol{u}, w) \ , \ M_{\alpha\beta} = B_{\alpha\beta\lambda\mu}\kappa_{\lambda\mu}(w) \ .$$

As previously, $\boldsymbol{N}, \boldsymbol{M}$ are the membrane forces tensor and bending moments tensor, respectively. In the absence of the obstacle the equilibrium equations are given by

$$N_{\alpha\beta,\beta} + p_\alpha = 0, \quad M_{\alpha\beta,\beta\alpha} + (N_{\alpha\beta} w,_\beta),_\alpha + p = 0 \quad \text{in} \quad \Omega .$$

We impose the following boundary conditions

$$\boldsymbol{u} = 0 \text{ on } \Gamma_0 , \quad w = \frac{\partial w}{\partial \boldsymbol{n}} = 0 \text{ on } \Gamma .$$

An appropriate space for displacements is

$$V_2 = \{(\boldsymbol{u}, w) \in H^1(\Omega)^2 \times H_0^2(\Omega) | \ \boldsymbol{u} = 0 \text{ on } \Gamma_0\} .$$

The functional of the total potential energy is given by

$$J_2(\boldsymbol{u}, w) = \frac{1}{2}\int_\Omega [A_{\alpha\beta\lambda\mu}(\varepsilon_{\alpha\beta}(\boldsymbol{u}) + \frac{1}{2}w,_\alpha w,_\beta)(\varepsilon_{\lambda\mu}(\boldsymbol{u}) + \frac{1}{2}w,_\lambda w,_\mu)$$
$$+B_{\alpha\beta\lambda\mu}\kappa_{\alpha\beta}(w)\kappa_{\lambda\mu}(w)]d\boldsymbol{x} - \int_\Omega (p_\alpha u_\alpha + pw)d\boldsymbol{x} - \int_{\Gamma_1} r_\alpha u_\alpha d\Gamma.$$

The nonlinear strain measure renders the functional J_2 nonconvex on $H^1(\Omega)^2 \times H_0^2(\Omega)$ and on V_2. This functional is weakly lower semicontinuous and bounded from below. For the obstacle problem the set of kinematically admissible fields is specified by

$$K_2 = \{(\boldsymbol{u}, w) \in V_2 | \ w(\boldsymbol{x}) + h \leq f(x_\alpha + u_\alpha(\boldsymbol{x}) - hw,_\alpha(\boldsymbol{x})) \text{ is satisfied for } \boldsymbol{x} \in \Omega\}.$$

We assume that $K_2 \neq \emptyset$. We can now formulate the obstacle contact problem.
Problem ($\boldsymbol{P_2}$)

Find

$$\inf\{J_2(\boldsymbol{u}, w) \mid (\boldsymbol{u}, w) \in K_2\}.$$

The existence of a solution to Problem (P_2) is ensured by the following result.
Theorem 2.1. The functional J_2 has at least one minimizer on the set K_2.

3. Augmented Lagrangians methods for a class of nonconvex problems

Ito and Kunisch [2, 3] elaborated Lagrangian methods applicable to convex problems, cf. also [4, 5] this section we propose an extension to a class of nonconvex problems.
Nonconvex set of constraints. Consider now the following minimization problem.

$$(P) \qquad \min\{\frac{1}{2}a(u, u) - l(u) | g(u) \leq 0, u \in B\}, \tag{3.1}$$

where $a(\cdot,\cdot)$ is a symmetric bilinear and continuous form on $V \times V$, $V-$elliptic; l is a continuous linear functional and $g : B \to H$ is a *nonconvex*, continuous, Gâteaux differentiable mapping. The notations are similar to those used in [2]. More precisely, B, V and H are Hilbert spaces, the same as in [2].

We set $g(u) = G(u) - G_1(u)$, where $G(G_1)$ is convex (nonconvex). Ito and Kunisch [2] approach can be extended by combining their augmented Lagrangian technique with an iterative procedure:

mth step.

$$G(u) \le G_1(u^{m-1}), \quad m = 1, 2, \ldots \tag{3.2}$$

Then the set

$$K_m = \{u \in B | G(u) \le G_1(u^{m-1})\} \tag{3.3}$$

is convex. At each step m we define a family of augmented Lagrangian problems by

$$(P)_{m,c,\lambda} \qquad \min_{u^m \in B} L_{m,c}(u^m, \lambda^m),$$

where

$$L_{m,c}(u^m, \lambda^m) = \frac{1}{2}a(u^m, u^m) - l(u^m) + \langle \lambda^m, \hat{g}(u^m, \lambda^m, c)\rangle + \frac{c}{2}\|\hat{g}(u^m, \lambda^m, c)\|_H^2,$$

and $\lambda^m \in H, c > 0, c \in \mathbb{R}^+$. Moreover

$$\hat{g}(u^m, \lambda^m, c) = \sup\left(g(u^m), -\frac{\lambda^m}{c}\right).$$

The augmented Lagrangian algorithm to solve $(\boldsymbol{P})_{m,c,\lambda}$ runs as follows.

The Algorithm

(1) Choose $\lambda_1^m \in H$, $\lambda_1^m \ge 0$, and $c > 0$,
(2) put $n = 1$,
(3) solve $(\boldsymbol{P})_{m,c,\lambda_n^m}$ for u_n^m,
(4) put $\lambda_{n+1}^m = \lambda_n^m + c\hat{g}(u_n^m, \lambda_n^m, c) = \sup(0, \lambda_n^m + cg(u_n^m))$
(5) put $n = n + 1$ and return to (3).

We observe that the parameter c may also depend on m.
Applying Ito and Kunisch [2] results we get

$$\nu \sum_{n=1}^{\infty} \|u_n^m - u^{*m}\|_V^2 \le \frac{1}{2c}\|\lambda_1^m - \lambda^{*m}\|_H^2 \le \sup_{m \in \mathbb{N}} \frac{1}{2c}\|\lambda_1^m - \lambda^{*m}\|_H^2 < \infty \tag{3.4}$$

since c may be taken sufficiently large, such that for each $m \in \mathbb{N}$ we have

$$\frac{1}{2c}\|\lambda_1^m - \lambda^{*m}\|_H^2 < C_1, \quad C_1 > 0. \tag{3.5}$$

Example 1. As we know, the set of constraints $\{u | g(u) \le 0\}$ identified with K, K_1 or K_2 is, in general, nonconvex. Define

$$K^m = \{(\boldsymbol{u}, w) \in H^1(\Omega)^2 \times H^2(\Omega) | w(\boldsymbol{x}) + h \leq f(\boldsymbol{x} + \boldsymbol{u}^{m-1}(\boldsymbol{x}) - h\nabla w^{m-1}(\boldsymbol{x})\}$$

The last set of constraints is convex.

Example 2. More general is the case where

$$a_m(u, u) = A(\Lambda u + N_1(u^{m-1}), \Lambda u + N_1(u^{m-1})).$$

Here $\Lambda(N_1)$ is a linear (nonlinear) operator.

Particularly, for von Kármán plates we have, cf. Sec. 2.

$$a_m(\boldsymbol{u}, w; \boldsymbol{u}, w) = \frac{1}{2}\int_\Omega [A_{\alpha\beta\lambda\mu}(\varepsilon_{\alpha\beta}(\boldsymbol{u}) + \frac{1}{2}w^{m-1}_{,\alpha} w^{m-1}_{,\beta})(\varepsilon_{\lambda\mu}(\boldsymbol{u}) + \frac{1}{2}w^{m-1}_{,\lambda} w^{m-1}_{,\mu})$$

$$+ \frac{1}{2}B_{\alpha\beta\lambda\mu}\kappa_{\alpha\beta}(w)\kappa_{\lambda\mu}(w)]\, dx.$$

Nonconvex extension of [3]. Consider a nonconvex minimization problem

$$(Q) \qquad \min\{f(u) + \varphi(N(u)) | u \in K\}.$$

Here N is a nonlinear mapping and K a nonconvex set. The remaining notations are the same as in [3]. Assume that $N(u) = \Lambda u + N_1(u)$ and

$$K = \{u | G(u) \leq G_1(u)\}.$$

Then one can apply the results due to Ito and Kunisch [3] to the *sequence of convex problems*

$$(Q_m) \qquad \min\{f(u) + \varphi(\Lambda u + N_1(u^{m-1})) | u \in K^m\}.$$

4. Example

We deal with an obstacle problem for a thin, axisymmetric clamped circular plate, of unit diameter. For simplicity, the contact condition is formulated on the mid-plane of the plate.

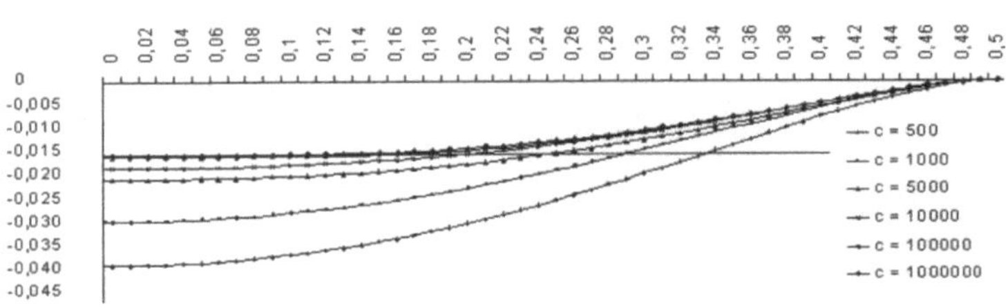

Fig. 4.1. The plate transverse displacements for step 1: $\lambda_1 = 1$, $c = 500$, 1000, 5000, 10000, 100000, 1000000; c - parameter in the augmented Lagrangian

To solve this problem, Ito-Kunisch [2] augmented Lagrangian method has been used. Obviously, the problem considered is convex.

The functional of the total potential energy of the clamped circular plate is given by

$$J(v) = \pi \int_0^{\frac{1}{2}} r\{[\frac{1}{r}\frac{d}{dr}(r\frac{dv}{dr})]^2 - 2fv(r)\}\, dr.$$

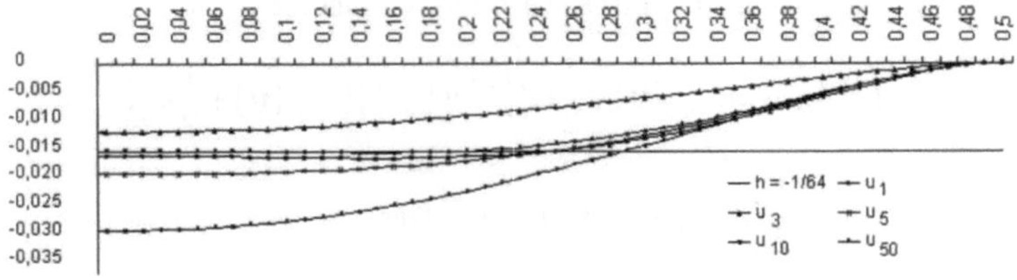

Fig. 4.2. The plate transverse displacement for steps 1, 3, 5, 10, 50; c=1000

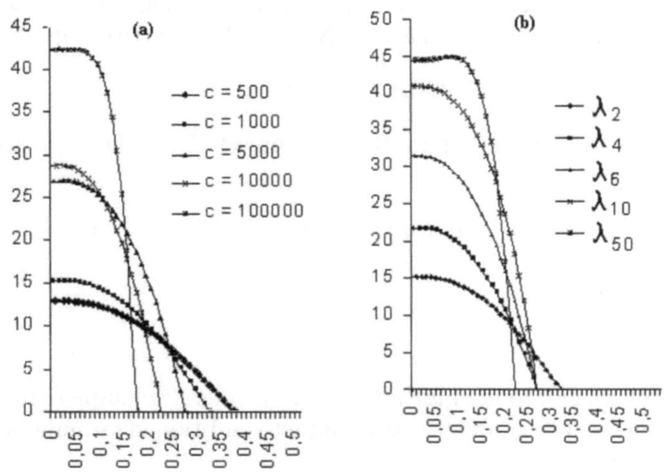

Fig. 4.3. Lagrangian multiplier λ_2 (a) step 1: $\lambda_1 = 1, c$ as in Fig. 4.1.
(b) $\lambda_2, \lambda_4, \lambda_6, \lambda_{11}, \lambda_{50}$

We formulate the convex minimization problem:
find $u \in K$ such that $J(u) = \inf_{v \in K} J(v)$, where

$$K = \{v \in H^2(0, \frac{1}{2}) \,|\, v(\frac{1}{2}) = 0,\ v'(\frac{1}{2}) = 0,\ v(r) \geq -h,\ r \in (0, \frac{1}{2})\}.$$

To apply the convex augmented Lagrangian method we set

$$g(v) = -v - h \leq 0, \text{ (since } v \geq -h).$$

The augmented Lagrangian has now the following form:

$$L_c(v, \lambda) = J(v) + \frac{\pi}{c} \int_0^{0.5} \{[\sup(0, \lambda + cg(v))]^2 - \lambda^2\} r \, dr.$$

We observe that the Lagrangian multiplier λ represents the contact forces. The results of calculation by using FEM are depicted in Figs 4.1-4.3. The augmented Lagrangian solutions approach to the problem with obstacle.

References

1. W. R. Bielski and J. J. Telega, Existence of solutions to obstacle problems for linear and nonlinear elastic plates, Mathematical and Comput. Modelling, **28**, 1998, 55–66.
2. K. Ito and K. Kunisch, An augmented Lagrangian technique for variational inequalities, Appl. Math. Optim., **21**, 1990, 223–241.
3. K. Ito and K. Kunisch, Augmented Lagrangian methods for nonsmooths, convex optimization in Hilbert spaces, in: Control of Partial Differential Equations and Applications, ed. by E. Casas, pp. 107–117, Marcel Dekker, 1995.
4. J. J. Telega, A. Gałka, Augmented Lagrangian methods for contact problems, optimal control and image restoration, in: From Convexity to Nonconvexity, ed. by P. D. Panagiotopoulos, Kluwer, in press.
5. J. J. Telega and A. Gałka, Augmented Lagrangian methods and applications to contact problems, in: Theoretical Foundationa of Civil Engineering, ed. by W. Szcześniak, pp. 335–348, Oficyna Wydawnicza Politechniki Warszawskiej, Warsaw

Longitudinal Wave Propagation in Conical Rods Subject to Impacts

P. Eberhard, B. Hu and W. Schiehlen

Institute B of Mechanics, University of Stuttgart, Germany

Summary. Numerical computation techniques are of increasing importance in the study of impact-contact problems. However, analytical solutions for simple impact systems are still of considerable utility since they provide revealing insights into the nature of the problem under investigation and furnish critical checks on the accuracy and efficiency of computational schemes. Besides, both analytical and numerical results should be validated by experiments as much as possible. Due to the interdisciplinary character of the research, good coordination and interaction between numerical, experimental and analytical approaches prove to be indispensable. In this paper, advances in the symbolical computation of longitudinal wave propagation in impacted rods are discussed and results from ongoing experiments are reported. The theoretical and experimental results are compared.

1. Introduction

The behavior of structural elements under impact loading is a subject of great interest in dynamical structural analysis. Since during impacts very high forces are applied to the colliding bodies over a very short period of time, transient impact responses should be considered in terms of wave propagation theory. Insight into stress wave propagation plays a fundamental role in various areas of modern technology, see Höschl, Okrouhlík, Červ and Beneš [1]. The simulation of transient impact responses in mechanical systems requires a suitable mechanical modeling. A survey of the models and the related time scales is presented in Schiehlen [2].

Rods are among the simplest engineering structures, but the propagation of waves in rods can be complicated due to some complex dynamical boundary conditions and wave reflections. In history, longitudinal wave propagation in rods with uniform cross sectional area along the length has been intensively investigated for more than hundreds years, see Szabó [3]. Vibrations resulting from longitudinal impacts were first examined by Bernoulli, Navier and Poisson. A detailed treatment was provided by St. Venant and later supplemented by Boussinesq and Donnel, see Goldsmith [4] and Timoshenko and Goodier [5]. In the form of Laguerre polynomials, some solutions were given in Nikitin [6]. Recently, this classical impact problem has been solved completely using a computer algebra system, see Hu, Eberhard and Schiehlen [7].

In contrast to uniform rods, however, only few investigations were performed for conical rods. Landon and Quinney [11] studied the problem analytically for an infinitely long cone of small conicity angle by solving the one-dimensional wave equation for a pulse moving away from the apex. For a conical rod with finite length where the wave reflection must also be considered, there do not exist to our knowledge analytical results for the wave propagation. An experimental investigation for stress wave propagation in a cylindrical bar with a conical end hitting a rigid wall was given by Suh [14]. Some results for the vibration of a conical rod can be found in Abrate [10] and Kumar and Sujith [12]. Recently, it was pointed out that the analysis of conical rods is important to the study of piles in deep foundation, see Meek and Wilf [13], and to the study of high velocity impact of projectiles hitting composite structures, see Abrate [9]. Having solved the longitudinal wave propagation in a uniform rod successfully using symbolical computations, in this paper we will show that the longitudinal wave propagation in a conical rod can be solved symbolically, too.

Experimental investigations of longitudinal wave propagations in uniform rods were reviewed by Al-Mousawi [8]. Using modern measurement instruments like Laser-Doppler-Vibrometers, we checked this classical impact system and found out that both, the classical Hertzian contact theory and St. Venant's contact theory, cannot explain the experimental results satisfactorily.

2. Symbolical computation for a conical rod

The conical rod under consideration is homogeneous and linearly elastic. It has Young's modulus E and density ρ. Its length is L and the radius of the cross-section of the conical rod $r(x) = r_0 + \alpha x$ is a linear function of the position x, see Fig. 2.1. It is assumed that the rod is slender so that its transverse motions can be neglected. The struck ground is rigid and remains at rest during the impact. The contact surface is supposed to be perfectly planar so that St. Venant's contact theory can be used. The resistant force of the air acting on the rod is neglected. According to these assumptions, the

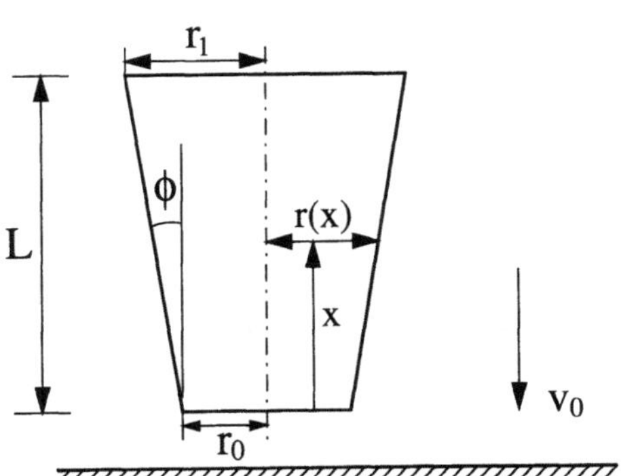

Fig. 2.1. Geometric description of a conical rod

governing equation for the longitudinal waves in the rod is an inhomogeneous partial differential equation with varying coefficients and reads as

$$\frac{\partial^2 u(x,t)}{\partial t^2} = c^2 \frac{\partial^2 u(x,t)}{\partial x^2} + \frac{2\alpha c^2}{r(x)} \frac{\partial u(x,t)}{\partial x} - g\,, \tag{2.1}$$

where the material parameter $c = \sqrt{E/\rho}$ is the wave propagation velocity and the constant g is the gravitational acceleration. At the beginning of the impact, which is denoted by the time $t = 0$, the rod has uniform displacement and velocity along its length, i.e.,

$$u(x,0) = 0\,, \qquad v(x,0) = -v_0\,. \tag{2.2}$$

Since the upper end of the rod is free, the stress at $x = L$ always vanishes and we have the boundary condition for the upper end of the rod

$$\frac{\partial u(L,t)}{\partial x} = 0\,. \tag{2.3}$$

For the lower end of the rod, we have different boundary conditions during the impact and during the free flight. During the impact, the displacement and the velocity at the lower end remain zero,

$$u(0,t) = 0\,, \qquad v(0,t) = 0\,, \tag{2.4}$$

and during the free flight the stress at the lower end vanishes,

$$\frac{\partial u(0,t)}{\partial x} = 0\,. \tag{2.5}$$

Now we have the wave equation (2.1), the initial conditions (2.2) and the boundary conditions (2.3) and (2.4) or (2.5). Therefore, the mathematical model for the impact problem of a conical rod striking the rigid ground is completely stated.

The wave equation (2.1) is a non-homogeneous linear partial differential equation with variable coefficients. It can be proven that the general solution of the wave equation (2.1) reads as

$$u(x,t) = \frac{L^2}{r(x)} \left(f_1\left(\frac{ct}{L} - \frac{x}{L}\right) + f_2\left(\frac{ct}{L} + \frac{x}{L}\right)\right) - \frac{1}{2} g t^2 \tag{2.6}$$

where two real functions f_1 and f_2 are determined from the initial conditions and boundary conditions, see Hu, Eberhard and Schiehlen [15]. Using a computer algebra system, we can obtain these two functions f_1 and f_2 much more easily than by paper and pencil. For their computation, two points should be emphasized. The first point is that due to the impact at the instant $t = 0$ and the succeeding wave reflections, the functions $f_1'(y)$ and $f_2'(y)$ are not continuous at points $y = 0, 1, 2, \ldots$, while the functions $f_1(y)$ and $f_2(y)$ at these points are still continuous. The other point is that the duration of contact is unknown. Since the lower end of the rod has different boundary conditions during the impact and during the free flight, the duration of contact t_c must be determined.

Since the functions $f_1'(y)$ and $f_2'(y)$ are not continuous at $y = 0, 1, 2, \ldots$, they must be computed piecewisely. We decompose the univariant variable y into a couple (n, z) with

$$z = y - [y] \in [0, 1) \quad \text{and} \quad n = [y] \in \mathbb{N}. \tag{2.7}$$

The mapping between y and the couple (n, z) is a one-to-one mapping. We can now also use the two variables n and z to express the functions $f_1(y)$ and $f_2(y)$,

$$f_1(y) = f_1^*(n, z) \quad \text{and} \quad f_2(y) = f_2^*(n, z)\,. \tag{2.8}$$

To determine the functions $f_1^*(n, z)$ and $f_2^*(n, z)$, we need the recurrence relations for the integer $n \in \mathbb{N}$. Using the boundary condition at the upper end of the rod, we obtain

$$f_2^*(n+2, z) = f_1^*(n, z) + [f_2^*(n+1, 1) - f_1^*(n, 0)]e^{bz} + 2be^{bz} \int_0^z e^{-bz} f_1^*(n, z)dz \tag{2.9}$$

where $b = \alpha L/r_0$. During the impact, we can use (2.4) to derive another recurrence relation to determine $f_1^*(n, z)$ from $f_2^*(n, z)$,

$$f_1^*(n, z) = -f_2^*(n, z) + \frac{1}{2}\alpha_0 \beta (n + z)^2 \tag{2.10}$$

with $\alpha_0 = r_0/L$ and $\beta = gL/c^2$. Using the initial condition (2.4) yields the functions $f_1^*(-1, z)$ and $f_2^*(0, z)$ for the start of the recursion

$$f_1^*(-1, z) = \frac{\nu}{2}\left(-\alpha_0(-1 + z) + \frac{\alpha}{2}(-1 + z)^2\right), \tag{2.11}$$

$$f_2^*(0, z) = -\frac{\nu}{2}\left(\alpha_0 z + \frac{\alpha}{2} z^2\right) \tag{2.12}$$

where $\nu = v_0/c$. By means of the start of the recursion (2.11) and (2.12) and the recurrence relations (2.9) and (2.10), we can determine all unknown functions $f_1^*(n, z)$ and $f_2^*(n, z)$ during the impact. However, the duration of contact t_c must be computed numerically since t_c is determined by a transcendental function, for which it is impossible to get an analytical solution. During the free flight, equation (2.4) is no longer true and (2.5) should be used instead. Similarly, we can compute the functions f_1 and f_2 during the free flight symbolically. More details can be found in Hu, Eberhard and Schiehlen [15].

As a numerical example, we consider three steel rods with the same material $E = 206\,GPa$, $\rho = 7900\,kg/m^3$, same length $L = 1\,m$ and same mass $m = 9.9274\,kg$. We investigate two conical rods with $\alpha = \pm 0.02$ and compare them with a cylindrical rod with $\alpha = 0$. The initial impacting velocity v_0 is set to $1\,m/s$. The contact forces and three-dimensional visualizations of the stress waves in the three rods are shown in Fig. 2.2, where $\tau = ct/L$ and

$\xi = x/L$ are the dimensionless time and position, respectively. It can be seen that the conical rods have a much more complicated impact behavior than the cylindrical rod. Among the three rods, the rod with $\alpha = -0.02$ has the largest contact force and the shortest duration of contact.

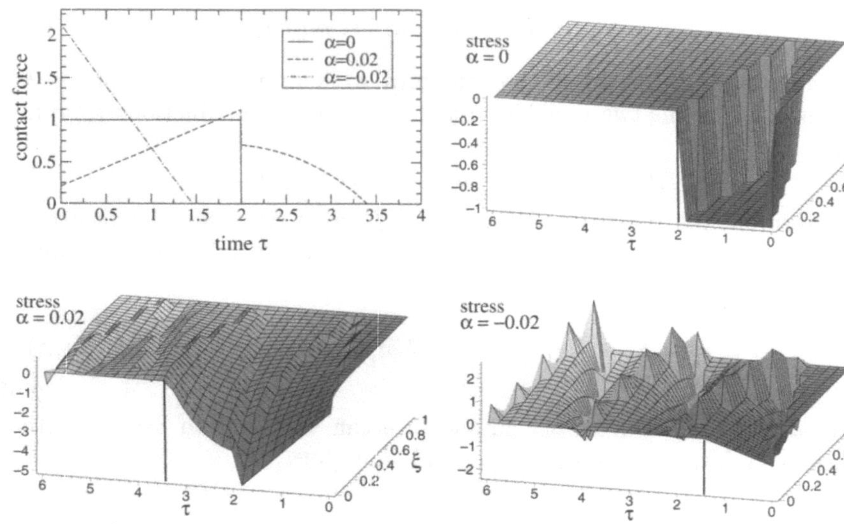

Fig. 2.2. Contact forces and stress waves in the three rods

3. Experimental investigation of longitudinal waves

For a long time, theories for longitudinal wave propagation problems were extensively developed. However, the experimental work was lagging behind the theory, and early investigations were limited to the measurement of overall effects such as the coefficient of restitution. The use of strain gauges from the forties onwards gave an important boost to the experimental work. By the end of the fifties many experimental investigations related to longitudinal wave propagation were carried out and the theory began to lag behind the experimental work. Nowadays the rapid development of digital computers made it possible to simulate complex transient impact responses numerically and with the advancement of laser techniques, the wave propagations can also be measured with Laser-Doppler-Vibrometers. In order to compare different impact-contact theories and different numerical methods, we reinvestigate the longitudinal wave propagation in a cylindrical rod struck by a rigid mass.

The experimental setup is shown schematically in Fig. 3.1. A ball made of steel with diameter $30\,mm$ and mass $m_2 = 0.11\,kg$ was used as a rigid body striking a brass rod. The impacting ball was suspended in a frame by thin wire. The brass rod is geometrically a uniform cylindrical rod with diameter

20 mm and length $L = 1.5\,m$. It was hung in two sets of thin wires at two positions with the same distance 0.3 m from the two ends of the rod. The strains at positions A and B with distances 0.5 m and 1 m from the struck end were measured using strain gauges and corresponding amplifiers. For the measurement of displacements and velocities, a Laser-Doppler-Vibrometer was used. The motion at the free end of the rod and the motion of the striking ball were measured by the vibrometer separately.

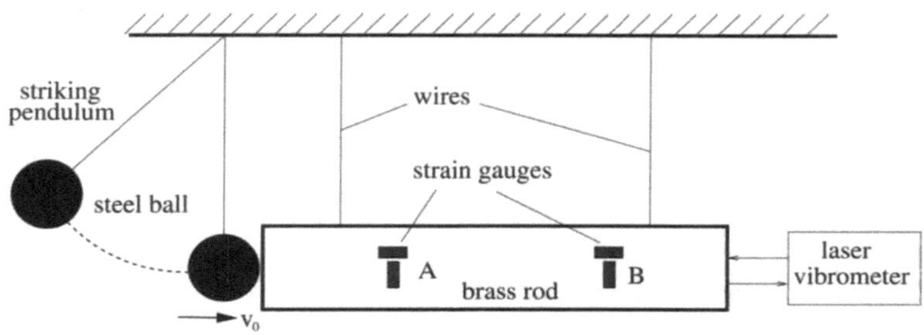

Fig. 3.1. Schematic description of the experimental setup

Typical measured strain signals at the positions A and B in the rod and the velocity and displacement signals at the free end of the rod are shown in Figs. 3.2 and 3.3. A statistical analysis of the measurement noise shows that the accuracy of the vibrometer is about ten times higher than the accuracy of the strain gauges and the amplifiers.

The measured results were compared with the classical Hertzian contact theory and St. Venant's contact theory. A comparison between theoretical and experimental results for the strain signal at position A and the velocity of the striking ball is shown in Fig 3.4. Hertzian theory can not predict the strains in the rod and St. Venant's theory predicts a totally wrong rebound

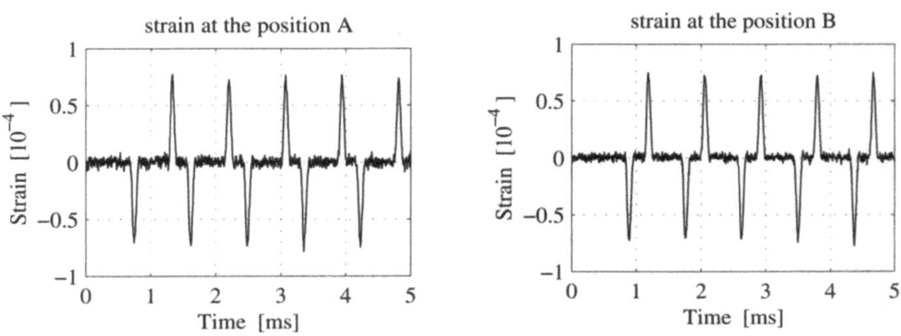

Fig. 3.2. Measured strain signals at the two positions in the rod

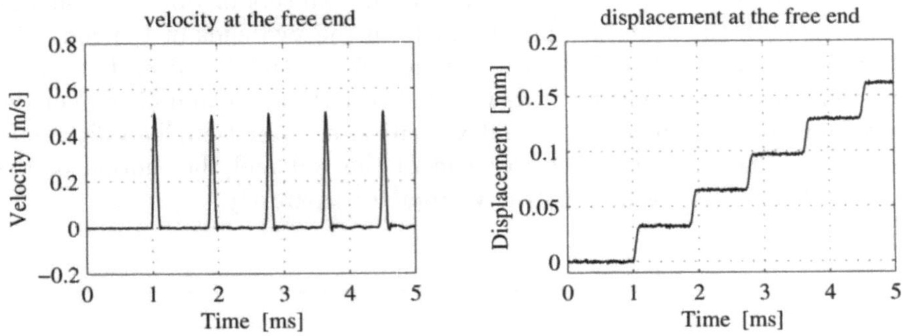

Fig. 3.3. Measured velocity and displacement at the free end of the rod

velocity for the striking ball. Both classical theories can not describe the experimental results satisfactorily. Therefore, improved contact theories will be developed. The developments play also a fundamental role for improved numerical computation methods.

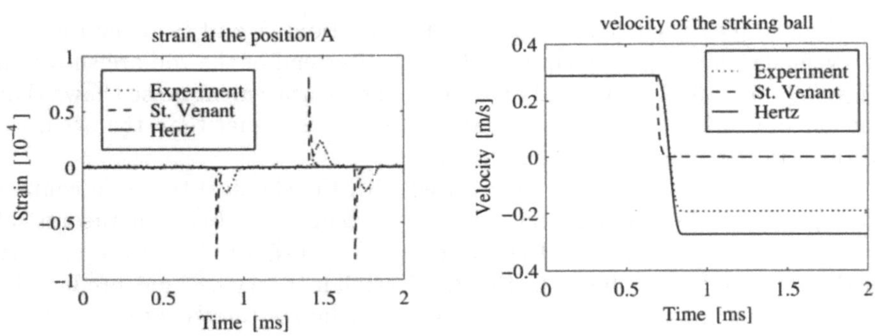

Fig. 3.4. Comparison between theoretical and experimental results

4. Conclusions

In this paper, it has been shown that the complicated impact problem of a conical rod striking the rigid ground can be solved by computer algebra. Since the given impact responses are exact, they may be used to validate numerical programs using finite element methods or boundary element methods. It is shown that even slightly conical shapes of colliding bodies play an important role in St. Venant's contact theory. Moreover, our ongoing experiments have shown that both, classical Hertzian and St. Venant's contact theories, can not predict experimental results satisfactorily. For an accurate simulation of the transient impact responses, new contact models are still required and currently developments are performed to improve the theory. Finally, we want to emphasize that the wave phenomena are very complex and they can only be tackled by a combination of analytical, numerical and experimental methods and by cooperation between scientists from different disciplines like mathematicians and engineers.

References

1. C. Höschl; M. Okrouhlík; J. Červ; J. Beneš: Analytical, computational and experimental investigations on stress wave propagation. *Appl. Mech. Rev.* **47** (1994), 77-99
2. W. Schiehlen: Unilateral contacts in machine dynamics. In F. Pfeiffer (ed.): IUTAM Symp. Unilateral Multibody Dynamics, Series: *Solid Mechanics and its Applications*, Vol. **72**. Dordrecht: Kluwer Academic Publishers, 1999.
3. I. Szabó: *Geschichte der mechanischen Prinzipien und ihrer wichtigsten Anwendungen*. Basel: Birkhäuser, 1977.
4. W. Goldsmith: *Impact: The Theory and Physical Behaviour of Colliding Solids.* London: Edward Arnold Ltd, 1960.
5. S.P. Timoshenko; J.N. Goodier: *Theory of Elasticity.* New York: McGraw Hill, 1970.
6. L. Nikitin: *Statics and dynamics of solids with external dry friction* (in Russian). Moscow: Moscow Liceum, 1998.
7. B. Hu; P. Eberhard; W.Schiehlen: Solving wave propagation problems symbolically using computer algebra. In V.I. Babitsky (ed.): *Dynamics of Vibro-Impact Systems*, pp. 231-240. Berlin: Springer, 1999.
8. M.M. Al-Mousawi: On experimental studies of longitudinal and flexural wave propagations: an annotated bibliography. *Appl. Mech. Rev.* **39** (1986), 853-864.
9. S. Abrate: Wave propagation during high velocity impacts on composite materials. Final Report for Summer Faculty Research Program, pp. 15.1-15.19, Air Forces Office of Scientific Research, 1993.
10. S. Abrate: Vibration of non-uniform rods and beams. *Journal of Sound and Vibration* **185** (1995), 703-716.
11. J.W. Landon; H. Quinney: Experiments with the Hopkinson pressure bar. *Proc. Royal Soc.* **A103** (1923), 622-643.
12. B.M. Kumar; R. Sujith: Exact solutions for the longitudinal vibration of nonuniform rods. *Journal of Sound and Vibration* **207** (1997), 721-729.
13. J.W. Meek; J.P. Wolf: Cone models for homogeneous soil I. *Journal of Geotechnical Engineering* **118** (1992), 667-685.
14. N.P. Suh: Stress-wave propagation in truncated cones against a "rigid" wall. *Experimental Mechanics* **7** (1967), 541-544.
15. B. Hu; P. Eberhard; W. Schiehlen: Symbolic wave propagation analysis of a falling conical rod impacting the rigid ground. Submitted to *Archive of Applied Mechanics*, 1999.

A Survey on Dynamic Contact Problems with Coulomb Friction

Ch. Eck and J. Jarušek

[1] Institute for Applied Mathematics, University Erlangen–Nürnberg, Germany,
[2] Mathematical Institute, Czech Academy of Sciences, Prague, Czech Republic.

Summary. A short survey on recent existence results for dynamic contact problems with Coulomb friction possibly including heat transfer is presented.

1. Introduction

Contact problems have many important applications in different fields of mechanics, e.g. in machine dynamics, civil engineering or in earthquake mechanics. Despite considerable efforts during the past decades, the analysis of these problems offers still many open questions, even in such fundamental topics as existence and uniqueness of solutions. This is particularly true for dynamic problems and for problems with Coulomb friction. In this contribution we survey recent results obtained by variational methods. In the first section the solvability of the dynamic contact problem with Coulomb friction and non–penetrability condition in displacement velocities for viscoelastic bodies is studied. In the second part a coupled thermoviscoelastic problem including a heat equation and a model for the heat generated by friction is investigated. We study two different models of heat transfer, a linearized one and a non–linear model including a specific constitutive law for the heat conductivity.

2. The dynamic contact problem with Coulomb friction

We consider a body occupying in a reference configuration a domain $\Omega \subset \mathbb{R}^N$ of dimension $N \in \{2, 3\}$. The deformation of the body is described by the displacement field u. The stresses are given from the displacements and the displacement velocities via the linearized strain tensor $e_{ij}(u) := \frac{1}{2}(\partial_j u_i + \partial_i u_j)$ by

$$\sigma_{ij}(u) = a^{(0)}_{ijk\ell} e_{k\ell}(u) + a^{(1)}_{ijk\ell} e_{k\ell}(\dot{u}). \tag{2.1}$$

Here, ∂_j denotes the derivative with respect to space variable x_j, the dot signifies the time derivative and the usual sum convention is employed. The coefficients $a^{(\iota)}_{ijk\ell} \in C^{1,1}(\Omega)$ are symmetric in the sense $a^{(\iota)}_{ijk\ell} = a^{(\iota)}_{jik\ell} = a^{(\iota)}_{k\ell ij}$ as well as bounded and elliptic, $a^{(\iota)}_0 \xi_{ij}\xi_{ij} \leq a^{(\iota)}_{ijk\ell}\xi_{ij}\xi_{k\ell} \leq A^{(\iota)}_0 \xi_{ij}\xi_{ij}$ for all symmetric tensors $\{\xi_{ij}\}$, $\xi_{ij} = \xi_{ji}$, with constants $0 < a^{(\iota)}_0 \leq A^{(\iota)}_0$. The boundary conditions consist of given boundary displacements $u = U$ on a part Γ_U, prescribed boundary tractions T_0 on a part Γ_T; on the remaining part Γ_C the body is in contact with a rigid foundation. The non–penetrability of the body and the rigid foundation is approximated by the Signorini–condition formulated in the displacement velocities, $\dot{u}_n \leq 0$, $T_n \leq 0$ and $T_n \dot{u}_n = 0$. The friction is given by the Coulomb law: $\dot{u}_t = 0 \Rightarrow |T_t| \leq \mathfrak{F}|T_n|$, $\dot{u}_t \neq 0 \Rightarrow T_t = -\mathfrak{F}|T_n|\frac{\dot{u}_t}{|\dot{u}_t|}$ with the coefficient of friction $\mathfrak{F}$ depending on x and possibly on $\dot{u}$. Here, n denotes the outer normal vector on the boundary Γ of Ω, the subscripts n and t signify the normal and tangential components of the corresponding vector fields and $T_i = \sigma_{ij}(u)n_j$ denotes the boundary traction. In order to complete the problem description, initial conditions $u(0,\cdot) = u_0$ and $\dot{u}(0,\cdot) = u_1$ are specified. The time interval to be considered is $I_{\mathfrak{T}} := (0, \mathfrak{T})$ with $\mathfrak{T} > 0$. Let $Q_{\mathfrak{T}} := I_{\mathfrak{T}} \times \Omega$ be the time–space domain and $S_{\mathfrak{T}} := I_{\mathfrak{T}} \times \Gamma$ be its lateral boundary consisting of the parts $S_{X,\mathfrak{T}} := I_{\mathfrak{T}} \times \Gamma_X$, $X = U, T, C$. For a measurable set M let $\langle \cdot, \cdot \rangle_M$ denote the $L_2(M)$ scalar product.

We will employ the usual notations $L_p(M)$, $H^\alpha(M)$ and $W_p^\alpha(M)$ for Lebesgue and Sobolev spaces. If α has two components, the first one refers to the time and the second to the space variables. Spaces of functions with values in $\mathbb{R}^N$ are denoted by bold letters. Let us introduce the sets

$$\boldsymbol{H}^{\frac{1}{2},1}_{S_{U,\mathfrak{T}}}(Q_{\mathfrak{T}}) := \left\{ v \in \boldsymbol{H}^{\frac{1}{2},1}(Q_{\mathfrak{T}})\,;\, v = 0 \text{ on } S_{U,\mathfrak{T}} \right\},$$
$$\mathfrak{C} := \left\{ v \in \boldsymbol{H}^{\frac{1}{2},1}(Q_{\mathfrak{T}})\,;\, v = \dot{U} \text{ on } S_{U,\mathfrak{T}} \text{ and } v_n \leq 0 \text{ on } S_{C,\mathfrak{T}} \right\}.$$

The frictional contact problem has a weak formulation in terms of the variational inequality (see e.g. [5], [3]):

Find a function $u : Q_{\mathfrak{T}} \to \mathbb{R}^N$ with $\dot{u} \in \mathfrak{C}$ and $\ddot{u} \in H^{\frac{1}{2}}(I_{\mathfrak{T}}, \boldsymbol{L}_2(\Omega))^*$ such that $u(0,\cdot) = u_0$, $\dot{u}(0,\cdot) = u_1$ and for each $v \in \mathfrak{C} \cap H^{\frac{1}{2}}(I_{\mathfrak{T}}, \boldsymbol{L}_2(\Omega))$ there holds

$$\begin{aligned} &\langle \ddot{u}, v - \dot{u} \rangle_{Q_{\mathfrak{T}}} + a^{(0)}(u, v - \dot{u}) + a^{(1)}(\dot{u}, v - \dot{u}) \\ &\quad + \langle \mathfrak{F}\,|T_n(u)|\,, |v_t| - |\dot{u}_t| \rangle_{S_{C,\mathfrak{T}}} \geq \mathfrak{L}(v - \dot{u}). \end{aligned} \tag{2.2}$$

Here $a^{(\iota)}(u,v) = \langle a_{ijk\ell} e_{ij}(u), e_{k\ell}(v) \rangle_{Q_{\mathfrak{T}}}$, $\iota = 0, 1$, are the bilinear forms of elastic and viscous energy and $\mathfrak{L}(v) := \langle f, v \rangle_{Q_{\mathfrak{T}}} + \langle T_0, v \rangle_{S_{T,\mathfrak{T}}}$ the linear functional of the outer forces.

A very fruitful approach to study this variational inequality is the penalty method. Replacing the non–penetrability condition by the non–linear boundary condition $T_n(u) = -\frac{1}{\delta}[\dot{u}_n]_+$, the penalized problem reads as:

Find $u : Q_{\mathfrak{T}} \to \mathbb{R}^N$ with $\dot{u} \in \dot{U} + \boldsymbol{H}^{\frac{1}{2},1}_{S_{U,\mathfrak{T}}}(Q_{\mathfrak{T}})$ and $\ddot{u} \in H^{\frac{1}{2}}(I_{\mathfrak{T}}, \boldsymbol{L}_2(\Omega))^*$ such that the initial conditions described above are satisfied and for each $v \in \dot{U} + \boldsymbol{H}^{\frac{1}{2},1}_{S_{U,\mathfrak{T}}}(Q_{\mathfrak{T}})$ there holds

$$\begin{aligned} &\langle \ddot{u}, v - \dot{u} \rangle_{Q_{\mathfrak{T}}} + a^{(0)}(u, v - \dot{u}) + a^{(1)}(\dot{u}, v - \dot{u}) + \left\langle \tfrac{1}{\delta}[\dot{u}_n]_+, v_n - \dot{u}_n \right\rangle_{S_{C,\mathfrak{T}}} \\ &\qquad + \left\langle \mathfrak{F}\tfrac{1}{\delta}[\dot{u}_n]_+, |v_t| - |\dot{u}_t| \right\rangle_{S_{C,\mathfrak{T}}} \geq \mathfrak{L}(v - \dot{u}). \end{aligned} \tag{2.3}$$

Problem (2.3) is much simpler to solve than the original problem, because here the non–monotone friction term is compact. The existence of a solution can be proved with the help of a suitable smoothing procedure for the absolute value in the friction term leading to a variational equation by solving the latter with the standard Galerkin method and passing to the limit for the smoothing parameter, see e.g. [5]. The existence result is proved under the following assumptions:

Assumption 1 *Let $\Gamma \in C^{0,1}$, let Γ_X, $X = U, T, C$, be measurable (w.r. to the surface topology). Let the assumptions for the coefficients $a^{(\iota)}_{ijk\ell}$ mentioned above be valid. Let $\mathfrak{F} \in L_\infty(\Gamma_C \times \mathbb{R}^N)$ such that $\mathfrak{F}(x,\cdot)$ is continuous for all $x \in \Gamma_C$ and $\mathfrak{F} \geq 0$. The given data satisfy $\mathfrak{L} \in \boldsymbol{H}^{\frac{1}{2},1}(Q_{\mathfrak{T}})^*$, $U, \dot{U} \in \boldsymbol{H}^1(Q_{\mathfrak{T}})$ and $U(0,\cdot) = u_0$, $\dot{U}(0,\cdot) = u_1$, $U = 0$ on $S_{C,\mathfrak{T}}$.*

The constants in the *a priori* estimates of the following theorems may depend on the geometry of the domain and on the given data. Their dependence on approximation parameters is explicitly indicated.

Theorem 2.1. *Under Assumption 1, the penalized contact problem (2.3) has a solution. Every solution satisfies the* a priori *estimate*

$$\|\dot{u}\|_{L_\infty(I_{\mathfrak{T}}, \boldsymbol{L}_2(\Omega))} + \|\dot{u}\|_{\boldsymbol{H}^{\frac{1}{2},1}(Q_{\mathfrak{T}})} + \left\|\tfrac{1}{\delta}[\dot{u}_n]_+^2\right\|_{L_1(S_{C,\mathfrak{T}})} \leq c.$$

If the coefficient of friction is solution–independent, $\mathfrak{F} = \mathfrak{F}(x)$, then the solution is unique.

The existence of a solution to the original contact problem is proved by letting the penalty parameter δ tend to zero. Then, however, the friction functional is no longer compact. This difficulty is tackled by the proof of a better regularity of the solution on the contact part of the boundary. Establishing such a regularity result is in fact the main step in the existence proof. This is done with the help of a localization and a straightening of the boundary and the application of a special shift–technique introduced by Fichera. Here the friction term is no longer of lower order, therefore it must be estimated by the quadratic form of the differential operator. This is possible, if the coefficient of friction is smaller than a definite bound $C_{\mathfrak{F}}$. This bound is calculated in [3] for the case of homogeneous, isotropic material in two dimensions and in [5] for general material. The result for the first case is depicted in Figure 2.1 in terms of the Young modulus E and the Poisson ratio ν of the *viscous* part of the strain–stress relation. In the second case the upper bound is given in terms of the value $z = \frac{\pi\sqrt{2a_1A_1}}{2c^2_{N-1}(\frac{1}{2})}$ with the constants a_1 and A_1 from the ellipticity and boundedness condition for the viscous part and with $c_{N-1}(\alpha) = c_1(\alpha)\int_{\mathbb{R}^{N-2}}(1+|s|^2)^{-\frac{N-1}{2}-\alpha}\,ds$, $c_1(\alpha) = 2^{2-2\alpha}\int_{\mathbb{R}}\frac{\sin^2(t)}{|t|^{1+2\alpha}}\,dt$ by the formula

$$C_{\mathfrak{F}} = \sqrt{\frac{a_1}{2A_1}} \begin{cases} \sqrt{\frac{z}{(z+1)(1+\sqrt{8z})}}, & z \geq 1, \\ \sqrt{\frac{\sqrt{z}}{2(1+\sqrt{8z})}}, & z \leq 1. \end{cases} \tag{2.4}$$

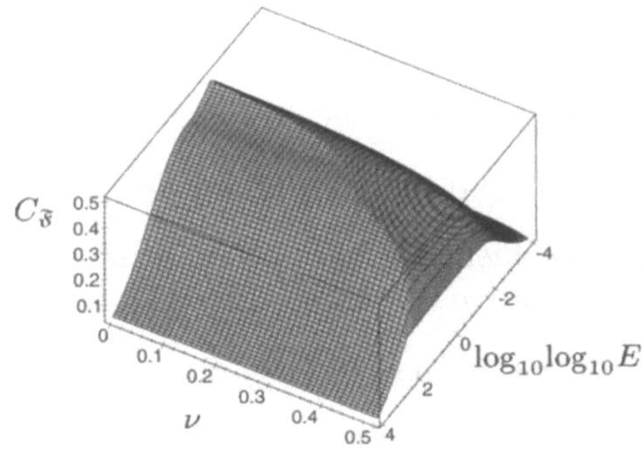

Fig. 2.1. Admissible coefficient of friction

Theorem 2.2. *Let Assumption 1 be valid, let $\Gamma_C \in C^{1,1}$, $\|\mathfrak{F}\|_{L_\infty(\Gamma_C\times\mathbb{R}^N)} < C_{\mathfrak{F}}$, $\operatorname{supp}\mathfrak{F} \subset \Gamma_{\mathfrak{F}}\times\mathbb{R}^N$ with $\Gamma_{\mathfrak{F}} \subset \Gamma_C$ having a* positive *distance to $\Gamma\backslash\Gamma_C$, $f \in \boldsymbol{H}^{\frac{1}{4},\frac{1}{2}}(Q_{\mathfrak{F},\mathfrak{T}})^*$ with $Q_{\mathfrak{F},\mathfrak{T}} := I_{\mathfrak{T}} \times \Omega_{\mathfrak{F}}$, where $\Omega_{\mathfrak{F}} \subset \Omega$ is a $C^{1,1}$ domain whose boundary contains $\Gamma_{\mathfrak{F}}$. Then the solution to the penalized problem satisfies*

$$\|u\|_{\boldsymbol{H}^{\frac{1}{2},1}(S_{\mathfrak{F},\mathfrak{T}})} + \left\|\tfrac{1}{\delta}[\dot{u}_n]_+\right\|_{L_2(S_{\mathfrak{F},\mathfrak{T}})} \leq c$$

with $S_{\mathfrak{F},\mathfrak{T}} := I_{\mathfrak{T}} \times \Gamma_{\mathfrak{F}}$ and c depending on the geometry of the domain, the coefficients $a^{(\iota)}_{ijk\ell}$ and the given data f, T_0, U, u_0, u_1. A suitable sequence u_k of solutions to the penalized problem with penalty parameter $\delta_k \to 0$ converges to a solution of the original contact problem (2.2).

Remark. The constant $C_{\mathfrak{F}}$ given here for the dynamic problem is smaller than that valid for the static case without viscosity, [2]. Exact trace and inverse trace theorems seem to be better for elliptic than for parabolic problems. Recently a similar result for the quasistatic problem was proved by L.E. Andersson [1] with $C_{\mathfrak{F}}$ dependent on the domain Ω.

3. Frictional contact problems with heat transfer

If the influence of heat generation and heat transfer shall be included into the model, it is necessary to add the heat equation for the temperature Θ,

$$\dot{\Theta} - \partial_i(c_{ij}\partial_j\Theta) + b_{ij}(\Theta)e_{ij}(\dot{u}) = 0 \text{ in } Q_{\mathfrak{T}}, \tag{3.1}$$

supplemented by an initial condition $\Theta(0,\cdot) = \Theta_0$ and by boundary conditions. For the latter we choose Dirichlet conditions $\Theta = \Theta_0$ on $S_{U,\mathfrak{T}}$ and the condition of heat exchange with heat exchange coefficient K and outer temperature Υ

$$c_{ij}\partial_j\Theta n_i = K(\Upsilon - \Theta) \text{ on } S_{T,\mathfrak{T}}. \tag{3.2}$$

On the contact part of the boundary the heat generated by friction is added,

$$c_{ij}\partial_j\Theta n_i = \mathfrak{F}|T_n||\dot{u}_t| + K(\Upsilon - \Theta) \text{ on } S_{C,\mathfrak{T}}. \tag{3.3}$$

In the viscoelastic contact problem, the heat deformation is described by an additional term in the constitutive law which then reads as

$$\sigma_{ij}(u,\Theta) = a^{(0)}_{ijk\ell}e_{k\ell}(u) + a^{(1)}_{ijk\ell}e_{k\ell}(\dot{u}) - b_{ij}(\Theta), \quad i,j = 1,\dots,N.$$

If the term $\mathfrak{F}|T_n||\dot{u}_t|$ describing the heat generated by friction is used in combination with *linear* constitutive material laws, then this term has a higher order of growth than the monotone volume energy terms. Since this term is also non–monotone, it is rather difficult to establish the required *a priori* estimates. A similar problem holds for the physically simplest law $b_{ij}(\Theta) = \widetilde{b}_{ij}\Theta$ for the heat exchange tensor. In order to avoid these problems, we consider here two simplified models:

In **Model 1** we use a linear constitutive law for the heat exchange tensor. Here $\{c_{ij}\}$ depends on the space variable only and satisfies the usual conditions $c_{ij} = c_{ji}$ and $c_0\xi_i\xi_i \le c_{ij}\xi_i\xi_j \le C_0\xi_i\xi_i$ for all $\{\xi_i\} \in \mathbb{R}^N$ with $0 < c_0 \le C_0 < +\infty$. The dependence of the heat exchange term on the temperature is neglected, $b_{ij}(\Theta) = \widetilde{b}_{ij}$, and the frictional heat $\mathfrak{F}|T_n||\dot{u}_t|$ is replaced by $J(\mathfrak{F}|T_n|,|\dot{u}_t|)$ with a function J satisfying the conditions $|J(x,y)| \le c_1(1+|x|+|y|)$ with $c_1 \in \mathbb{R}$ and $J(f_n,g_n) \rightharpoonup J(f,g)$ in $L_\alpha(S_{C,\mathfrak{T}})$ for $f_n \rightharpoonup f$, $g_n \to g$ in $L_2(S_{C,\mathfrak{T}})$ with $\alpha > \frac{8}{5}$. If the convergence of f_n is strong, then the convergence of $J(f_n,g_n)$ must be strong, too.

In **Model 2** the original form of the frictional heat $J(x,y) = xy$ and the heat exchange term $b_{ij}(\Theta) = \widetilde{b}_{ij}\Theta$ are kept. In order to control the growth of these terms, a nonlinear constitutive law for the heat equation is chosen. The heat conductivity depends locally Lipschitz–continuously on the *temperature gradient* and satisfies the growth condition

$$c_0\left(1+|\nabla\Theta|^2\right)\xi_i\xi_i \le c_{ij}(\nabla\Theta)\xi_i\xi_j \le C_0\left(1+|\nabla\Theta|^2\right)\xi_i\xi_i$$

for all $\{\xi_i\} \in \mathbb{R}^N$ with $0 < c_0 \le C_0$, the monotonicity condition

$$\begin{aligned}\langle c_{ij}(\nabla\Theta)\partial_j\Theta - c_{ij}(\nabla\Psi)\partial_j\Psi, \partial_i(\Theta-\Psi)\rangle_{Q_\mathfrak{T}} \\ \ge c_1\|\nabla(\Theta-\Psi)\|^4_{L_4(Q_\mathfrak{T})} + c_2\|\nabla(\Theta-\Psi)\|^2_{L_2(Q_\mathfrak{T})}\end{aligned}$$

for all $\Theta,\Psi \in L_4\big(I_\mathfrak{T};W^1_4(\Omega)\big)$ with $c_1,c_2 > 0$, and the continuity condition

$$c_{ij}(\nabla\Theta^{(k)})\partial_j\Theta^{(k)} \to c_{ij}(\nabla\Theta)\partial_j\Theta \text{ in } \boldsymbol{L}_{\frac{4}{3}}(Q_\mathfrak{T})$$

for $\Theta^{(k)} \to \Theta$ in $L_4\big(I_{\mathfrak{T}}; W_4^1(\Omega)\big)$. An example for such a coefficient tensor is $c_{ij}(\Xi) = \delta_{ij}\big(\gamma_0(x) + \gamma_1(x)|\Xi|^2\big)$ with $0 < c_3 \le \gamma_0, \gamma_1 \le c_4 < +\infty$ and the Kronecker symbol δ_{ij}. In Model 2 we also require mes $\Gamma_U \neq \emptyset$. In both models we assume $\widetilde{b}_{ij}$ to be Lipschitz, bounded and symmetric.

The different models for the heat conductivity lead to different sets of admissible functions for the variational formulation of the thermo–viscoelastic contact problem:

$$\mathfrak{V} := \begin{cases} \{\varphi \in H^{\frac{1}{2},1}(Q_{\mathfrak{T}}) \,|\, \varphi = 0 \text{ on } S_{U,\mathfrak{T}}\} & \text{for Model 1,} \\ \{\varphi \in L_4\big(I_{\mathfrak{T}}; W_4^1(\Omega)\big) \,|\, \varphi = 0 \text{ on } S_{U,\mathfrak{T}}\} & \text{for Model 2.} \end{cases} \tag{3.4}$$

The weak formulation of the problem is given by

Find (u, Θ) *with* $\dot{u} \in \mathfrak{C}$, $\ddot{u} \in H^{\frac{1}{2}}(I_{\mathfrak{T}}; \boldsymbol{L}_2(\Omega))^*$, $\Theta \in \Theta_0 + \mathfrak{V}$, $\dot{\Theta} \in \mathfrak{V}^*$, $u(0,\cdot) = u_0$, $\dot{u}(0,\cdot) = u_1$ *and* $\Theta(0,\cdot) = \Theta_0$ *such that for all* $v \in \mathfrak{C}$

$$\langle \ddot{u}, v - \dot{u}\rangle_{Q_{\mathfrak{T}}} + a^{(0)}(u, v - \dot{u}) + a^{(1)}(\dot{u}, v - \dot{u}) - \langle b_{ij}(\Theta), e_{ij}(v - \dot{u})\rangle_{Q_{\mathfrak{T}}} + \langle \mathfrak{F}|T_n|, |v_t| - |\dot{u}_t|\rangle_{S_{C,\mathfrak{T}}} \ge \mathfrak{L}(v - \dot{u}) \tag{3.5}$$

and for all $\varphi \in \mathfrak{V}$

$$\langle \dot{\Theta}, \varphi\rangle_{Q_{\mathfrak{T}}} + \langle c_{ij}\partial_j\Theta, \partial_i\varphi\rangle_{Q_{\mathfrak{T}}} + \langle b_{ij}(\Theta)e_{ij}(\dot{u}), \varphi\rangle_{Q_{\mathfrak{T}}} + \langle K(\Theta - \Upsilon), \varphi\rangle_{S_{\mathfrak{T}}} = \langle J(\mathfrak{F}|T_n|, |\dot{u}_t|), \varphi\rangle_{S_{C,\mathfrak{T}}}. \tag{3.6}$$

The solvability of the coupled problem is studied by fixed point arguments. Let $\mathfrak{S}_u\Theta^{(0)}$ denote the solution of the contact problem with given temperature $\Theta^{(0)}$ and let $\mathfrak{S}_\Theta(u)$ denote the solution of the heat problem with given displacement field u. The operators $\mathfrak{S}_u$ and $\mathfrak{S}_\Theta$ are well defined, if the corresponding problems have unique solutions. Then every solution of the coupled problem (3.5), (3.6) is also a fixed point of the operator $\Phi := \mathfrak{S}_\Theta \circ \mathfrak{S}_u : \mathfrak{X} \to \mathfrak{X}$ mapping a set $\mathfrak{X}$ of admissible temperatures into itself. Unfortunately, the uniqueness of the solution to the contact problem is not clear. Therefore we apply this method to the *penalized problem*, consisting of the penalized contact problem (2.3) with the modified linear functional $\mathfrak{L}_\Theta(v) := \mathfrak{L}(v) + \langle b_{ij}(\Theta), e_{ij}(v)\rangle_{Q_{\mathfrak{T}}}$ and the heat problem (3.6) with $J(\mathfrak{F}|T_n|, |\dot{u}_t|)$ replaced by $J(\mathfrak{F}\frac{1}{\delta}[\dot{u}_n]_+, |\dot{u}_t|)$. Then the following two results are true, [4], [6]:

Theorem 3.1. *In addition to the assumptions of Theorem 2.2 and the assumptions on* b_{ij} *of Model 1 or Model 2 let* $\Theta^{(0)} \in H^{\frac{1}{4},\frac{1}{2}}(Q_{\mathfrak{T}})$ *and let* $\mathfrak{F}$ *be independent of the solution. Then the penalized contact problem (2.3) with* $\mathfrak{L}$ *replaced by* $\mathfrak{L}_\Theta$ *has a unique solution* $u = \mathfrak{S}_u(\Theta^{(0)})$. *This solution satisfies*

$$\|u\|_{\boldsymbol{H}^{\frac{1}{2},1}(Q_{\mathfrak{T}})} \le c_1\|\Theta^{(0)}\|_{L_2(Q_{\mathfrak{T}})} + c_2, \tag{3.7}$$

$$\|u\|_{\boldsymbol{H}^{\frac{1}{2},1}(S_{\mathfrak{F},\mathfrak{T}})} + \big\|\tfrac{1}{\delta}[\dot{u}_n]_+\big\|_{L_2(S_{\mathfrak{F},\mathfrak{T}})} \le c_3\|\Theta^{(0)}\|_{H^{\frac{1}{4},\frac{1}{2}}(Q_{\mathfrak{T}})} + c_4. \tag{3.8}$$

The solution operator $\mathfrak{S}_u : L_2(Q_{\mathfrak{T}}) \to \boldsymbol{H}^{\frac{1}{2},1}(Q_{\mathfrak{T}})$ *is Lipschitz.*

Theorem 3.2. *Let the above assumptions for c_{ij}, b_{ij} and J for Model 1 or Model 2 be valid, let $\Theta_0(0,\cdot) \in L_2(\Omega)$, $\Theta_0 \in H^{\frac{1}{2},1}(Q_{\mathfrak{T}})$, $\Upsilon \in L_2(S_{\mathfrak{T}})$, $K \in L_\infty(\Gamma)$, $K \geq 0$ and $u \in \boldsymbol{H}^{\frac{1}{2},1}(Q_{\mathfrak{T}})$. In the case of Model 2 we also assume $\Theta_0 \in L_4\big(I_{\mathfrak{T}}, W_4^1(\Omega)\big)$. Then problem (3.6) has a unique solution Θ. In the case of Model 1 the solution satisfies*

$$\|\Theta\|_{H^{\frac{1}{2},1}(Q_{\mathfrak{T}})} \leq c_1\|\dot{u}\|_{L_2(I_{\mathfrak{T}};\boldsymbol{H}^1(\Omega))} + c_2\left\|\tfrac{1}{\delta}[\dot{u}_n]_+\right\|_{L_2(S_{\mathfrak{F}})} + c_3 \tag{3.9}$$

and $\mathfrak{S}_\Theta : u \to \Theta$ is continuous from $\boldsymbol{H}^{\frac{1}{2},1}(Q_{\mathfrak{T}})$ to $H^{\frac{1}{2},1}(Q_{\mathfrak{T}})$. For Model 2 the following a priori *estimate is valid for every $\alpha < \frac{1}{2}$,*

$$\|\Theta\|^2_{H^{\alpha,1}(Q_{\mathfrak{T}})} + \|\Theta\|^3_{L_4(I_{\mathfrak{T}},W_4^1(\Omega))} \leq c_1(\alpha,\delta)\|\dot{u}\|^2_{L_2(I_{\mathfrak{T}};\boldsymbol{H}^1(\Omega))} + c_2(\alpha,\delta), \tag{3.10}$$

and the continuity holds from $\boldsymbol{H}^{\frac{1}{2},1}(Q_{\mathfrak{T}})$ to $L_4\big(I_{\mathfrak{T}}; W_4^1(\Omega)\big) \cap H^{\frac{1}{4}}\big(I_{\mathfrak{T}}; L_2(\Omega)\big)$.

Now we apply the Schauder fixed point theorem to the operator $\Phi := \Theta^{(0)} \mapsto \Theta$ on a subset $\mathfrak{B}$ of a Banach space $\mathfrak{X}$. In the case of Model 1 we choose $\mathfrak{X} := L_2(Q_{\mathfrak{T}})$ and $\mathfrak{B} := \big\{\varphi \in L_2(Q_{\mathfrak{T}}) \,\big|\, \|\varphi\|_{L_2(Q_{\mathfrak{T}})} \leq R\big\}$ with $R > 0$. Then it is possible to prove

$$\|\Theta\|^2_{L_2(Q_{\mathfrak{T}})} \leq \mathfrak{T}\left(c_1\big\|\Theta^{(0)}\big\|^2_{L_2(Q_{\mathfrak{T}})} + c_2\right)$$

with c_1, c_2 dependent on the domain and the given data and c_1 *independent* of the initial data. For a sufficiently small time interval and an appropriate constant R, the Schauder theorem yields the existence of a solution to the coupled problem. By iteration this result is extended to the whole time interval $I_{\mathfrak{T}}$. In the case of Model 2 let $\mathfrak{X} := H^{\frac{1}{4}}(I_{\mathfrak{T}}; L_2(\Omega))$ and

$$\mathfrak{B} := \big\{\varphi \in H^{\frac{1}{4}}(I_{\mathfrak{T}}; L_2(\Omega)) \,\big|\, \|\varphi\|_{H^{\frac{1}{4}}(I_{\mathfrak{T}};L_2(\Omega))} + \|\varphi\|^2_{L_2(Q_{\mathfrak{T}})} \leq R\big\}.$$

If $\Theta^{(0)} \in \mathfrak{B}$, then there holds

$$\|\Theta\|_{H^{\frac{1}{4}}(I_{\mathfrak{T}};L_2(\Omega))} + \|\Theta\|^2_{L_2(Q_{\mathfrak{T}})} \leq c_1(\delta)R^{\frac{2}{3}} + c_2(\delta).$$

Hence, for a suitable choice of R, the assumptions of the Schauder fixed point theorem are again satisfied. Consequently the following result is obtained:

Theorem 3.3. *Under the assumptions of Theorems 3.1 and 3.2 there exists a solution to the penalized thermoviscoelastic contact problem. This solution satisfies the estimates*

$$\|u\|_{\boldsymbol{H}^{\frac{1}{2},1}(Q_{\mathfrak{T}})} + \|u\|_{\boldsymbol{H}^{\frac{1}{2},1}(S_{\mathfrak{F},\mathfrak{T}})} \leq c_1 \text{ and}$$

$$\begin{cases} \|\Theta\|_{H^{\frac{1}{2},1}(Q_{\mathfrak{T}})} \leq c_2 & \text{for Model 1,} \\ \|\Theta\|_{H^{\alpha,1}(Q_{\mathfrak{T}})} + \|\Theta\|_{L_4(I_{\mathfrak{T}},W_4^1(\Omega))} \leq c_3(\alpha,\delta) & \text{for Model 2 with } \alpha < \frac{1}{2}. \end{cases}$$

The solvability of the original coupled problem is now proved by performing the limit $\delta \to 0$ for the penalty parameter. This is a standard procedure in the case of Model 1. However, for Model 2 we need an *a priori* estimate, uniform with respect to the penalty parameter. Here, the treatment of the frictional heat is very delicate. With rather tricky estimates employing embedding, interpolation and trace results for non–Hilbertian Sobolev–Slobodetskii spaces based mainly on [9], it is possible to prove in the case $N \leq 3$

$$||\Theta||_{H^{\alpha,1}(Q_{\mathfrak{T}})} + ||\Theta||_{L_4(I_{\mathfrak{T}},W_4^1(\Omega))} + ||\Theta||_{L_p(S_{\mathfrak{T}})} \leq c \tag{3.11}$$

for all $\alpha < \frac{1}{2}$, $p < 6$ with c independent of δ. The application of the interpolation technique requires $\Gamma \in C^{1,1}$. Using this estimate, it is possible to pass to the limit $\delta \to 0$ of the penalty parameter and to prove the following result:

Theorem 3.4. *Let the assumptions of Theorem 3.3 and, in the case of Model 2, $\Gamma \in C^{1,1}$ as well as $N \leq 3$ be valid. Then there exists a sequence $\delta_k \to 0$ of penalty parameters and a corresponding sequence (u_k, Θ_k) of solutions to the penalized problem such that the latter converges to a solution of the original thermoviscoelastic contact problem (3.5), (3.6).*

Acknowledgement. The support of this work by the Grant Agency of the Czech Academy of Sciences under grant No. A1075707 is gratefully acknowledged.

References

1. L.E. Andersson. Existence results for quasistatic contact problems with Coulomb friction. Preprint.
2. C. Eck, J. Jarušek: Existence results for the static contact problem with Coulomb friction. *Math. Mod. Meth. Appl. Sci.* **8**(3), 445–468, 1998.
3. C. Eck, J. Jarušek: Existence of solutions for the dynamic frictional contact problem for isotropic viscoelastic bodies. Preprint 242, Institute for Applied Mathematics, University Erlangen–Nürnberg, 1998.
4. C. Eck, J. Jarušek: The solvability of a coupled thermoviscoelastic contact problem with small Coulomb friction and linearized growth of frictional heat. *Math. Meth. Appl. Sci.* **22**, 1221–1234, 1999.
5. J. Jarušek, C. Eck: Dynamic contact problems with small Coulomb friction for viscoelastic bodies. Existence of solutions. *Math. Models Meth. Appl. Sci.* **9**(1), 11–34, 1999.
6. J. Jarušek, C. Eck: Existence of solutions to a nonlinear coupled thermoviscoelastic contact problem with Coulomb friction. Preprint No. 236, Institute of Applied Mathematics, University Erlangen–Nürnberg, 1998.
7. J.L. Lions, E. Magenes: *Problèmes aux limites non-homogènes et applications*, Dunod, Paris 1968.
8. J. Nečas: *Les méthodes directes en équations elliptiques*, Academia, Praha 1967.
9. H. Triebel: *Interpolation Theory, Function Spaces, Differential Operators.* North Holland, Amsterdam - New York - Oxford 1978.

Numerical Simulation of Noise Radiation from Rolling Tires

L. Gaul, U. Nackenhorst and B. Nolte

Institute A of Mechanics, University of Stuttgart, Institute of Mechanics and Institute of Mathematics, University of the Federal Armed Forces Hamburg, Germany.

Summary. The noise radiation from rolling bodies is studied numerically by using a sequential finite element–boundary element approach of the field equations. The equations of motion for the rolling wheel are developed in the frame of an Arbitrary Eulerian Lagrangian description which provides a time independent formulation for steady state rolling and a spatial description of vibrations. The noise radiation caused by the vibration modes of a steady state rolling wheel is computed by a boundary element approach. An outlook on the simulation of acoustic radiation due to tire–road interaction is given.

1. Introduction

With increasing traffic density the reduction of traffic noise becomes more and more important. Success in reducing sound sources in cars has been achieved. This holds for example for the main engine noise. Thus, nowadays the sound of the rolling tires is the main source of noise radiation. This effect is dominating when cars exceed a speed of 25 mph and trucks a speed of 40 mph.

The rolling noise is mainly caused by the vibrations of the tires [1]. These vibrations are excited by the physical contact of the rolling tires with the road surface. As a result airborne sound is generated.

The special excitation mechanisms are deterministic on the one hand but as well can be stochastic. E.g. an excitation can be caused by the periodical contact of the tire treads and the roughness of the road surface. Thus, one step to overcome the noise problem is to optimize the tire tread design in an acoustical sense. Modern treads are designed with uneven spacing generating a broader and lower excitation spectrum. This technique may be used to decrease the sound pressure level up to 5 dB [2]. However, this technique has been tested to the extent of its limits. Further reduction techniques may include the following:

a) The structural dynamic behavior of the wheel or the tires needs to be improved acoustically. In this context it has to be ensured that other important tire characteristics, including security or economical aspects will not be negatively influenced.
b) The whole tire–road system needs to be optimized because the characteristic quality of the road surface implies a great potential for noise reduction [3].

Due to the complicated structure of pneumatic tires with strongly non–linear mechanical response the exact mechanisms of all noise sources are not well understood yet. Of course, the sound pressure level of a passing vehicle is measured easily, but these measurements at a few fixed positions are not significant for the entire noise emission. More complete measurements of the noise radiation of rolling tires are possible only in the laboratory [4]. Another way to investigate the sound noise origin is to use numerical simulation methods. Besides, the numerical simulation enables the observation of single design–parameter influences on the noise radiation as well as of safety and economy aspects.

2. The simulation procedure

To explain the technique for the numerical simulation of rolling noise in this paper an illustrative way has been chosen rather than a detailed mathematical description for which is referred to [7]. The technique splits into three main steps:

1. The computation of the strongly nonlinear motion of the steady state rolling wheel by a finite element approach,
2. A modal analysis of the vibrational behavior of the rolling wheel in its stationary rolling contact state,
3. The computation of the noise radiation based on the results of step 2.

In the first step the equilibrium of a steady state rolling wheel is computed by use of a special finite element approach which is described in more details in the next section. This problem is strongly nonlinear due to the nonlinear material behavior of tire composites, large displacements and the contact problem itself. Therefore, an incremental – iterative procedure is necessary for this analysis, expressed by the equations

$$ {}^{t}\mathbf{K}_{\text{eff}}\left({}^{t}\boldsymbol{\phi}\right)\,\Delta\boldsymbol{\phi} \;=\; {}^{t+\Delta t}\mathbf{f}^{(e)} - {}^{t}\mathbf{f}^{(i)} + {}^{t}\mathbf{f}^{(c)}\,, \qquad (2.1)$$

$$ {}^{t+\Delta t}\boldsymbol{\phi} \;=\; {}^{t}\boldsymbol{\phi} + \Delta\boldsymbol{\phi}\,. \qquad (2.2)$$

Herein ${}^{t}\mathbf{K}_{\text{eff}}\left({}^{t}\boldsymbol{\phi}\right)$ is the effective stiffness matrix which depends on the actual state of deformation ${}^{t}\boldsymbol{\phi}$ and the contact state. The target of the Newton–Raphson like procedure is to find the equilibrium of the applied external forces ${}^{t+\Delta t}\mathbf{f}^{(e)}$, the internal forces due to the divergence of the stress tensor ${}^{t}\mathbf{f}^{(i)}$ and the contact forces ${}^{t}\mathbf{f}^{(c)}$. For a realistic modelling of the tire behavior more than 50.000 unknowns have to be solved. This is done on modern unix–workstations in a reasonable amount of time. However, this is only possible by using a special relative kinematics formulation described in the next section. Additionally, stable and efficient contact algorithms are needed.

After the configuration $\bar{\boldsymbol{\phi}}$ of steady state rolling is investigated a modal–analysis is performed in this deformed configuration. By doing this, small linear vibrations φ are superimposed on the large elastic deformation $\bar{\phi}$, going along with the relative velocities $\dot{\varphi}$. In addition, the contact conditions are kept fixed for this eigen–analysis. For this an eigenvalue problem

$$\left\{\lambda_i \begin{bmatrix} \mathbf{M} & \mathbf{0} \\ \mathbf{0} & \mathbf{I} \end{bmatrix} + \begin{bmatrix} \mathbf{G} & \mathbf{K}_{\text{eff}}(\bar{\phi}) \\ -\mathbf{I} & \mathbf{0} \end{bmatrix}\right\} \begin{bmatrix} \dot{\varphi} \\ \varphi \end{bmatrix}_i = \mathbf{0} \tag{2.3}$$

has to be solved. It is meaningful that the eigenvalues λ_i and eigenvectors φ_i are complex valued quantities due to the gyroscopic effects expressed by the skew–symmetric matrix **G**. Furthermore, **M** is the mass matrix and **I** is the identity matrix in eq. (2.3).

From the computed relative velocities $\dot{\varphi}$ at the wheel surface, e.g. from the calculated eigenmodes or a modal superposition of these, the relative velocity components perpendicular to the wheel $\mathbf{v}_n$ are extracted. These velocity components are the Neumann input data for the third step, the computation of the noise radiation by use of the boundary element method [6]. This computation is performed in two steps. In the first step the pressure distribution on the surface of the wheel and the road is computed by

$$\mathbf{F}\,\mathbf{p}_n = \mathbf{J}\,\mathbf{v}_n\ , \tag{2.4}$$

where **F** and **J** are the corresponding boundary element matrices and $\mathbf{p}_n$ decribes the sound pressure at the nodes on the boundary Γ. Then the sound analysis at arbitrary points **x** in the surrounding air is computed with the nodal data by the evaluation of surface integrals, e.g. the sound pressure is computed from

$$p(\boldsymbol{\xi}) = \int_{\Gamma_x} g(\mathbf{x},\boldsymbol{\xi})\, \mathrm{D}_n\left(p_n(\mathbf{x})\right) \mathrm{d}\Gamma_x - \int_{\Gamma_x} p_n(\mathbf{x})\, \mathrm{D}_n\left(g(\mathbf{x},\boldsymbol{\xi})\right) \mathrm{d}\Gamma_x\ , \tag{2.5}$$

where g is the so–called Green's function and D_n is the directional derivative normal to the surface Γ [9].

In a first approach the road surface influence can be bounded as soft–walled or hard–walled, which is efficiently implemented by mirroring techniques as discussed in section 5. A more realistic description of the reflection of sound waves on the road surface can be modeled by prescribed impedances, but then the road needs to be discretized as well. The discritization has to be truncated at a certain distance from the wheel.

3. Numerical simulation of stationary rolling wheels

For the description of rolling in relative kinematics an Arbitrary Lagrangian Eulerian (ALE) formulation has been chosen, which is depicted schematically in fig. 3.1. The motion Ψ is decomposed into the rigid body motion χ which is described in Eulerian coordinates, i.e. the mesh is fixed in space, and into the deformation ϕ described in Lagrangian coordinates. The material time derivative of the motion,

$$\frac{\mathrm{d}\boldsymbol{\Psi}}{\mathrm{d}t} = \frac{\partial \boldsymbol{\phi}}{\partial t} + \mathrm{Grad}\boldsymbol{\phi} \cdot \dot{\boldsymbol{\chi}} \,, \tag{3.1}$$

then splits into the relative velocity and into the convective velocity. In the steady state case the relative velocity vanishes. Thus, the stationary rolling state is modeled independent of the time and no expensive time discretization is necessary. Furthermore, the fine discretization in space necessary for the contact analysis can be concentrated in the contact region only and the mesh can be coarse over the complete circumference. And moreover, the relative dynamics is described in space immediately, as it is needed for the computation of the sound radiation.

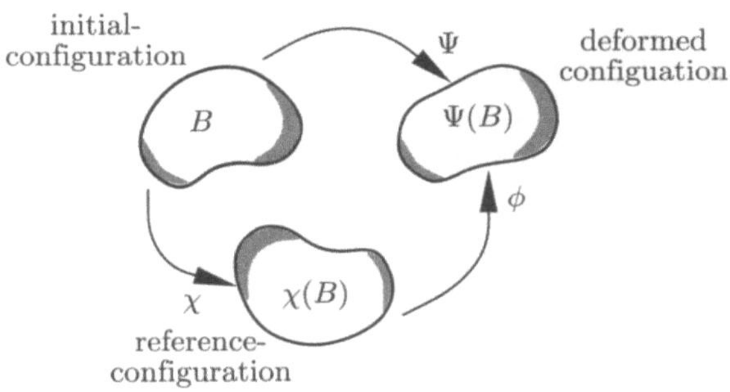

Fig. 3.1. The ALE mapping of motion.

For illustration, the ALE–description can be interpreted as the introduction of an observer riding on the axis of the rolling wheel. The ALE–observer monitors the motion of an arbitrary particle which at the moment is located at the fixed point **x** but it does not collect the history of the material particles. Therefore, special formulations and algorithms for the treatment of the contact conditions, especially for the computation of the contact shear tractions, have to be worked out. For details the reader is referred to e.g. [5].

4. Numerical analysis of sound radiation

For the analysis of sound radiation from vibrating structures the boundary element method has the following decisive advantages:

a) The sound waves can radiate without reflections from artificial boundaries. The Sommerfeld radiation condition is fulfilled inherently.
b) Only the acoustically relevant boundaries, i.e. sound generating or reflecting surfaces, have to be discretized. This reduces the dimension of the problem by one. Therefore, the number of unknowns is much smaller than in the case of a comparable finite element discretization.

c) And furthermore, no discretization of the infinite road surface is necessary as long as it can be modeled soft– or hard–walled.

The basic field equations of the acoustical problem are the wave equations for the sound pressure and for the particle velocity. A transformation into the frequency domain leads to the Helmholtz equation. For deriving a boundary integral formulation, the field equations are mapped onto the relevant surfaces by weighting with a test function g and integration over the entire domain. With a suitable choice of the test function for a source free domain a complete mapping on the boundary is achieved by double partial integration and application of the divergence theorem. The test function

$$g = \frac{1}{4\pi r} e^{-ikr} \tag{4.1}$$

is called the Fundamental Solution or the Green's Function for the full space [9]. Herein r is the Euclidian distance between the field point ξ and the so called load point x as shown in fig. 4.1. Because the Sommerfeld radiation condition is fulfilled a priori by use of the test function eq. (4.1), no waves are reflected from the boundary Γ_∞ at infinity. Additionally, by the use of the modified test functions

$$g_s(r, r') = \frac{1}{4\pi}\left(\frac{1}{r}e^{-ikr} + \frac{1}{r'}e^{-ikr'}\right) \tag{4.2}$$

$$g_h(r, r') = \frac{1}{4\pi}\left(\frac{1}{r}e^{-ikr} - \frac{1}{r'}e^{-ikr'}\right) \tag{4.3}$$

soft– and hard–walled surfaces of the road respectively are implemented without discretization of the road surface. This can be visualized as a superposition of two sound fields, the actual scattering sound field generated by the radiator and a reflected wave from the road surface as illustrated in fig. 4.1. This is called the mirror technique. In conclusion only the surface of the tire has to be discretized.

For a more detailed description of the acoustical behavior of the road surface impedance boundary conditions need to be described. But this requires a discretization of the road surface too, going along with additional numerical effort and a truncation error of the finite discretization.

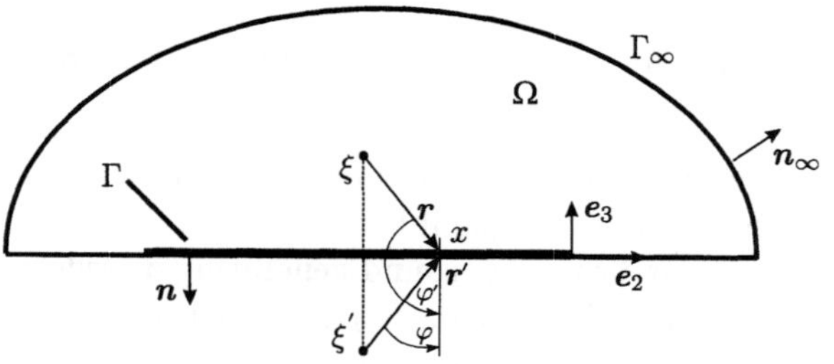

Fig. 4.1. The mirror technique.

5. Example: A simple wheel

For demonstrating the computation procedure the noise radiation caused by fundamental vibration modes of a simple wheel in rolling contact has been analyzed. In the first step the wheel is set up into rotation at a speed of 25 mph and pressed against the rigid road surface. In this particular deformed state of stationary rolling the computation of the eigenfrequencies and the eigenvectors is carried out by eq. (2.3). Gyroscopic effects have been neglected here, thus only real eigensolutions have been computed. From the eigenvectors the relative velocities have been obtained and their components normal to the surface have been extracted. These normal components of the relative velocity are the Neumann boundary data of the subsequent BEM–analysis for computing the sound pressure on the surface by the linear system in eq. (2.4). Finally, at arbitrary field points in the domain the sound pressure has been computed by solving eq. (2.5).

In fig. 5.1 the results for at typical eigenmode at 1424 Hz are depicted. In this example the road surface has been modeled as hard–walled. On the left of the figure the modeshape is shown at the top and the sound pressure on the wheels surface below. On the right the sound pressure level in the surrounding air is shown at two field–point meshes. The upper field–point mesh is perpendicular to the symmetry plane of the wheel and the second one has been chosen in this plane. Because the sound pressure calculation in this example is based on normalized modeshapes the absolute values of the sound pressure is not given. The pressure values are scaled to the maximum value p_{max} found near the contact area on the wheels surface.

In this example the radiation is dominant in the plane perpendicular to the wheel. The sound radiation in the symmetry plane of the wheel is negligible. It should be noticed that the highest sound pressure level is located in the surrounding of the wheels contact zone.

To demonstrate the influence of the road a different boundary condition namely a soft-walled surface has been analyzed too. The results are shown in fig. 5.2. In comparison with the hard–walled surface (fig. 5.1) an evidently different sound radiation characteristic is obtained and the maximum sound pressure level is only half of that in the case of the hard–walled road surface.

6. Conclusions

The aim of the presented research activities is the development of numerical simulation strategies for noise generation analysis. In this paper a first step for a better understanding of rolling noise has been described. Based on eigenmodes, which are determined in the deformed state of the stationary rolling wheel by finite element techniques, sound radiation analysis is carried out by the boundary element method. This allows the simulation of the radiation characteristics of single mode shapes. The computation strategy has

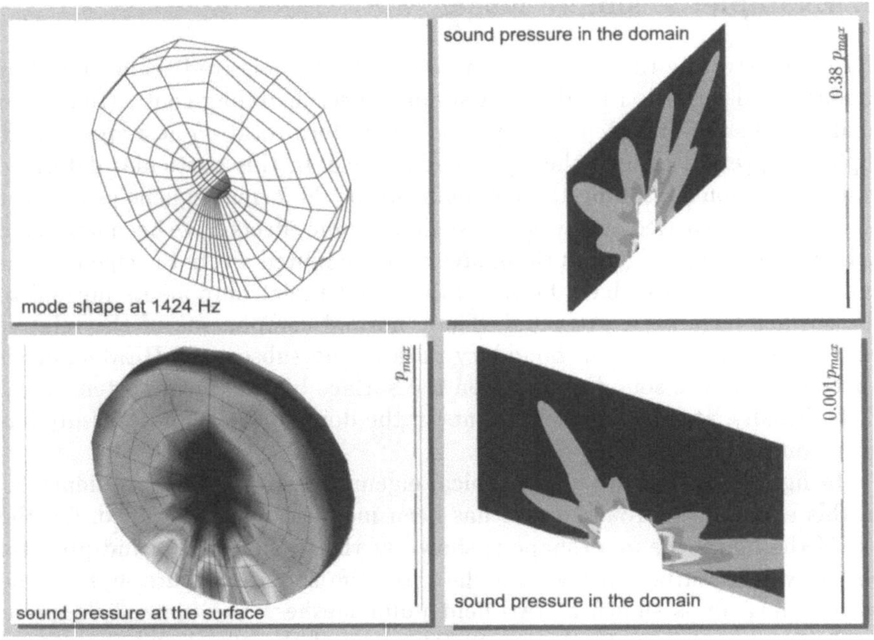

Fig. 5.1. Noise radiation at 1424 Hz, hard–walled road surface ($p_{max} = 2\bar{p}$).

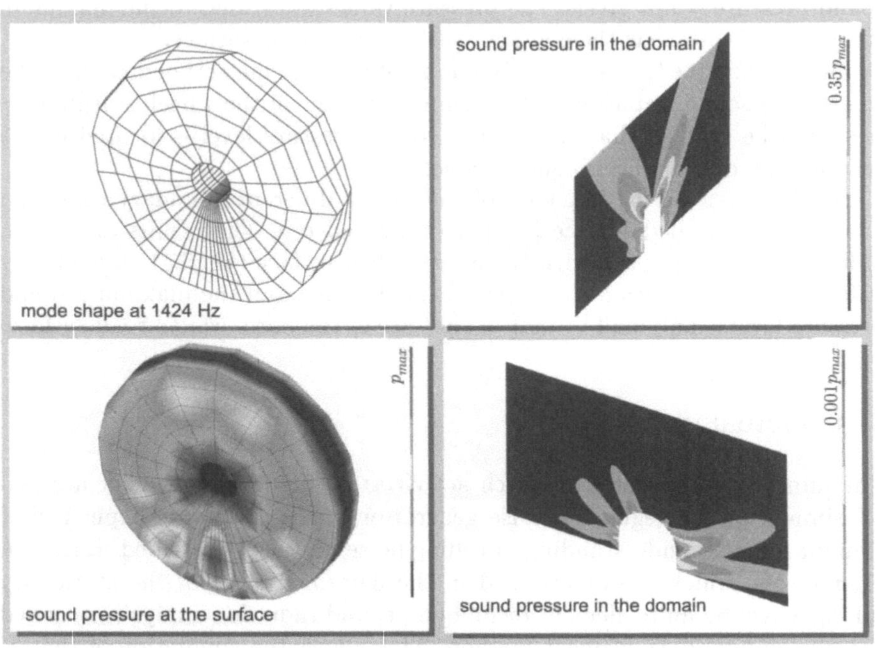

Fig. 5.2. Noise radiation at 1424 Hz, soft–walled road surface ($p_{max} = \bar{p}$).

been demonstrated by the analysis of a simple model wheel. Furthermore, the influence of different boundary conditions, e.g. soft– and hard–walled road surfaces, has been demonstrated. First results with a realistic tire model have been reported in [8].

The next step will be the examination of excitation spectra from the road surface roughness and the simulation of the real vibration behavior of rolling wheels by a modal superposition approach. Simultaneously, the reflection behavior of the rough road surface has to be measured and taken into account by impedance boundary conditions in the simulation.

References

1. Bschorr, O.: Reduktion von Reifenlärm. Automobil–Industrie, 6/86:721–729, 1986.
2. Tonhauser, J.: Einfluß des Reifen-/Fahrbahngeräusches auf das Außengeräusch von PKW; Stand der Technik. Proceedings Innovative Ideen zur Minderung der Kfz-Rollgeräusche, München, 10./11.06.1996.
3. Stenschke, R. und Jäcker, M.: Einfluß von Reifen und Fahrbahnen auf das Reifen/Fahrbahn–Geräusch von Kraftfahrzeugen und administrative Maßnahmen zur Verminderung. Proceedings Innovative Ideen zur Minderung der Kfz-Rollgeräusche, München, 10./11.06.1996.
4. Saemann, E.-U.: Schallquellenortung am Reifen mittels akustischer Holographie im Zeitbereich. In Reifen–Fahrbahn–Fahrwerk, VDI–Berichte 1350, VDI–Verlag, Düsseldorf, 1997.
5. Nackenhorst, U.: On the finite element analysis of steady state rolling contact. In Aliabadi, M.H. and Brebbia, C.A.(eds.): Contact Mechanics, Computational Techniques. Computational Mechanics Publications, Southampton, 1993.
6. Nolte, B.: Randelementberechnungen und Nahfeldmessungen zur akustischen Fluid-Struktur-Interaktion. PhD-Thesis, Institute of Mechanics, University of the Federal Armed Forces Hamburg, 1998.
7. Nackenhorst, U. and Nolte, B.: Numerische Simulation von Rollgeräuschen, in Nixdorff, K. (ed.), Anwendungen der Akustik in der Wehrtechnik, Proc. of the 11th Conference in Meppen, 22.–24. 09. 1998.
8. Nackenhorst, U., Herster, P., Mahmoud, K. und Rainer, G. Ph.: Ein Beitrag zur akustischen Optimierung von Reifen mittels numerischer Simulation, in: Reifen — Fahrbahn — Fahrwerk, VDI–Berichte 1350, VDI–Verlag, Düsseldorf, 1997.
9. Gaul, L. and Fiedler, C.: Methode der Randelemente in Statik und Dynamik, Vieweg Braunschweig, 1997.

Stress Singularities in Composites

A.-M. Sändig

Mathematical Institute A, University of Stuttgart, Germany

Summary. It is well known that stress singularities appear in linear elastic composites near geometrical and structural pecularities as corner of inclusions and crossing points, where the interfaces intersect the exterior boundary or cross one another. We investigate the regularity properties of linear elastic and harmonic fields via Mellin techniques. Furthermore, we study the behaviour of solutions of general semilinear transmission problems in two-dimensional composites. In the nonlinear case we use the implicit function theorem in appropriate Sobolev spaces. We characterize a class of semilinear problems whose local solutions have a similar behaviour as the solutions of the linearized problems.

1. Introduction

The theory of general linear elliptic problems in domains with a piecewise smooth boundary is well developed. In papers by Kondrat'ev [10] and Maz'ya, Plamenevsky [13, 14] the Fredholm property of the corresponding linear operators in domains with conical points is investigated in several scales of functional spaces and the solutions u are decomposed into a linear combination of some singular functions S_j and a more regular remainder

$$u = \sum c_j S_j + u_{reg}.$$

Many mathematicians e.g. Dobrowolski[7], Kellogg[9], Costabel and Dauge[3], Costabel and Stephan[5] as well as mechanicians e.g. Williams[22], Bogy[2], Dempsey and Sinclair[6], Yosibash and Szabo[23] have studied two-dimensional linear interface problems for isotropic, orthotropic and anisotropic elastic composites. The extension and adaption of the general theory for linear boundary value problems to general interface problems was done in [17, 18, 19]. Similar as above the solutions can be decomposed in compound structures, e.g. near a crossing point of two-dimensional subdomains of different materials we have

$$\mathbf{u} = \sum c_j r^{\lambda_j} \mathbf{s}_j(r,\omega) + \mathbf{u}_{reg}.$$

Here $\mathbf{u} = (\mathbf{u}^1, \dots, \mathbf{u}^t)^\top, \mathbf{s}_j = (\mathbf{s}_j^1, \dots, \mathbf{s}_j^t)^\top$ and $\mathbf{u}^m, \mathbf{s}_j^m$ denote the restriction of the corresponding fields to the subdomain Ω_m of the structure, $m = 1, \dots, t$; (r, ω) are the polar coordinates with respect to a crossing point; λ_j are real or complex numbers. Vectors and the corresponding spaces are written bold.

In addition to the linear problems we consider a system of semilinear second

order equations together with boundary and transmission conditions in composites. We investigate the local solvability in a neighbourhood of $\mathbf{u} = \mathbf{0}$ and the regularity of the solutions with the help of the implicit function theorem. The main idea is to work in appropriate function spaces, namely in Sobolev spaces with detached asymptotics, where the operator of the semilinear problem is Fréchet differentiable. Such spaces are used also in [15]. We adapt and extend ideas of the paper[1], where semilinear boundary value problems in homogeneous nonsmooth domains are studied, to composites.
The paper is organized as follows: in Section 2 we consider two-dimensional boundary transmission problems in composites for the Laplacian and the Lamé system and describe the singular behaviour of their solutions near crossing points. Then we show mapping properties of the operators generated by general linear problems and acting in Sobolev spaces with detached asymptotics in Section 3. Semilinear problems are formulated in Section 4. The Fréchet differentiability of the semilinear operators in Sobolev spaces with detached asymptotics is handled in Section 5. In the last Section 6 the asymptotic behaviour of the local solution of the semilinear problem is described.

2. Linear examples: Laplacian and Lamé system

Let $\Omega \subset R^2$ be a bounded domain consisting of t subdomains Ω_m, that means $\overline{\Omega} = \bigcup_{i=1}^{t} \overline{\Omega}_m$. We denote by $\partial\Omega = \overline{\Gamma}^D \cup \overline{\Gamma}^N$ the exterior boundary of Ω and by $\Gamma_{ml} = \partial\Omega_m \cap \partial\Omega_l \neq \emptyset$ the interfaces between the corresponding subdomains.
The boundary transmission problem for the Laplacian can be formulated in the subdomains Ω_m

$$\begin{array}{rcll} \mu^m \Delta u^m & = & f^m & \text{in} \quad \Omega_m \\ u^m & = & 0 & \text{on} \quad \partial\Omega_m \cap \Gamma^D \\ \mu^m \frac{\partial u^m}{\partial n^m} & = & b^m & \text{on} \quad \partial\Omega_m \cap \Gamma^N \\ u^m - u^l & = & 0 & \text{on} \quad \Gamma_{ml} \\ \mu^m \frac{\partial u^m}{\partial n^m} + \mu^l \frac{\partial u^l}{\partial n^l} & = & b^{ml} & \text{on} \quad \Gamma_{ml}. \end{array} \tag{2.1}$$

The corresponding boundary transmission problem for the Lamé equation system reads

$$\begin{array}{rcll} \mu^m \Delta \mathbf{u}^m + (\lambda^m + \mu^m)\nabla \mathrm{div} \mathbf{u}^m & = & \mathbf{f}^m & \text{in} \quad \Omega_m, \\ \mathbf{u}^m & = & \mathbf{0} & \text{on} \quad \partial\Omega_m \cap \Gamma^D, \\ \sigma_m(\mathbf{u}^m)\mathbf{n}_m & = & \mathbf{b}^m & \text{on} \quad \partial\Omega_m \cap \Gamma^N, \\ \mathbf{u}^m - \mathbf{u}^l & = & \mathbf{0} & \text{on} \quad \Gamma_{ml}, \\ \sigma_m(\mathbf{u}^m)\mathbf{n}_m + \sigma_l(\mathbf{u}^l)\mathbf{n}_l & = & \mathbf{b}^{ml} & \text{on} \quad \Gamma_{ml}, \end{array} \tag{2.2}$$

where the components of the stress tensor $\sigma_m(\mathbf{u}^m)$ are given by

$$\sigma_m^{k,l}(\mathbf{u}^m) = \lambda^m(\sum_{j=1}^{2} \partial_j u_j^m)\delta_{kl} + \mu^m(\partial_k u_l^m + \partial_l u_k^m).$$

Here $\mathbf{n}_m$ is the exterior normal vector, $\mathbf{u}^m = (u_1^m, u_2^m)^\top$ denotes the displacement vector in Ω_m. Note that problems with non-vanishing Dirichlet data can be transformed to the above problems, provided some compatibility conditions are satisfied on $\partial\Omega_m \cap \Gamma^D \cap \Gamma_{ml}$ [18].

Vertex singularities

Here, a vertex P is a crossing point of interfaces, a corner point of an inclusion or a point, where an interface intersects the exterior boundary. The set of the vertices is denoted by $\mathcal{P}$. A vertex singularity has the form, discarding the logarithmic terms

$$r^\lambda \Phi_\lambda(\omega) = r^\lambda \begin{pmatrix} \Phi_1(\lambda,\omega) \\ \Phi_2(\lambda,\omega) \\ \cdots \\ \Phi_t(\lambda,\omega) \end{pmatrix}. \tag{2.3}$$

Note, that only q nonvanishing components occur in (2.3), where $q \le t$ denotes the number of subdomains adjacent to the vertex.

The functions (2.3) have to be a solution of the homogeneous system (2.1) or (2.2) in an infinite q-tuple-cone C_P, whose closure coincides with $\overline{\Omega}_{n_1} \cup \overline{\Omega}_{n_2} \cup \cdots \cup \overline{\Omega}_{n_q}$ in a neighbourhood of P. Writing our problem (2.1) or (2.2) in polar coordinates with respect to P, it is equivalent to say that Φ_λ is solution of

$$\mathcal{A}_P(\lambda)\Phi_\lambda = 0, \tag{2.4}$$

where $\mathcal{A}_P(\lambda)$ is the operator pencil, obtained by inserting (2.3) into the homogeneous boundary transmission problems. Using the Mellin transform

$$\hat{\mathbf{u}}(\lambda,\omega) := \int_0^\infty r^{-\lambda}\mathbf{u}(r,\omega)\frac{dr}{r},$$

which maps rD_r into the complex parameter λ, $\mathcal{A}_P(\lambda)$ is simply the Mellin transformed operator of our system (2.1) or (2.2), localized in a neighbourhood of P. We desribe now the behaviour of a weak solution of (2.1) or (2.2) near a vertex $P \in \mathcal{P}$.

Theorem 2.1. *Let $\mathbf{f}^m \in \mathbf{L}_2(\Omega_m)$, $\mathbf{b}^m \in \mathbf{H}^{\frac{1}{2}}(\partial\Omega_m \cap \Gamma^N), \mathbf{b}^{ml} \in \mathbf{H}^{\frac{1}{2}}(\Gamma_{ml})$. Then a weak solution $(\mathbf{u}_1, \mathbf{u}_2, \cdots, \mathbf{u}_t) \in \mathbf{H}^1(\Omega_1) \times \mathbf{H}^1(\Omega_2) \cdots \times \mathbf{H}^1(\Omega_t)$ of problem (2.1) or (2.2) admits the following decomposition:*

$$\mathbf{u} = \mathbf{u}^{reg} + \sum_{P\in\mathcal{P},\gamma\in\Lambda_P} \eta_P c_\gamma^P \mathbf{v}_\gamma^P, \tag{2.5}$$

where $\Lambda_P = \{\gamma = (\lambda,\mu,\kappa) : \lambda$ is an eigenvalue of $\mathcal{A}_P(\lambda)$ in the strip $0 < Re\lambda \le 1, \mu = 1,\ldots,I_\lambda^P, \kappa = 0,\ldots M_{\lambda,\mu}^P\}$, I_λ^S denotes the geometrical multiplicity of λ, $\{\Phi_{\lambda,\mu,\kappa}^P\}_{\mu,\kappa}$ is a canonical system of Jordan chains of $\mathcal{A}_P(\lambda)$ with *respect to* λ, $M^P_{\lambda,\mu} + 1$ *are the lengths of the Jordan chains,* η_P *are cut-off functions equal to 1 near* P,

$$\mathbf{v}^P_\gamma = \mathbf{v}^P_\gamma(r,\varphi) = r^\lambda \sum_{q=0}^{\kappa} \frac{(\ln r)^q}{q!} \Phi^P_{\lambda,\mu,\kappa-q}(\varphi),$$

c^P_γ *are constants and finally* $\mathbf{u}^{reg} \in \mathbf{H}^2(\Omega_1) \times \mathbf{H}^2(\Omega_2) \times \cdots \times \mathbf{H}^2(\Omega_t)$.

Let be Re λ_0 the smallest real part of the eigenvalues of $\mathcal{A}_P$ in the strip $0 <$ Re $\lambda \leq 1$. The following estimates are sharp taking in account all material parameters:

Estimates from below for Re λ_0 for the Laplacian [11, 4, 18]

		1 material	2 materials	3 materials	4 materials
convex, D or N	boundary crossing point	$1\ (\frac{\pi}{\omega_0})$	$1/2\ (\frac{\pi}{2\omega_0})$	0	0
polygon, D or N		1/2	1/4	0	0
convex, D and N	boundary crossing point	1/2	0	0	0
polygon, D and N		1/4	0	0	0
	interior crossing point		1/2	1/4	0

For pure Dirichlet or Neumann data, 2 materials and $\mu_2 > \mu_1 > 0$, we have material depending estimates [11]

$$\text{Re } \lambda_0 > \frac{\arcsin\sqrt{\frac{\mu_1}{\mu_1+\mu_2}} + \arctan\left(-\sqrt{\frac{\mu_2}{\mu_1}}\right) + \pi}{\omega_0};$$

for mixed boundary data, 2 materials and $\mu_2 > \mu_1 > 0$, we get [11]

$$\text{Re } \lambda_0 > \frac{\arcsin\sqrt{\frac{\mu_1}{\mu_1+\mu_2}} + \arctan\sqrt{\frac{\mu_1}{\mu_2}}}{\omega_0}.$$

Estimates from below for Re λ_0 for the Lamé system
As mentioned in the introduction there are several methods known for the computation of singular terms of elastic fields in two-dimensional composites. But, to our knowledge, estimates from below for Re λ_0 for the Lamé system are only known for homogeneous materials near boundary corner points (here are the eigenvalues λ the roots of transcendent equations [8, 20]) and for convex domains consisting of two materials near boundary crossing points for pure Dirichlet or Neumann conditions on the exterior boundary [18]. In these cases we have the same values as in the first column of the table

(1material) and also the value 1/2 at the first place of the second column (2materials). We expect that estimates as in the above table are valid for the Lamé operator also in the other cases.

3. Linear elliptic boundary transmission problems in Sobolev spaces with detached asymptotics

We consider more general elliptic boundary transmission problems

$$\begin{array}{rclll} -\partial_j[a^m_{ij\alpha\beta}(x)\partial_i u^m_\alpha] + b^m_{i\alpha\beta}(x)\partial_i u^m_\alpha + c^m_\alpha(x)u^m_\alpha = f^m_\beta & & & \text{in} & \Omega_m, \\ \left[a^m_{ij\alpha\beta}(x)\partial_i u^m_\alpha\right]\nu^m_j & = & b^m_\beta & \text{on} & \partial\Omega_m \cap \Gamma^N, \\ u^m_\beta & = & 0 & \text{on} & \partial\Omega_m \cap \Gamma^D, \\ \left[a^m_{ij\alpha\beta}(x)\partial_i u^m_\alpha\right]\nu^m_j + \left[a^l_{ij\alpha\beta}(x)\partial_i u^l_\alpha\right]\nu^l_j & = & b^{ml}_\beta & \text{on} & \Gamma_{ml}, \\ u^m_\beta - u^l_\beta & = & 0 & \text{on} & \Gamma_{ml}, \end{array} \tag{3.1}$$

where $\beta = 1, \ldots, k$; $m, l = 1, \ldots, t$. The summation convention is used for repeated indices $i, j = 1, 2$; $\alpha = 1, \ldots, k$.
Let be $\Gamma_S = \bigcup\limits_{m,l} \Gamma_{ml}$ the interface set.

Assume:
(E) *The problem (3.1) is elliptic and has a unique weak solution* $\mathbf{u} \in \mathbf{W}^1_2(\Omega, \Gamma^D) \cap \mathbf{W}^1_p(\Omega)$ *for* $f^m_\beta \in L_p(\Omega_m)$, $b^m_\beta \in W_p^{1-1/p}(\partial\Omega_m \cap \Gamma^N)$, $b^{ml}_\beta \in W_p^{1-1/p}(\Gamma_{ml})$, $p = 2 + \varepsilon$. *Here is* $\mathbf{W}^1_2(\Omega, \Gamma^D) = \{\mathbf{u} \in \mathbf{W}^1_2(\Omega),\ \mathbf{u} = \mathbf{0}$ *on* Γ^D *in the trace sense*$\}$.

From the general theory [17, 14, 16] it follows that the restrictions of the weak solution $\mathbf{u}|_{\Omega_m} = \mathbf{u}^m$ allow a decomposition

$$\mathbf{u^m} = \sum_{P \in \mathcal{P}} \sum_{j=1}^{l} c^P_j \mathbf{w}^m_j(P) + \tilde{\mathbf{u}}^m, \tag{3.2}$$

where $\mathbf{w}^m_j(P)$ are singular functions to the vertex P as in (2.5), c^P_j are constants and $\tilde{\mathbf{u}}^m \in \mathbf{W}^2_p(\Omega_m)$. Furthermore, the regularity assumption in (E) is always satisfied for sufficiently small ε.
The asymptotics (3.2) motivates the introduction of Sobolev spaces with detached asymptotics [15]. Denoting by $\mathbf{w}_j(P) = (\mathbf{w}^1_j(P), \ldots, \mathbf{w}^t_j(P))^\top$ the singular functions in Ω we choose a basis $\{\mathbf{z}_1, \ldots, \mathbf{z}_s\}$ of the finite dimensional linear space generated by the functions $\{\mathbf{w}_1(P), \ldots, \mathbf{w}_l(P), P \in \mathcal{P}\}$. We define

$$\mathbf{D}^2_p(\Omega) := (\text{span}\langle \mathbf{z}_1, \ldots, \mathbf{z}_s \rangle \otimes \mathbf{W}^2_p(\Omega)) \cap \mathbf{W}^1_2(\Omega, \Gamma^D) \tag{3.3}$$

with the norm

$$\|\mathbf{u}\|_{D_p^2(\Omega)} = \sum_{j=1}^{s} |c_j| + \|\tilde{\mathbf{u}}\|_{\mathbf{W}_p^2(\Omega)}. \tag{3.4}$$

The follwing theorem can be proved as in [1]:

Theorem 3.1. *Assume that the condition (E) is satisfied. Then the operator L defined by problem (3.1)*

$$L : \mathbf{D}_p^2(\Omega) \to \mathbf{L}_p(\Omega) \times \mathbf{W}_p^{1-1/p}(\partial\Gamma^N) \times \mathbf{W}_p^{1-1/p}(\Gamma_S) \tag{3.5}$$

is an isomorphism.

4. Semilinear problems

We consider the following boundary transmission problem for a system of k semilinear equations

$$\begin{array}{rcll}
-\partial_j[a^m_{ij\alpha\beta}(x)\partial_i u^m_\alpha] + b^m_{i\alpha\beta}(x,\mathbf{u^m})\partial_i u^m_\alpha + c^m_\beta(x,\mathbf{u^m}) &=& f^m_\beta & \text{in } \Omega_m, \\
\left[a^m_{ij\alpha\beta}(x)\partial_i u^m_\alpha\right]\nu^m_j &=& b^m_\beta & \text{on } \partial\Omega_m \cap \Gamma^N, \\
u^m_\beta &=& 0 & \text{on } \partial\Omega_m \cap \Gamma^D, \\
\left[a^m_{ij\alpha\beta}(x)\partial_i u^m_\alpha\right]\nu^m_j + \left[a^l_{ij\alpha\beta}(x)\partial_i u^l_\alpha\right]\nu^l_j &=& b^{ml}_\beta & \text{on } \Gamma_{ml}, \\
u^m_\beta - u^l_\beta &=& 0 & \text{on } \Gamma_{ml},
\end{array} \tag{4.1}$$

where $\beta = 1,\ldots,k;\ m,l = 1,\ldots,t.$
We use the implicit function theorem to investigate the behaviour of the solutions near vertex points.

Theorem 4.1. *[24, Theorem 4.B] Let X, Y be two Banach spaces. Suppose $N : \mathcal{U}(x_0) \subset X \to Y$ is a mapping of $\mathcal{U}(x_0)$ into Y and $y_0 = Nx_0$. We assume that N is Fréchet differentiable in $\mathcal{U}(x_0)$ and continuous at x_0 and that the Fréchet derivative $N'(x_0) : X \to Y$ is an isomorphism. Then there exist a unique mapping N^{-1} defined in a neighbourhood $\mathcal{V}(y_0) \subset Y, N^{-1} : \mathcal{V}(y_0) \to X$, such that $NN^{-1}y = y$ for all $y \in \mathcal{V}(y_0)$.*

The formal linearization at $\mathbf{u} = \mathbf{0}$ of problem (4.1) reads

$$\begin{array}{rcll}
-\partial_j[a^m_{ij\alpha\beta}(x)\partial_i u^m_\alpha] + b^m_{i\alpha\beta}(x,\mathbf{0})\partial_i u^m_\alpha + \langle \partial c^m_\beta(x,\mathbf{0})\mathbf{u}^m &=& f^m_\beta & \text{in } \Omega_m, \\
\left[a^m_{ij\alpha\beta}(x)\partial_i u^m_\alpha\right]\nu^m_j &=& b^m_\beta & \text{on } \partial\Omega_m \cap \Gamma^N, \\
u^m_\beta &=& 0 & \text{on } \partial\Omega_m \cap \Gamma^D, \\
\left[a^m_{ij\alpha\beta}(x)\partial_i u^m_\alpha\right]\nu^m_j + \left[a^l_{ij\alpha\beta}(x)\partial_i u^l_\alpha\right]\nu^l_j &=& b^{ml}_\beta & \text{on } \Gamma_{ml}, \\
u^m_\beta - u^l_\beta &=& 0 & \text{on } \Gamma_{ml},
\end{array} \tag{4.2}$$

where $\beta = 1,\ldots,k;\ m,l = 1,\ldots,t.$

Theorem 3.1 motivates the choice of appropriate spaces X and Y. Assume that the condition (E) is satisfied for (4.2). Then the corresponding operator

$$L : X = \mathbf{D}_p^2(\Omega) \to Y = \mathbf{L}_p(\Omega) \times \mathbf{W}_p^{1-1/p}(\partial\Gamma^N) \times \mathbf{W}_p^{1-1/p}(\Gamma_S) \qquad (4.3)$$

is an isomorphism. It remains to show that the operator N generated by (4.1) is indeed Fréchet differentiable and $N'(\mathbf{0}) = L$.

5. Continuity and Fréchet–differentiability of the semilinear operator

We write the problem (4.1) in the form of an operator equation

$$N\mathbf{u} = (\mathbf{f}, \mathbf{b}|_{\Gamma^N}, \mathbf{b}|_{\Gamma^S}) \qquad (5.1)$$

Using properties of composition operators proved by (Marcus/Mizel [12], Valent [21]) we can show similar to [1]:

Theorem 5.1. *Suppose that the coefficients of (4.1)* $a_{ij\alpha\beta} \in C^1(\overline{\Omega}), b_{i\alpha\beta}$, $c_\beta \in C^2(\overline{\Omega} \times R^2)$ *and* $p > 2$. *Then the operator* $N : \mathbf{D}_p^2(\Omega) \to \mathbf{L}_p(\Omega) \times \mathbf{W}_p^{1-1/p}(\partial\Gamma^N) \times \mathbf{W}_p^{1-1/p}(\Gamma_S)$ *is continuously Fréchet differentiable and its Fréchet derivative at* $\mathbf{u} = \mathbf{0}$ *coincides with the operator* L *defined by (4.2).*

6. Asymptotic behaviour of the solution of the semilinear problem

Now, we are in position to summarize the assumptions which we need in order to apply Theorem 4.1 to the nonlinear problem (4.1):

A1 $a_{ij\alpha\beta} \in C^1(\overline{\Omega}), b_{i\alpha\beta}, c_\beta \in C^2(\overline{\Omega} \times R^2)$.
A2 $c_\beta(x; 0) = 0$.
A3 $\mathbf{f} \in \mathbf{L}_p(\Omega)$, $\mathbf{b}|_{\Gamma^N} \in \mathbf{W}_p^{1-1/p}(\Gamma^N)$, $\mathbf{b}|_{\Gamma^S}) \in \mathbf{W}_p^{1-1/p}(\Gamma^S)$ $\beta = 1, \ldots, k$, with sufficiently small norms and $p = 2 + \varepsilon$ with a small positive ε.
A4 The linearized problem (4.2) satisfies the assumption (E).

We formulate the main result:

Theorem 6.1. *Let the assumptions* A1-A4 *be satisfied. Then the semilinear problem (4.1) has a unique solution* $\mathbf{u} \in \mathbf{D}_p^2(\Omega)$ *in a neighbourhood* $\mathbf{u} = \mathbf{0}$.

We remark, that a paper about the singular behaviour of solutions of quasilinear elliptic systems in domains with corners and edges is in preparation where the implicit function theorem in weighted Sobolev spaces with detached asymptotics is used.

References

1. M. Bochniak, A.–M. Sändig, *Local solvability and regularity results for a class of semilinear elliptic problems in nonsmooth domains,* Mathematica Bohemica 2–3, 124 (1999) 245-254.
2. D. B. Bogy *Two edge–bonded elastic wedges of different materials and wedge angles under surface tractions,* Trans. ASME J. appl. Mech. **38** (1971) 377–386.
3. M. Costabel and M. Dauge, *Computation of corner singularities in linear elasticity,* Lecture notes in pure and applied mathematics **167** (1995) 59–69.
4. M. Costabel, M. Dauge and S.Nicaise *Singularities of Maxwell interface problems* Math. Modelling and Numerical Analysis, **33** (1999) 627-649.
5. M. Costabel and E. P. Stephan, *A direct boundary integral equation method for transmission problems,* J. Math. Anal. Appl. **106** (1985) 376–413.
6. J. P. Dempsey and G. B. Sinclair, *On the singular behavior at the vertex of a bi–material wedge.* J. Elast. **11** (1981) 317–327.
7. M. Dobrowolski, *Numerical approximation of elliptic interface and corner problems.* Habilitationsschrift, Bonn (1981).
8. P. Grisvard, *Problémes aux limites les polygones, Mode d'emploi* EDF Bull. Rech. Sér.C Math.Inform. No 1, (1986) 21-59.
9. R. B. Kellogg, *On the Poisson equation with intersecting interfaces.* Appl. Anal. **4** (1975) 102–129.
10. V. A. Kondrat'ev, *Boundary problems for elliptic equations in domains with conical or angular points,* Trans. Moscow Math. Soc. 16 (1967) pp. 209–292.
11. T. Kühn *Die Regularität der Lösungen von Interface-Problemen in Gebieten mit singulären Punkten,* Diplomthesis, University Rostock, 1992.
12. M. Marcus, V.J. Mizel, *Complete characterization of functions which act, via superposition, on Sobolev spaces,* Trans. AMS 251 (1979) pp. 187–218.
13. V. G. Maz'ya, B. A. Plamenevsky, *On the coefficients in the asymptotics of solutions of elliptic boundary value problems in domains with conical points,* Math. Nachr. 76 (1977) pp. 29–60.
14. V. G. Maz'ya, B. A. Plamenevsky, *Weighted spaces with nonhomogeneous norms and boundary value problems in domains with conical points,* Amer Math. Soc. Transl. 123 (1984) pp. 89–107.
15. S.A. Nazarov, *On the two–dimensional aperture problem for Navier–Stokes equations,* C.R. Acad. Sci Paris, t. 323, Série I (1996) pp. 699–703.
16. S. A. Nazarov, B. A. Plamenevsky, *Elliptic Problems in Domains with Piecewise Smooth Boundaries,* Walter de Gruyter, Berlin, 1994.
17. S. Nicaise and A.–M. Sändig, *General interface problems I/II,* Math. Meth. in the Appl.Sci. **17** (1994) 395–450.
18. S. Nicaise, A.-M. Sändig, *Transmission problems for the Laplace and elasticity operators: Regularity and boundary integral formulation,* Mathematical Models and Methods in Applied Sciences, Vol.9, No.6 (1999) 855-898
19. A.-M. Sändig *The Shapiro-Lopatinskij condition for boundary and transmission problems for the Lamè system,* preprint 98/14 SFB 404, University Stuttgart
20. A.-M. Sändig, U. Richter and R. Sändig, *Regularity of boundary value problems for the Lamé equations in a polygonal domain,* Rostocker Mathematisches Kolloquium 36 (1989) 21-50
21. T. Valent, *Boundary Value Problems of Finite Elasticity,* Springer, 1988
22. M. L. Williams, *The stresses around a fault or crack in dissimilar media,* Bull. Seismol. Soc. America **49** (1959).
23. Z. Yosibash and B. Szabo, *Numerical analysis of singularities in two-dimensions I,* Int. J. Numer. Methods Eng. **39** (1996) 409–434.
24. E. Zeidler, *Nonlinear Functional Analysis and its Applications I: Fixed–Point Theorems,* Springer–Verlag, New York, 1986.

MIX
Papier aus verantwortungsvollen Quellen
Paper from responsible sources
FSC® C105338

If you have any concerns about our products,
you can contact us on
ProductSafety@springernature.com

In case Publisher is established outside the EU,
the EU authorized representative is:
Springer Nature Customer Service Center GmbH
Europaplatz 3, 69115 Heidelberg, Germany

Printed by Libri Plureos GmbH
in Hamburg, Germany